Real-Time Robotics With IoT and Edge AI:

Methods, Metrics, and Applications

Mohammed Aquil Mirza
https://orcid.org/0000-0001-7756-4363
The Hong Kong Polytechnic University, Hong Kong

Wu Cong
https://orcid.org/0000-0002-8652-3501
Liaoning Technical University, China

IGI Global
Scientific Publishing
Publishing Tomorrow's Research Today

Vice President of Editorial	Melissa Wagner
Director of Acquisitions	Mikaela Felty
Director of Book Development	Jocelynn Hessler
Production Manager	Mike Brehm
Cover Design	Jose Rosado

Published in the United States of America by
IGI Global Scientific Publishing
701 East Chocolate Avenue
Hershey, PA, 17033, USA
Tel: 717-533-8845 | Fax: 717-533-7115
Website: https://www.igi-global.com E-mail: cust@igi-global.com

Library of Congress Cataloging-in-Publication Data

LCCN: 2026021637 (CIP Data Pending)
ISBN13: 979-8-2600-0701-3
Isbn13Softcover: 979-8-2600-0702-0
EISBN13: 979-8-2600-0703-7
British Cataloguing in Publication Data
A Cataloguing in Publication record for this book is available from the British Library.

Table of Contents

Detailed Table of Contents

Edge AI is reshaping industrial robotics by moving perception, inference, and coordination closer to sensors, actuators, and motion controllers operating under strict timing constraints. This chapter examines latency optimization in real-time robotic control as an architectural and benchmarking problem rather than a hardware-selection issue alone. It develops a layered reference architecture linking sensing, middleware, inference runtimes, control execution, safety envelopes, and edge-cloud orchestration. It also proposes a benchmarking framework centered on end-to-end latency, jitter, deadline misses, control-age, energy efficiency, and thermal stability. Drawing on literature in edge computing, ROS 2, embedded AI, deterministic networking, and industrial robotics, the chapter synthesizes design trade-offs and presents literature-grounded industrial case studies in machine tending, vision-guided handling, and mobile robotics, concluding with implementation guidance and a research agenda.

Edge AI swarm robotics is reshaping supply chain business intelligence by moving perception, coordination, and decision support closer to operational events. This chapter examines how decentralized robot collectives, embedded sensing, and edge analytics can create real-time visibility for warehouses, fulfillment centers, ports, and cold-chain networks. It synthesizes literature on edge computing, swarm robotics, digital twins, and supply chain analytics to develop an architectural and managerial framework for low-latency intelligence. The discussion integrates benchmarking metrics, governance requirements, and literature-grounded industrial case analyses. Attention is given to latency budgets, model deployment patterns, task allocation, resilience, cybersecurity, and human oversight. The chapter concludes with implementation recommendations and a future research agenda for trustworthy, scalable, and economically viable swarm-enabled intelligence in contemporary global multi-echelon supply networks under uncertainty.

Autonomous mobile robots face critical perception failures due to severe motion blur during rapid rotation. Standard deblurring models introduce geometric artifacts when applied to planar yaw motion. To resolve this, this chapter details a real-time, hardware-accelerated perception pipeline developed in ROS 1 for the memory-constrained NVIDIA Jetson Orin Nano. The authors propose a Cartesian Cascaded Refinement Network (CAR-Net) that eliminates polar coordinate biases, corrects linear yaw blur, and achieves a 29.52 dB PSNR. Additionally, an optimized YOLOv8 obstacle detection model was generated using feature extraction trained on unmodified data to prevent knowledge loss. Freezing specific layers preserved robust object tracking. Physical stress tests utilizing high-intensity steering (up to 1.2 rad/s) demonstrated that the TensorRT-optimized pipeline reduced frame latency from 32 to 23 frames. This edge-based solution maintains a stable 20Hz throughput, significantly enhancing autonomous obstacle avoidance capabilities.

Airlines discard an estimated $4 billion in untouched food annually, yet biosecurity regulations mandate immediate incineration of cabin waste, precluding post-flight auditing. This chapter presents SkyWaste, a resilient edge-cloud IoT architecture that transforms crew tablets into waste sensors producing item-level telemetry without cloud connectivity during flight. The architecture comprises three planes: an Edge Plane producing structured telemetry (~300 bytes/tray) within 67 ms, a Cloud Plane with six modular serverless stacks, and an MLOps Plane for active learning. A store-and-forward protocol with crash-resilient session linkage ensures data integrity. Manifest integration uses embedding similarity to bridge catering vocabulary and detection taxonomy. Validated across 489 tests and backed by an airline partner agreement.

The Hong Kong public medical system is facing a critical manpower crisis while frontline nurses are spending up to one-third of their time on manual visitor management. This project proposes a Smart Door Automation System leveraging an AIoT architecture to streamline real-time access control. The system integrates a React.js self-service portal with a decentralized edge-computing module powered by a Raspberry Pi 4. To ensure patient privacy, identity verification—incorporating Tesseract OCR for HKID extraction and ArcFace-based facial matching, which is executed locally at the ward entrance. Experimental results have demonstrated an end-to-end latency of 13.47 seconds and an OCR accuracy of about 90%. Automating verification and maintaining secure audit trails can alleviate the administrative burden on staff while providing safeguarding services to patients and a privacy-aware blueprint for future Smart Hospitals.

A simple misremembering can trigger household food waste: people can forget about food that is already stored, they can have something that was lost somewhere behind other goods, and they cannot see the printed expiry dates before it is too late. In this project, an in-depth AIoT smart-fridge prototype is introduced that considers the problem as a workflow and not as a single recognition task. An ESP32-S3 is used to sense door openings, a Jetson Orin to sense items, their position, and expiry zone, a Raspberry Pi to compare inventory records with predicted date parts, and a mobile application to display the retained value over a local MQTT-based network. The contribution does not claim to resolve all the cases of fridges. Instead, it is a practical integration of sensing, edge vision, local messaging, persistence, and readable inventory display for a small set of expirable items.

This report presents a local smart-fridge prototype that connects a door-trigger node, a Jetson Orin vision service, a Raspberry Pi backend, and a Vue-based web interface through MQTT. Each fridge interaction is treated as one session and is linked across the full pipeline by a shared door_session_id. In the current prototype, the Orin records while the door is open and processes the session after the door closes. The Pi stores results in SQLite, updates the inventory state, and serves the user interface. The web application shows inventory, alerts, review tasks, and an optional recipe feature. The evaluation uses real session artifacts, backend logs, database rows, and an MQTT door-node validation archive. The results show add-item and remove-item handling, safe unknown-item fallback, empty-session handling, and a support-image refinement example for egg classification. The main contribution is a traceable local workflow that avoids writing uncertain detections into inventory until the user confirms them.

Smart homes that rely on the cloud often have problems with high transmission latency, privacy risks, and a lot of extra bandwidth. This project proposes a decentralized edge-computing automation system that incorporates an Arduino Uno, an ESP8266 Wi-Fi module, a local Python edge gateway, and mobile rule engines like iOS Shortcuts to address these limitations. The local gateway takes care of heavy computational tasks, such as parsing complex JSON and logging data all the time. This lets the resource-constrained microcontroller work as a very optimized hardware abstraction layer. To get around strict hardware limits (2KB SRAM and 32KB Flash), advanced memory management techniques were used. The proposed system successfully achieved sub-second latency, eliminated hardware servo jitter, and enabled highly customizable, local automation routines without reliance on proprietary cloud ecosystems, proving the viability of decentralized, low-cost IoT computing.

Falls and toileting-related accidents remain a serious patient-safety concern in hospital wards, particularly in privacy-sensitive toilet environments. This project developed a Smart Ward AIoT Toilet Safety Monitoring System that integrates an ESP32 entrance module, MQTT communication, relational event logging, dynamic threshold prediction, and a web interface with AI-assisted support. To preserve privacy, the system avoids camera surveillance and replaces fixed timeout rules with patient-specific safety thresholds. A prototype was implemented and evaluated as an integrated end-to-end system. Experimental results showed that the XGBoost-based configuration achieved the strongest performance, demonstrating the feasibility of the proposed privacy-aware monitoring approach.

This project presents a smart home automation system based on edge computing to address the limitations of mainstream IoT systems, which rely on cloud-based architectures. The project is divided into two systems. The first system is an online simulation in Wokwi to validate the feasibility with multiple sensors controlling coloured LEDs for appliance operation and circuit structure simulation. The second system is a physical system using the ESP32 microcontroller. The system integrates with Wi-Fi and a web server for real-time monitoring and generates historical charts using Chart.js. The user can use iOS Shortcuts for HTTP-based control to achieve remote control. Edge computing ensures that all data processing and decision-making are completed locally, solving the problems of current smart home systems on the market. The simulation shows how edge computing can deliver privacy-preserving, resilient and efficient smart automation suitable for real-world environments. Finally, this project further proposes using MQTT to address the limitations of current systems.

Preface

INTRODUCTION

Real-Time Robotics With IoT and Edge AI: Methods, Metrics, and Applications gathers a family of studies and system designs that advance the capability to sense, reason, and act in time-critical robotic contexts. At its core, the book investigates how edges; local processing at the device, gateway, or nearby server; coupled with AI techniques, can reduce latency, increase reliability, and enable robust decision making in dynamic environments. Across the chapters, readers encounter a common throughline: the shift from centralized cloud-centric intelligence to distributed, latency-aware architectures that fuse perception, communication, and control in real time. The volume foregrounds practical engineering choices, benchmarking practices, and deployment considerations, while also exploring theoretical constructs such as closed-loop performance, end-to-end latency budgets, and the interplay between sensing fidelity, computational constraints, and network dynamics. By threading together edge AI, IoT protocols, and human-centric interfaces, the book offers a coherent lens on how modern robots perceive, decide, and act within the constraints of real-world systems.

Today's engineered world increasingly demands systems that respond promptly to changing conditions, even when connectivity is intermittent or bandwidth is limited. Industrial automation, healthcare, transportation, smart buildings, and logistics are all increasingly populated by autonomous agents that must operate with minimal latency and maximal resilience. Edge AI and IoT enable this by pushing computation closer to the data source, reducing round-trips to distant servers, and exploiting local context for faster and more reliable decisions. The COVID-era acceleration of remote monitoring, the rise of Industry 4.0 concepts, and the growing ubiquity of sensor-rich devices have created a vast ecosystem where latency optimization, adaptive control, and secure, privacy-preserving data flows are not luxuries but prerequisites. In many settings, from hospital corridors and smart homes to factories and supply chains, stakeholders seek systems that can translate streams of sensory input into timely actions, while maintaining safety, explainability, and accountability. The chapters in this book illuminate how researchers and practitioners are meeting these demands through pragmatic designs, rigorous benchmarking, and thoughtful consideration of regulatory, ethical, and operational constraints.

The intended audience comprises researchers, practitioners, and decision-makers who design, implement, or evaluate real-time intelligent systems at the edge. Graduate students and postdocs working at the intersection of robotics, AI, embedded systems, and IoT will find both foundational concepts and cutting-edge approaches applicable to their research agendas. Engineering managers and system integrators tasked with deploying edge-enabled robotics in domains such as industrial automation, healthcare, aviation, logistics, and smart infrastructure will benefit from the collected lessons on architecture, latency budgeting, system reliability, and performance evaluation. Finally, policymakers and standards developers interested in privacy, safety, and interoperability in AIoT ecosystems will discover cross-cutting themes about how real-time systems can meet societal needs without compromising ethical norms. Below is a chapter by chapter summary to help supplement your reading.

CHAPTER SUMMARIES

Chapter 1: Edge AI-Driven Latency Optimization in Real-Time Robotic Control

This chapter presents a systems-oriented view of latency as a primary design variable in real-time robotics. It emphasizes end-to-end pathways from perception to actuation, discusses architectural partitioning across robot, edge, and cloud layers, and offers benchmarking methods that capture closed-loop performance. The narrative integrates architectural decisions with practical evaluation strategies, illustrating how to bound deadline misses, manage jitter, and ensure safe operation under load.

Chapter 2: Edge AI Swarm Robotics for Real-Time Supply Chain Intelligence

Here, the focus is on distributed multi-robot systems in supply chains, where edge-enabled perception, coordination, and analytics enable near-instantaneous responses. The chapter articulates how swarm behaviors, local decision rules, and edge processing combine to deliver real-time visibility and adaptive operational intelligence across a network of moving agents, with attention to scalability, fault tolerance, and auditability.

Chapter 3: ROS-Based Mobile Robot for Environmental Monitoring and Data Analysis

This chapter details a mobile robotic platform that integrates ROS-based navigation with on-board sensing and real-time data analytics for environmental monitoring. It highlights the challenges and solutions for synchronized data collection in dynamic indoor and outdoor environments, emphasizing reproducibility, modularity, and the role of ROS in rapid prototyping and deployment.

Chapter 4: SkyWasteTracker – An Edge-Cloud Architecture for Connectivity-Constrained Aviation Environments

SkyWasteTracker investigates edge-centric designs for aviation environments characterized by intermittent connectivity and stringent regulatory constraints. It explores how on-device processing and opportunistic cloud synchronization can support waste tracking, ESG reporting, and data integrity in a highly constrained setting, drawing lessons that extend to other regulated, connectivity-limited domains.

Chapter 5: Real-Time IoT and Edge AI Framework for Hospital Visitor Access Control and Smart Door Automation

This chapter surveys a safety- and privacy-conscious IoT/edge framework enabling secure access control and automated environmental adaptation in hospital corridors. It discusses architectural choices, latency budgets, and human-in-the-loop interfaces that balance operational efficiency with patient safety and regulatory compliance.

Chapter 6: Real-Time Edge AIoT Fridge System for Inventory Tracking and Expiry-Aware Monitoring

Focusing on domestic and small-business environments, this chapter presents an edge AIoT system for inventory management and product expiry monitoring. It emphasizes local processing, privacy-preserving data handling, and resilient operation in scenarios with limited connectivity and variable network availability.

Chapter 7: Session-Based Inventory Tracking on ESP32 with MQTT and Edge AI

This chapter extends edge AI concepts to resource-constrained hardware, illustrating how session-based inventory tracking can be implemented on microcontrollers with MQTT-enabled communication and lightweight onboard analytics. It addresses practical considerations for energy efficiency, reliability, and real-time responsiveness in constrained devices.

Chapter 8: Real-Time Edge-Enabled Smart Home Automation with Arduino, MQTT, and Rule-Based Interfaces

A pragmatic chapter that blends simple, accessible hardware with edge-oriented processing and MQTT-based messaging. It emphasizes rule-based control, local decision-making, and scalable home automation patterns suitable for educational settings and rapid prototyping.

Chapter 9: Real-Time Access Control via Smart Doors and Windows Using AIoT

This chapter explores secure and responsive access control mechanisms leveraging AIoT principles. It discusses perception, decision logic, and communication protocols necessary to ensure reliable entry/exit management, while addressing privacy, safety, and redundancy concerns.

Chapter 10: Smart Automation Using Arduino UNO and iOS Shortcuts for Efficient Edge Computing

Concluding the collection, this chapter demonstrates how widely available off-the-shelf hardware and consumer platforms (Arduino UNO and iOS Shortcuts) can cohere into an agile, edge-enabled automation workflow. It emphasizes accessibility, developer-friendly design patterns, and practical demonstrations of edge computing in daily life and small-scale deployments.

CONCLUSION

Real-Time Robotics With IoT and Edge AI: Methods, Metrics, and Applications offers a timely, substantive contribution to the field by synthesizing architectural patterns, benchmarking practices, and application-specific lessons across a spectrum of real-time robotic systems. By foregrounding edge computing as a core engine for latency-aware perception, decision making, and actuation, the book provides both a conceptual framework and a practical toolkit for researchers and practitioners. It clarifies how end-to-end latency budgets can be defined, measured, and optimized, and it highlights the critical interplay between hardware constraints, software architecture, and network dynamics in achieving robust, real-time performance. The diversified chapter set demonstrates that edge AI is not a single-venue solution but a multi-domain enabler; capable of transforming environmental sensing, healthcare, aviation, smart homes, and industrial automation. As such, the book equips readers with the knowledge to design systems that are safer, more efficient, privacy-preserving, and capable of operating under real-world constraints. In doing so, it contributes to a field that increasingly requires transparent, auditable, and scalable real-time intelligent systems, helping to bridge the gap between theoretical advances and their responsible, impactful deployment.

Chapter 1
Edge AI-Driven Latency Optimization in Real-Time Robotic Control

Mohammad Al Khaldy
http://orcid.org/0009-0009-7502-4668
University of Petra, Amman, Jordan

Hewa Majeed Zangana
http://orcid.org/0000-0001-7909-254X
Duhok Polytechnic University, Iraq

ABSTRACT

Edge AI is reshaping industrial robotics by moving perception, inference, and coordination closer to sensors, actuators, and motion controllers operating under strict timing constraints. This chapter examines latency optimization in real-time robotic control as an architectural and benchmarking problem rather than a hardware-selection issue alone. It develops a layered reference architecture linking sensing, middleware, inference runtimes, control execution, safety envelopes, and edge-cloud orchestration. It also proposes a benchmarking framework centered on end-to-end latency, jitter, deadline misses, control-age, energy efficiency, and thermal stability. Drawing on literature in edge computing, ROS 2, embedded AI, deterministic networking, and industrial robotics, the chapter synthesizes design trade-offs and presents literature-grounded industrial case studies in machine tending, vision-guided handling, and mobile robotics, concluding with implementation guidance and a research agenda.

DOI: 10.4018/979-8-2600-0701-3.ch001

1. INTRODUCTION

Industrial robotics is reaching an inflection point where the utility of AI isn't about the model itself but rather whether the AI can run in an envelope that has a time limit. For many workstations in a factory, a great model that arrives too late is not nearly as important as an average model that always arrives on time. This makes latency a first order design variable for robotic control. The designing question is no longer how to get AI on to robots but how to place sensing, inference, communication and control on distributed embedded controllers, edge computers and cloud services such that the close loop is responsive, predictable, safe and economical. Edge computing has become a key enabler of this vision because it reduces the distance of data to actuation. Without this reduction, all these demands of a time-critical close-loop would have been impossible (Shi et al., 2016; Satyanarayanan, 2017; Chen & Ran, 2019).

This problem matters because robots in industrial settings are taking an increasingly diverse set of roles. They are no longer performing isolated, repetitive motion with fixed logic. In flexible cells, collaborative workspaces, lineless assembly systems, intra-logistics networks and smart machine tools, there is a need to quickly mix vision, path planning, quality inspection and contextual coordination. Each scenario has heterogeneous latency budget requirements. A high-speed pick-and-place cell may only allow tens of milliseconds between image capture and motor update. A mobile manipulator traversing an aisle can tolerate slightly longer inference times but will be quite sensitive to network jitter and synchronization drift. Meanwhile, a fleet-level orchestration platform may take hundreds of milliseconds or even seconds to plan a better route but depends on lower-layer control and middleware that behave deterministically on much shorter timescales. So, research on robotics at the edge emphasizes not just lowering average latency but also constructing the full stack so that its variability, deadline misses and scheduling interactions are bounded (Tahir et al., 2025; Macenski et al., 2022; Tipsuwan & Chow, 2003).

A related challenge is that latency is rarely caused by one component. It is an emergent property of the full robotic pipeline: sensor exposure and preprocessing, message transport, middleware serialization, executor wake-up, model loading and inference, memory transfer, control computation, actuation dispatch, and even safety interlocks. In ROS 2-based systems, for example, latency can be shaped by executor design, callback scheduling, middleware selection, thread isolation, and tracing visibility as much as by processor speed (Maruyama et al., 2016; Park, Delgado, & Choi, 2020; Ye et al., 2023). In embedded AI workloads, model architecture, quantization, accelerator mapping, thermal throttling, and batch semantics influence responsiveness and determinism alongside raw throughput (Sze et al., 2017; Cantero et al., 2022; Jayanth, Gupta, & Prasanna, 2024). As a result, real-time optimization cannot be reduced to buying a faster edge device. It requires an architectural approach.

This chapter argues that latency optimization in edge AI-driven robotics should be treated as a systems design and benchmarking problem. Architecturally, designers must decide which functions remain on the robot, which are delegated to nearby edge servers, and which can tolerate cloud-level delay. They must also define control hierarchies, fallback modes, synchronization strategies, and observability mechanisms so that the system continues to behave safely under load, packet loss, or inference degradation. From a benchmarking standpoint, teams need metrics that reflect closed-loop behavior rather than only stand-alone inference speed. A model that delivers high frames per second on a test image stream may still produce poor robotic performance if communication latency, queueing delay, or actuation jitter dominate the real loop. For industrial readers, this means benchmark protocols must connect platform-level metrics to task-level consequences such as path tracking, grasp timing, cycle time, reject rate, or queueing bottlenecks.

The literature base for this chapter spans four streams. The first is edge computing and edge intelligence, which defines why distributed compute placement matters and how edge resources differ from cloud environments in constraints and opportunities (Shi et al., 2016; Satyanarayanan, 2017; Murshed et al., 2021; Wang et al., 2025). The second is real-time robotic middleware and systems engineering, especially work around ROS 2, executor design, tracing, and distributed timing analysis (Macenski et al., 2022; Choi, Xiang, & Kim, 2021; Bédard, Lütkebohle, & Dagenais, 2022; Puck et al., 2021). The third is embedded AI benchmarking and model optimization, which contributes methods for comparing accelerators, quantization strategies, and object-detection pipelines under power and memory limits (Cantero et al., 2022; Kang & Somtham, 2022; Lema, Usamentiaga, & García, 2024). The fourth is industrial deployment work, including manufacturing, mobile robotics, digital twins, and 5G/TSN integration, which grounds architectural claims in applied settings (Kehl et al., 2022; Zhang et al., 2023a; Bergs et al., 2024; Mtowe et al., 2025).

The chapter makes four contributions. First, it synthesizes technical foundations for latency-aware edge AI in robotic control and clarifies why end-to-end delay should be understood in control-theoretic and systems terms rather than as a single communication statistic. Second, it develops a layered reference architecture that relates edge inference, middleware, deterministic networking, safety constraints, and orchestration across industrial robotics deployments. Third, it proposes a benchmarking framework based on latency decomposition, control-age, jitter, deadline miss ratio, energy use, and thermal stability, supported by repeatable experimental design. Fourth, it presents literature-grounded industrial case studies that translate the framework into machine tending, vision-guided handling, and mobile robot coordination scenarios. Throughout, tables and figures are introduced in the text before placement.

2. TECHNICAL AND THEORETICAL FOUNDATIONS OF EDGE AI FOR REAL-TIME ROBOTIC CONTROL

The conceptual foundation of low-latency edge AI in robotics lies at the intersection of edge computing, networked control systems, robotic middleware, and efficient neural inference. Edge computing literature provides the baseline rationale. Shi et al. (2016) define edge computing as moving computation, storage, and services closer to data sources, while Satyanarayanan (2017) emphasizes ultra-low-latency use cases as a decisive driver of this migration. In robotics, that shift matters because the costs of delay are not only user-perceived responsiveness or cloud bandwidth, but also control instability, degraded path quality, unsafe interactions, and lost production time. Chen and Ran (2019) extend the argument by showing that deep learning at the edge is motivated by the need to localize inference under resource, privacy, and responsiveness constraints. Murshed et al. (2021) and Wang et al. (2025) further show that edge intelligence is shaped by limited power, constrained memory, accelerator diversity, and dynamic workload placement, all of which are highly visible in robotic systems.

Robotic control, however, is more demanding than generic edge AI because it is a feedback problem. In a networked control system, delay is not just a service metric; it alters the age and relevance of feedback signals used to compute the next action. Tipsuwan and Chow (2003) show that network-induced delay, packet loss, and jitter directly influence control performance, while Gautam et al. (2021) trace the evolution of networked control systems toward architectures in which communication and computation must be designed together. For AI-enabled robots, this means perception delay, inference latency, and transport delay must be interpreted through the loop they participate in. A 20 ms inference delay is benign if it supports supervisory replanning at 5 Hz, but problematic if it feeds a fine manipulation policy that should react within a 10 ms control interval. The useful design abstraction is therefore not "fast AI" in general but "task-matched latency budgets" tied to loop frequency, actuator dynamics, and safety constraints.

The middleware layer has become especially important because it often determines whether the rest of the stack can exploit edge placement. ROS 2 is now a dominant production-oriented robotics middleware, but its real-time behavior depends strongly on executor design, DDS configuration, operating-system settings, and message-flow structure. Macenski et al. (2022) describe ROS 2 as a scalable, modular architecture intended for deployment outside research laboratories, which explains its growing industrial relevance. Yet early work by Maruyama et al. (2016) already showed that transport latency in ROS 2 varies significantly by use case, data size, configuration, and DDS vendor. Subsequent empirical research has made this more precise. Park et al. (2020) quantify the real-time characteristics of ROS 2 in multiagent robotic systems and show that system load, middleware choice, and kernel configuration materially affect timing behavior. Ye et al. (2023) demonstrate that Linux real-time optimizations and system-level tuning can meaningfully reduce jitter and improve predictability, which is essential for high-precision motion equipment.

The executor and scheduling problem is equally significant. ROS 2 applications are often built as chains of callbacks with data dependencies, and timing can degrade when callbacks compete for shared executors or are awakened in non-priority-aware ways. Yang and Azumi (2020) analyze executor behavior in ROS 2 and show why default execution semantics are not always sufficient for real-time applications. Choi et al. (2021) respond with PiCAS, a priority-driven, chain-aware scheduler that improves timeliness by aligning callback execution with end-to-end chains rather than local event order alone. Kronauer et al. (2021) examine latency in multi-node ROS 2 systems, clarifying that distributed pipelines must be understood at the chain level. Puck et al. (2021) then move from software-only analysis to time-synchronized distributed networks for robotic control, showing how externally measured timing can expose bottlenecks that would otherwise remain hidden. This family of work suggests that latency optimization in robotic systems is irreducibly end-to-end: it requires simultaneous attention to middleware semantics, network timing, and workload partitioning.

Observability tools are indispensable in this context because complex chains make it difficult to identify where delay actually originates. Bédard et al. (2022) show that low-overhead tracing for ROS 2 enables detailed timing analysis without overwhelming real-time workloads, while Li et al. (2022) extend tracing into Autoware-related performance analysis for complex applications. Bédard et al. (2023) further show how message-flow analysis with causal links can reveal the timing of distributed chains, providing much richer information than isolated node-level logs. These tools matter for industrial practice because optimization without tracing often degenerates into guesswork: teams may swap accelerators or compress a model while the dominant latency remains in message serialization, image acquisition, or callback wake-up.

At the inference layer, efficient deep neural processing becomes a second theoretical pillar. Sze et al. (2017) explain why throughput alone is a misleading comparison for neural accelerators; memory movement, dataflow choices, model architecture, and quantization strategy can matter more than arithmetic intensity. This perspective aligns with applied benchmarking studies. Cantero et al. (2022) benchmark object-detection models on embedded devices and show that deployment feasibility depends on the interaction among model family, quantization level, runtime conversion, and target hardware. Galliera and Suri (2022) compare off-the-shelf deep-learning-capable edge devices and accelerators, while Kang and Somtham (2022) and Lema et al. (2024) evaluate modern edge accelerators and object-detection models under realistic embedded constraints. Jayanth et al. (2024) add a heterogeneous platform perspective, emphasizing that CPUs, GPUs, and NPUs have distinct latency and throughput profiles depending on workload shape and parallelism. For robotic control, the critical implication is that "best" hardware is workload-dependent. Motion-conditioned vision tasks, lightweight classification, multimodal fusion, or recurrent prediction may each favor different accelerator mixes and deployment runtimes.

Networking research supplies a third foundational piece. Edge robotics operates over Ethernet, Wi-Fi, 5G, TSN, or hybrid links, each with different latency variance and coordination affordances. Seol et al. (2021) survey time-sensitive networking and explain why deterministic Ethernet matters for bounded delay and interoperability in industrial settings. Zhang et al. (2024b) extend this logic specifically to industrial automation, where time-critical traffic from controllers, sensors, and actuators must coexist with less critical data on shared networks. In robotics, the significance of TSN is not merely lower average delay; it is the reduction of uncertainty and queueing behavior that complicate control-age guarantees. Kehl et al. (2022) demonstrate a 5G-plus-TSN prototype for edge-controlled mobile robotics, suggesting how deterministic wired and advanced wireless technologies can be integrated when robots move across heterogeneous connectivity domains. Dechouniotis et al. (2022) similarly frame edge robotics experimentation within next-generation IIoT testbeds, highlighting the need to evaluate networking, compute placement, and robot behavior together rather than separately.

Cloud and orchestration research adds a fourth element: scale. Low-latency control often begins with a single robot, but industrial value is increasingly tied to cells and fleets. Zhang et al. (2023b) propose KubeROS for automated and scalable deployment of ROS2-based multi-robot applications, while Corsaro et al. (2024) present Zenoh as a unified communication, storage, and computation fabric spanning cloud, edge, and microcontroller layers. Zhang et al. (2024a) and Chovet et al. (2025) show that middleware choices differ under Ethernet, Wi-Fi, 4G, and mesh conditions, which means architecture must anticipate scale-out and mobility. This is where edge design departs from pure embedded control. The key problem is not simply keeping one inference loop fast, but preserving latency guarantees while multiple robots, services, and data products share compute and network resources.

Finally, industrial case literature indicates that low-latency edge AI is becoming operationally relevant beyond experimental prototypes. Kapusi et al. (2022) show how deep learning can be integrated with an industrial SCARA machine for real-time object detection. Ozer et al. (2025) present a fully automated robotic stacking system with deep-learning-based visual perception. Zhang et al. (2023a) describe a multi-access edge computing framework for knowledge-sharing intelligent machine-tool swarms, while Bergs et al. (2024) demonstrate edge computing for robotic adaptivity in line-less mobile assembly systems. Mtowe et al. (2025) bring a digital twin perspective to low-latency multi-robot control, and Shata et al. (2024) connect real-time robotic part repair to 5G-cloud manufacturing. These studies do not converge on one universal architecture, but they do converge on a common principle: latency optimization is a multi-layer design issue involving control horizons, communication semantics, deployment topology, and application-specific safety margins.

3. REFERENCE ARCHITECTURES FOR LATENCY OPTIMIZATION IN EDGE ROBOTIC SYSTEMS

Table 1 shows that end-to-end robotic latency is a composite of sensing, transport, middleware, inference, control computation, and actuation delay. This motivates a layered reference architecture rather than ad hoc tuning at one point in the stack. Figure 1 presents the proposed reference architecture, it organizes edge AI robotic control into six interacting layers: physical interaction and sensing, time synchronization and deterministic data transport, middleware and message management, edge inference and runtime optimization, local control and safety supervision, and orchestration across the edge-cloud continuum. The intent is not to prescribe one vendor stack but to clarify where latency should be measured, budgeted, and governed.

Table 1. Principal latency sources in edge AI robotic control loops.

Latency source	Where it appears	Why it matters	Representative mitigation
Sensor acquisition and exposure	Camera, lidar, encoder, IO front-end	Sets the oldest possible data age; irregular sampling propagates through the loop	Hardware triggering, deterministic timestamping, lower-resolution crops
Transport and synchronization	Ethernet, Wi-Fi, 5G, TSN, clock distribution	Adds delay, jitter, and misalignment across distributed nodes	PTP/TSN, QoS classes, bounded wireless paths, clock audits
Middleware serialization and dispatch	DDS/RMW, brokers, callback executors	Queueing and wake-up behavior can dominate end-to-end timing	Executor isolation, tuned QoS, middleware selection, reduced copies
Inference runtime	CPU/GPU/NPU pipeline	Determines model response time and thermal headroom	Quantization, runtime compilation, accelerator-aware scheduling

continued on following page

Table 1. Continued

Latency source	Where it appears	Why it matters	Representative mitigation
Control and post-processing	Filtering, optimization, MPC, trajectory update	Transforms AI outputs into actuation-ready commands	Task-specific simplification, fixed-horizon solvers, local controllers
Actuation and safety gating	Robot controller, PLC, servo interface	Late or stale commands can be rejected or destabilizing	Priority channels, local safety envelopes, deadline-aware fallbacks

Figure 1. Layered edge AI reference architecture for real-time robotic control.

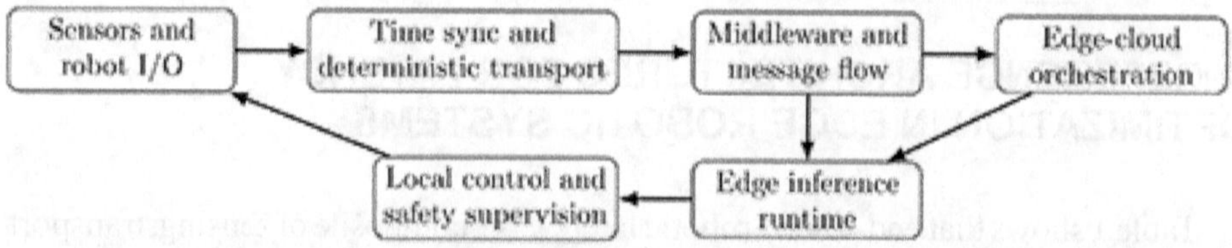

At the lowest level sit sensors, actuators, and robot controllers. This layer includes cameras, force-torque sensors, lidars, encoders, and IO devices as well as motion controllers, PLCs, and servo drives. Its design problem is acquisition determinism. A system cannot achieve bounded reaction time if exposure, timestamping, or encoder sampling are already inconsistent. For many industrial cells, the most reliable strategy is to keep high-frequency servo control on native robot controllers or hardened real-time processes while allowing AI-assisted perception and supervisory control to run on a nearby edge compute node. This preserves the most stringent loop inside the lowest-latency environment while still enabling adaptive behavior. Laurenzi et al. (2023) illustrate the importance of mixed real-time and non-real-time architectural separation in robotics middleware, and Puck et al. (2021) show why external timing validation remains useful even when internal software timing seems acceptable.

The second layer concerns time synchronization and deterministic transport. Synchronization is fundamental because distributed timing analysis is meaningless when clocks drift or event ordering cannot be trusted. In practice, IEEE 1588 Precision Time Protocol, TSN profiles, or carefully managed time bases are required when multiple hosts, controllers, and sensors contribute to the same control chain. Seol et al. (2021) and Zhang et al. (2024b) show that TSN matters in industrial automation because it allows traffic classes with bounded latency and reduced jitter to coexist with less critical flows. Kehl et al. (2022) extend this into mobile robotics by combining 5G and TSN for edge-controlled systems, demonstrating that networking architecture can materially shape robot responsiveness. The design implication is that edge AI should not be discussed independently from the communication substrate. A highly optimized model running behind a non-deterministic network remains a weak real-time solution.

The third layer is middleware and message management. Here the main functions are publish-subscribe communication, serialization, queueing, executor dispatch, and inter-process coordination. ROS 2 dominates this layer in many contemporary deployments, but the choice of RMW implementation, brokering approach, and executor topology has measurable consequences. Maruyama et al. (2016) identified these trade-offs early. Park et al. (2020), Ye et al. (2023), and Yang and Azumi (2020) show that middleware overhead and executor semantics can dominate message age, especially when callbacks are chained across nodes. Zhang et al. (2024a) compare DDS, MQTT, and Zenoh combinations across edge-to-edge and edge-to-cloud communication paths, while Chovet et al. (2025) evaluate ROS2 middlewares over mesh networks. Their combined lesson is that middleware should be selected per workload and network condition rather than by default habit. In high-frequency control chains, serialization size, discovery behavior, security overhead, and multicast assumptions can all change tail latency. For edge AI robotics, the criterion should be end-to-end chain timeliness, not merely message throughput under isolated load.

The fourth layer is edge inference and runtime optimization. This is the layer most commonly labeled "edge AI," yet it delivers value only when embedded within the others. It includes model loading, input preprocessing, accelerator scheduling, tensor runtime selection, quantization, batching policy, and thermal management. Chen and Ran (2019) explain why edge inference must account for resource limits and model partitioning, while Sze et al. (2017) show how memory movement and dataflow dominate efficiency. Applied studies by Cantero et al. (2022), Galliera and Suri (2022), Kang and Somtham (2022), and Lema et al. (2024) demonstrate that the same model can behave very differently across Jetson-class GPUs, edge TPUs, NPUs, and CPU-only targets. Jayanth et al. (2024) reinforce that heterogeneity is not a nuisance but a design opportunity: different compute units excel on different workload structures. For robotics, the correct question is therefore not "Which accelerator is fastest?" but "Which platform meets the tail-latency, power, thermal, and software-integration requirements of this control chain?"

A design mistake frequently observed in industrial proof-of-concepts is the use of throughput-oriented settings that increase average performance while worsening control relevance. Batch processing, asynchronous buffering, or oversized image pipelines may raise frames per second but inflate control-age because decisions reflect older sensor states. A perception loop serving a moving robot should generally prioritize bounded single-sample latency over aggregate throughput. This distinction is central to benchmark design and to architectural placement. In some tasks, moving image preprocessing to the camera interface or using fixed-region crops can save more relevant milliseconds than changing the model. In others, quantization or runtime compilation produces the dominant benefit. Architecture should therefore expose the whole timing path so that engineering effort is directed toward the true bottleneck.

The fifth layer is local control, coordination, and safety supervision. This layer integrates AI outputs into actionable robot behavior. It includes control-law execution, trajectory update, supervisory logic, constraint handling, and fallbacks to safe states or nominal policies. From a control perspective, this is where stale inference becomes dangerous: a good detector that arrives too late can still degrade manipulation or navigation. A sound edge architecture keeps a clear boundary between advisory AI outputs and safety-critical control primitives. For example, object localization may be produced by an edge GPU process, but final trajectory execution may remain under a controller that enforces velocity, collision, and workspace limits. This approach is evident in several applied systems, including SCARA-based object handling and automated stacking, where deep learning enhances perception but the robot still executes bounded control behaviors under deterministic local supervision (Kapusi et al., 2022; Ozer et al., 2025). It is also visible in digital-twin-based coordination, where an edge-hosted state model supports planning while local controllers preserve immediate safety envelopes (Mtowe et al., 2025).

The sixth layer is orchestration across the edge-cloud continuum. This layer handles lifecycle management, fleet analytics, model rollout, historical storage, remote monitoring, and slower optimization functions. It should not sit in the critical path of fast loops unless the network and service guarantees are strong enough. Instead, it should provide support functions such as policy distribution, trace collection, benchmark storage, digital twin synchronization, and workload reconfiguration. KubeROS (Zhang et al., 2023b) illustrates how orchestration platforms can simplify deployment of ROS2-based multi-robot applications, while Zenoh (Corsaro et al., 2024) shows one way to unify communication and computation across different layers. In industrial practice, this layer is where maintainability, multi-cell scalability, and IT/OT convergence become manageable. It is also where governance enters: version control, cyber security, model rollback, and workload admission policies determine whether a low-latency prototype can become a robust production system.

Table 2 summarizes the architecture in terms of layer functions, principal latency concerns, and typical optimization levers. The table is placed here because the remainder of the chapter repeatedly returns to these layers when discussing benchmarking and case design.

Table 2. Layered reference architecture for latency-optimized edge robotic systems.

Layer	Primary responsibility	Main latency concern	Typical artifacts
Physical sensing and actuation	Acquire environment state and execute motion	Sampling determinism and actuation delay	Cameras, encoders, PLCs, robot controllers
Time synchronization and transport	Move data with bounded uncertainty	Jitter, packet delay, synchronization skew	PTP, TSN, 5G, industrial Ethernet
Middleware and message management	Coordinate distributed software components	Serialization, queue buildup, callback dispatch	ROS 2, DDS, Zenoh, MQTT
Edge inference runtime	Run AI models under resource limits	Tail latency, warm-up, thermal throttling	Tensor runtimes, GPUs, NPUs, quantized models
Local control and safety supervision	Convert intelligence into bounded actions	Control-age, deadline misses, fallback response	Trajectory update, MPC, safety monitors
Edge-cloud orchestration	Lifecycle management and fleet services	Shared-resource contention, rollout risk	Containers, observability, model registry, digital twin services

Beyond layering, successful architectures share three cross-cutting principles. The first is hierarchical timing: fast loops stay local, slower loops may use richer edge intelligence, and cloud services remain off the hard real-time path. The second is graceful degradation: the robot should retain a safe fallback behavior if inference is delayed, stale, or unavailable. The third is observability by design: time stamps, tracing hooks, and causal flow visibility must be present from the outset. Without them, teams cannot distinguish communication delay from inference delay, nor average latency from rare but critical deadline misses.

Figure 2 deepens this architectural view by showing the end-to-end latency decomposition loop, the figure emphasizes that responsiveness is governed by both cumulative delay and data age, not only by any single hop. This matters because industrial robotics often tolerates occasional long analytical jobs, but it does not tolerate silent age inflation in the main control path.

Figure 2. End-to-end latency decomposition and control-age loop.

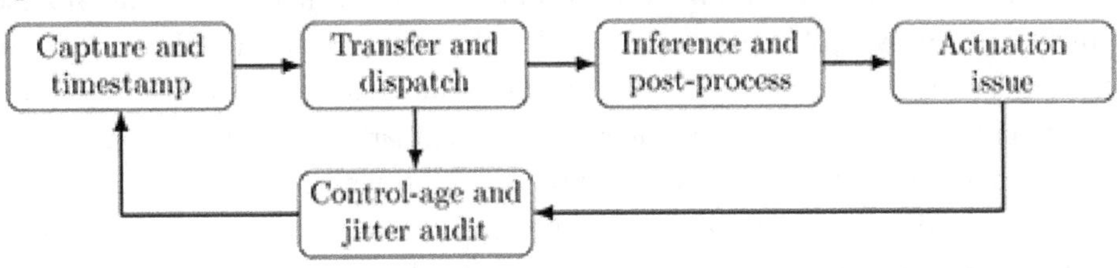

Architectural discussion also benefits from differentiating three deployment patterns. The first is on-robot edge AI, where perception and control execute on the robot or its immediate controller computer. This yields the lowest communication latency, but power, thermal, and upgrade constraints are tighter. The second is cell-edge AI, where a nearby industrial PC or edge server supports one robot or one workstation. This often offers the best trade-off for fixed industrial cells because it provides stronger compute capacity while maintaining predictable wired connectivity. The third is shared edge AI, where several robots or services rely on a common edge node. This can reduce cost and simplify management, but it requires stronger admission control and more careful tail-latency governance. Bergs et al. (2024), Zhang et al. (2023a), and Mtowe et al. (2025) all illustrate variants of this continuum in assembly, machining, and multi-robot coordination contexts.

A final architectural point concerns digital twins. In robotics discourse, digital twins are sometimes described at a strategic level, but for latency optimization their immediate value is operational: they help maintain a timely state estimate of robot, environment, and network conditions that can be used for prediction, safety, and scheduling. Mtowe et al. (2025) show how edge-enabled digital twins can support multi-robot collision avoidance and remote control. Zhang et al. (2023a) show a related idea in knowledge-sharing machine-tool swarms. In both cases, the twin is useful only if its update and synchronization latency remain appropriate for the decisions it supports. Designers should therefore treat the twin as another workload with its own timeliness budget rather than as an unlimited background service.

4. BENCHMARKING METRICS AND EVALUATION METHODOLOGY

Improving architecture systematically requires latency benchmarks that reflect real robotic behavior, not isolated software events. In edge robotics, benchmarking should therefore begin with clear control objectives and workload definitions rather than hardware datasheets. A meaningful benchmark must answer six key questions: What task is the robot performing? What timing constraints matter? Which pipeline segments are measured? Under what load and environmental conditions? What level of degradation is acceptable? And how repeatable are the results across runs, devices, and modes? Without these questions, benchmarks often produce optimistic averages that fail to predict actual deployment performance.

Figure 3 presents the proposed benchmarking workflow — from workload definition through trace instrumentation and controlled experiments to repeatable reporting that directly links system metrics to robotic outcomes. It serves as a practical template for fair architectural comparisons.

Figure 3. Benchmarking workflow for latency-aware robotic evaluation.

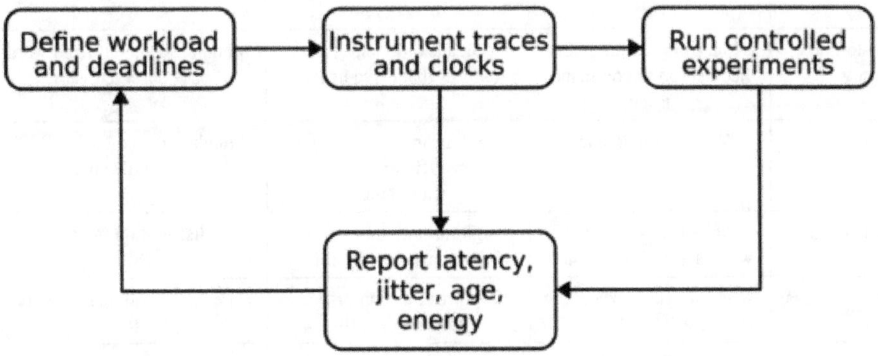

A meaningful benchmark usually requires at least three classes of metrics. The first class covers timing metrics internal to the control chain. These include sensing latency, transport delay, executor wake-up latency, inference time, post-processing time, control computation time, and actuation dispatch time. Puck et al. (2021), Kronauer et al. (2021), and Bédard et al. (2022) show that distributed tracing and external synchronization are useful for exposing these components. The second class covers end-to-end control metrics. These include reaction time, response time, data age or control-age, deadline miss ratio, and jitter percentiles. Real-time systems research increasingly recognizes that average latency is a weak descriptor because the worst system behavior is often governed by tail events and age accumulation rather than by mean values alone. The third class covers platform sustainability metrics: energy per inference, thermal stability, memory footprint, and CPU/GPU/NPU utilization. These matter because embedded systems that initially meet timing budgets may later violate them due to thermal throttling or resource contention.

Table 3 synthesizes a benchmarking metric set appropriate for edge AI robotic control. The intention is to give readers a compact vocabulary for comparing papers, prototypes, and industrial deployments using criteria that capture both speed and robustness.

Table 3. Benchmark metrics for real-time edge AI robotic control.

Metric	Definition	Why averages are insufficient	Recommended reporting
End-to-end latency	Time from measurement acquisition to command issue	Means hide long-tail failures that drive unsafe behavior	Median, 95th, 99th, and worst-case observed values
Jitter	Variation in latency across samples	Equal means can still yield very different control smoothness	Standard deviation plus high-percentile spread
Control-age	Age of the data used when action is computed	High throughput can still produce stale decisions	Age distribution by task mode
Deadline miss ratio	Share of cycles exceeding a task deadline	Captures mission-critical failures directly	Per workload, per network mode, per thermal phase
Energy and thermal stability	Power use and heat behavior under sustained load	Fast devices may throttle and lose timing later	Power traces, temperature traces, long-run latency drift
Task-level outcome	Operational effect of timing quality	Infrastructure metrics alone do not show industrial value	Cycle time, grasp error, path deviation, defect-detection timeliness

Several metrics deserve special emphasis. The first is jitter, which is the variability around latency rather than its nominal value. In robotic control, two systems with the same average latency can behave very differently if one has tightly bounded jitter and the other exhibits occasional spikes. Jitter matters for servo smoothness, synchronization with exteroceptive sensing, and coordinated multi-robot behavior. Ye et al. (2023) and Park et al. (2020) both show that ROS 2 timing quality depends strongly on kernel and middleware configuration, which often appears more clearly in jitter and deadline-miss distributions than in average delay.

The second is control-age, sometimes described as the age of information used for action. A perception result may be computed quickly, but if the camera frame was queued, copied, and post-processed for too long before inference, the control command can still be based on stale reality. This is especially relevant in machine vision tasks involving moving conveyors, mobile bases, or collaborative workspaces with humans. Throughput-optimized pipelines can inadvertently increase control-age when batching, buffering, or asynchronous handoff grows faster than immediate decision quality. For this reason, benchmarks should record both inference latency and end-to-end age from measurement acquisition to actuation issue.

The third is percentile reporting. Tail latency at the 95th, 99th, or 99.9th percentile is often more informative than the mean when control deadlines matter. Industrial systems frequently operate under variable backgrounds, thermal conditions, and message contention; a design that fails only one percent of the time may still be unacceptable if the missed deadlines coincide with high-speed motion or safety-sensitive interactions. Jayanth et al. (2024), Cantero et al. (2022), and Lema et al.

(2024) all reinforce the need for benchmark transparency about workload structure and resource contention because different devices may exchange average performance for very different tail behaviors.

The fourth is synchronization accuracy. Distributed robotic cells often blend camera streams, force signals, controller states, and middleware traces from multiple hosts. Without consistent clock alignment, measured latency segments are difficult to interpret. Puck et al. (2021) show the value of time-synchronized distributed networks in robotic control evaluation, while Kehl et al. (2022) highlight the role of synchronized industrial networking in edge-controlled mobile robotics. A benchmarking framework should therefore report on how clocks are synchronized and what timestamp resolution is available.

Experimental design is as important as metric choice. A robust benchmark should specify the robotic workload, input rate, model version, precision format, runtime, thermal state, and network conditions. It should also disclose whether the benchmark is run in isolation or alongside background processes such as mapping, logging, or orchestration services. Inference timing measured on a bare device can be useful, but it is not a substitute for application-level measurement in a realistic execution graph. This is why Bédard et al. (2023) and Li et al. (2022) are so useful: they help connect low-level measurements to actual message flows and node chains. In industrial evaluation, synthetic microbenchmarks and application-level benchmarks should be treated as complementary. The former help isolate runtime or accelerator behavior; the latter determine whether the whole system meets operational objectives.

Networking and middleware benchmarks require similar care. It is common to report mean message latency under ideal Ethernet settings, but industrial systems may rely on Wi-Fi, 5G, mesh, or mixed topologies with varying payload sizes and discovery behavior. Zhang et al. (2024a) and Chovet et al. (2025) demonstrate why middleware comparisons should be run across multiple networks and application patterns. Results that are strong under Ethernet may reverse under Wi-Fi or mesh routing. Likewise, enabling security features can impose latency and CPU penalties that should be measured rather than assumed. For industrial deployments, a realistic benchmark should therefore include network modes, security configurations, and loss or contention scenarios that resemble the intended operating environment.

Table 4 translates this benchmarking logic into optimization levers by stack layer. The table links common bottlenecks to interventions such as quantization, callback isolation, TSN scheduling, runtime pinning, image-pipeline reduction, and edge placement changes. It is positioned here because benchmarking is valuable only if it leads to actionable design changes.

Fair comparison also requires workload taxonomies. For this chapter, four workload categories are especially useful. The first is perception-dominant workloads, such as object detection, pose estimation, or defect segmentation, where inference and image transfer dominate. The second is middleware-dominant workloads, such as multi-node coordination or multi-robot message exchange, where serialization and scheduling dominate. The third is control-dominant workloads, such as model-predictive control or high-frequency servo adaptation, where computation and synchronization must be tightly bounded. The fourth is orchestration-dominant workloads, such as fleet coordination and edge-offloading policies, where local control remains real-time but shared infrastructure influences service quality. A good benchmark reports which category it belongs to and avoids drawing broad conclusions from only one type.

Table 4. Optimization levers across the edge AI robotic stack.

Stack layer	Typical bottleneck	Optimization lever	Validation signal
Image and sensor pipeline	Oversized frames or copies	Cropping, lower bit depth, zero-copy paths	Lower control-age without loss of task success
Middleware and executors	Callback contention and queueing	Priority-aware scheduling, node isolation, tuned QoS	Reduced 99th percentile latency and miss ratio
Inference runtime	Large models, poor accelerator mapping	Quantization, pruning, TensorRT/NPU compilation	Stable single-sample latency and thermal headroom
Networking	Contention or roaming variability	TSN classes, wired offload, 5G slicing, local autonomy	Lower jitter and handover-safe behavior
Control computation	Heavy post-processing or solver load	Simplified filters, bounded-horizon optimization	Bounded compute time within loop budget
Deployment and lifecycle	Shared-resource interference	CPU pinning, container quotas, staged rollouts	Repeatable timing under sustained mixed workloads

Benchmarking should end with task-level relevance, not only platform-level statistics. In vision-guided manipulation, the endpoint may be successful grasp rate under moving inputs, cycle time, or pick error as a function of perception age. In mobile robotics, it may be path-tracking error, collision-free replanning success, or stop distance under delayed updates. In machine-tool or repair applications, it may be tracking precision, defect-detection timeliness, or rework reduction. This is where industrial case studies are valuable: they show how seemingly abstract timing metrics influence outcomes that operations teams actually care about. It is also where benchmark honesty matters most. A chapter or paper that reports only hardware speed without control impact is incomplete for real-time robotics.

Finally, benchmark reporting should be reproducible. Researchers and practitioners should disclose software versions, model variants, operating-system parameters, tracing methods, thermal conditions, and workload traces. Where possible, benchmark suites should include both median and percentile metrics, warm-up behavior, and long-run stability under sustained load. Thermal and energy metrics are particularly important for edge devices deployed in enclosures or near industrial heat sources. Many industrial disappointments occur not because a design failed immediately, but because it degraded after fifteen or thirty minutes of sustained operation. Latency optimization in edge AI is therefore inseparable from thermal design, deployment context, and runtime discipline.

5. INDUSTRIAL CASE STUDIES

Architecture and benchmarking become most useful when translated into industrial scenarios. This section therefore presents three literature-grounded case studies that illustrate how latency optimization changes with task structure. The aim is not to claim one universal best architecture, but to show how the chapter's framework can be adapted to different control horizons, sensing modalities, and operational constraints. Table 5 summarizes the cases before they are discussed in detail.

Table 5. Literature-grounded industrial case studies for latency-aware edge robotics.

Case	Operational setting	Dominant latency challenge	Architecture emphasis	Key evaluation outcome
Vision-guided handling	SCARA pick-and-place and automated stacking	Perception age inside short cycle times	Dedicated cell-edge inference with bounded handoff to controller	Cycle time, pick success, tail-latency stability
Adaptive machine tending	Inspection, repair, and line-less assembly	Coordination across perception, machine state, and planning	Hybrid edge node with local safety loops and shared analytics	Inspection timeliness, handoff delay, rework reduction
Mobile robot coordination	AMRs or mobile manipulators over heterogeneous networks	Variable network quality and distributed state freshness	Layered autonomy with edge-assisted coordination and local fallbacks	Path error, continuity under handover, collision-free replanning

The first case concerns vision-guided handling in a high-throughput workcell, represented by industrial SCARA pick-and-place and stacking applications. Kapusi et al. (2022) show how deep learning can be deployed with an industrial SCARA machine for real-time object detection, while Ozer et al. (2025) demonstrate a fully automated stacking system with deep-learning-based perception. In these settings, the decisive latency challenge is that visual inference must be synchronized with fast and repeated manipulation cycles. The robot can often execute motion deterministically once a pose or class decision is available, but the overall cycle time suffers if image capture, transfer, and inference are too slow or too variable. A practical architecture for this class of application keeps camera ingestion and inference on a nearby cell-edge computer, uses fixed image regions where possible, avoids large asynchronous buffers, and feeds only the minimal decision variables back to the motion controller. Benchmarking should emphasize single-frame control-age, 95th and 99th percentile inference latency, pick success rate under varying part flow, and long-run thermal stability of the edge device. In this class of work, a seemingly modest latency reduction can materially increase cell throughput because the effect compounds over every cycle.

The second case concerns adaptive machine tending, line-less assembly, and intelligent machining environments. Bergs et al. (2024) show how edge computing can support robotic adaptivity in line-less mobile assembly systems, and Zhang et al. (2023a) propose a multi-access edge computing framework for knowledge-sharing intelligent machine-tool swarms. Shata et al. (2024) add a computer-vision-based robotic part-repair scenario in advanced manufacturing. These applications differ from the first case because they often combine several slower but more heterogeneous loops: vision, task planning, quality inspection, and machine-state integration. Here the key latency problem is not only fast perception but also bounded coordination among digital twins, machine controllers, and knowledge services. Edge resources are valuable because they enable local reasoning near the production equipment while still supporting data sharing and digital twin synchronization. A strong design pattern is to keep immediate machine and robot safety loops local, place perception and analytics on a shared cell-edge node, and reserve cloud or central services for model updates, longer-horizon optimization, and historical knowledge accumulation. Benchmarking should therefore combine end-to-end loop latency with machine-quality outcomes such as inspection timeliness, handoff delay, repair-cycle variability, or tool-state update freshness. Tail latency and synchronization accuracy are especially important because these systems often rely on several data-producing entities that must agree on state within bounded time.

The third case concerns mobile robots and distributed multi-robot coordination. This class includes edge-controlled mobile robotics over 5G and TSN (Kehl et al., 2022), digital-twin-based multi-robot coordination (Mtowe et al., 2025), edge robotics experimentation in IIoT environments (Dechouniotis et al., 2022), and middleware studies for mesh-connected multi-robot systems (Chovet et al., 2025). The fundamental issue here is that motion, network quality, and compute availability vary together. Unlike fixed cells, robots move through changing radio conditions and may require coordination across several edge nodes or network segments. As a result, optimization is less about minimizing a single stationary pipeline and more about preserving bounded control-age under varying conditions. A useful architecture combines local autonomy for fast safety-critical reactions with edge-assisted coordination for path planning, shared state, or fleet optimization. Kehl et al. (2022) show how integrated networking can support this, while Mtowe et al. (2025) show the role of edge-enabled digital twins in avoiding collision and supporting remote control. Benchmarking in this class should include not only message and inference latency but also handover effects, path-tracking deviation under network variation, synchronization drift across nodes, and resilience when one communication mode degrades. In other words, the metric of interest is often continuity of acceptable behavior rather than raw minimum latency.

These cases highlight a broader point: industrial latency budgets are nested. A robot may tolerate one latency budget for servo control, another for perception-assisted correction, another for local coordination, and a fourth for cloud or fleet-level adaptation. The architecture must therefore prevent slower layers from silently contaminating faster ones. This is especially important in software stacks that make asynchronous composition easy. Queue buildup, discovery traffic, logging, or orchestration calls can slowly leak into critical paths unless resource isolation is deliberate. Here middleware and observability matter as much as model compression. Studies on ROS 2 tracing, callback scheduling, and middleware comparison are therefore not peripheral to industrial AI; they are central to whether industrial AI behaves acceptably under real workloads (Bédard et al., 2022; Choi et al., 2021; Zhang et al., 2024a).

The cases also show why benchmarking must align with economic context. In a pick-and-place cell, milliseconds can accumulate directly into cycle-time gains. In machine tending, more value may come from avoiding quality-related latency spikes than from lowering the mean. In mobile robotics, robustness to variable connectivity can matter more than absolute minimum delay. These differences should change what is optimized. A benchmarking program that treats all robotic tasks as equivalent will often reward the wrong platform or the wrong architectural intervention.

Another cross-case lesson concerns shared versus dedicated edge resources. Dedicated cell-edge resources often simplify latency governance and are well suited to fixed workcells with stable workloads. Shared edge resources can be attractive in mobile robotics or digital-twin-heavy systems because they enable coordination and centralized maintenance, but they introduce admission-control and contention problems. KubeROS (Zhang et al., 2023b) is relevant here because deployment automation becomes a timing issue once multiple robotic services share compute resources. Likewise, Zenoh (Corsaro et al., 2024) becomes attractive when a deployment needs communication flexibility across robot, edge, and cloud layers without forcing one transport model onto every workload.

The case studies thus support a balanced conclusion. Edge AI is most effective in industrial robotics when it is deployed close enough to preserve the relevant timing envelope, but abstracted enough to remain maintainable, observable, and updatable. That balance varies by case. High-speed cells favor tightly bounded local designs. Flexible manufacturing and machine-tool ecosystems favor hybrid edge architectures that keep hard loops local while sharing knowledge and analytics at the edge. Mobile robots favor layered autonomy in which local behaviors absorb short-term variability and edge services enhance coordination and situational intelligence whenever communication quality allows.

6. IMPLEMENTATION CHALLENGES, GOVERNANCE, AND DESIGN TRADE-OFFS

Even well-designed edge AI architectures confront recurring implementation barriers. The first is architectural over-centralization. Because edge servers often provide more compute than on-robot hardware, teams are tempted to move too much functionality away from the robot. This can create a brittle dependency on transport quality and service availability. The literature on edge computing consistently warns that latency reduction is only one advantage of the edge; availability, context proximity, and graceful degradation are equally important (Satyanarayanan, 2017; Tahir et al., 2025). For robotic control, this means that hard real-time or safety-critical primitives should not be casually externalized merely because an edge server is faster.

The second barrier is opaque software timing. Industrial teams often know the nominal frequency of each node or controller, but not the causal timing of the whole chain. Without tracing and message-flow analysis, optimization efforts can focus on the wrong layer. Bédard et al. (2022, 2023) and Li et al. (2022) show that low-overhead tracing is feasible even in demanding ROS 2 systems. Yet many deployments still instrument only inference runtimes or network ping measurements. The result is misleading: ping latency may be low while executor wake-up dominates end-to-end delay, or model inference may be fast while image preprocessing silently expands control-age. Architecture review should therefore mandate end-to-end observability as a production requirement, not as an optional debugging convenience.

The third barrier is benchmarking inconsistency. Edge AI papers often report different data sizes, batch modes, warm-up rules, clocks, and environmental conditions, making cross-study comparison difficult. This problem is visible in the embedded object-detection literature, where device comparisons can reverse depending on precision format, framework conversion, or image size (Cantero et al., 2022; Galliera & Suri, 2022; Kang & Somtham, 2022). Industrial teams face the same issue internally when procurement, controls, and AI groups each benchmark systems differently. A standardized benchmark protocol tied to control-age, percentile latency, and task-level outcomes is therefore a governance asset as much as a research method.

A fourth challenge is the conflict between flexibility and determinism. Platforms such as ROS 2, Kubernetes-based orchestration, and heterogeneous message fabrics create enormous modularity and deployment agility, but they can also add variability if resources are not isolated carefully. KubeROS (Zhang et al., 2023b) and middleware comparisons by Zhang et al. (2024a) and Chovet et al. (2025) show the value of scalable orchestration and configurable communication, yet they also imply a governance burden: workload placement, container overhead, security settings, and middleware discovery must be evaluated with timing objectives in mind. In industrial environments, elasticity is valuable only if it does not undermine bounded behavior in critical paths.

A fifth barrier concerns networking realism. Many robotic prototypes assume ideal Ethernet or stable Wi-Fi, while factories and warehouses present interference, roaming, variable uplink conditions, and mixed criticality. Seol et al. (2021), Zhang et al. (2024b), and Kehl et al. (2022) collectively show that deterministic or semi-deterministic networking is central to trustworthy robotic timing. Yet full TSN deployment can be expensive and operationally complex, and wireless segments often remain unavoidable for mobile platforms. The solution is not to avoid wireless robotics, but to design latency-aware fallback strategies. Local autonomy, deadline-aware message priorities, and bounded degradation modes are essential when network quality fluctuates.

A sixth challenge is thermal and energy behavior. Edge accelerators are attractive precisely because they offer substantial local inference capacity, but industrial deployment environments may include heat, dust, enclosure limits, or sustained duty cycles that alter device behavior. Jayanth et al. (2024) and Lema et al. (2024) remind us that heterogeneous platforms differ not only in peak speed but in how they trade power, thermal load, and latency. Long-duration benchmarks are therefore necessary before industrial deployment claims are made. A device that meets a 20 ms inference target for five minutes but slips to 45 ms after thermal saturation is not a valid real-time platform for sustained production use.

Security and maintainability form a seventh challenge. Macenski et al. (2022) note that ROS 2 adoption in production environments is closely tied to operational robustness, and that in turn requires disciplined lifecycle management. Edge AI adds model versioning, runtime dependencies, accelerator drivers, and sometimes remote orchestration platforms to the maintenance burden. Cybersecurity hardening may add overhead or complicate discovery and communication paths. The correct response is architectural separation and staged hardening: keep critical loops minimal, isolate noncritical services, and validate the latency cost of security measures rather than assuming it away.

To manage these challenges, organizations need governance loops rather than one-time optimization campaigns. Figure 4 therefore presents a continuous optimization and governance cycle, it connects measurement, diagnosis, architecture change, redeployment, and re-benchmarking, emphasizing that latency optimization is a recurring operational discipline.

Figure 4. Continuous optimization and governance loop for industrial edge AI robotics.

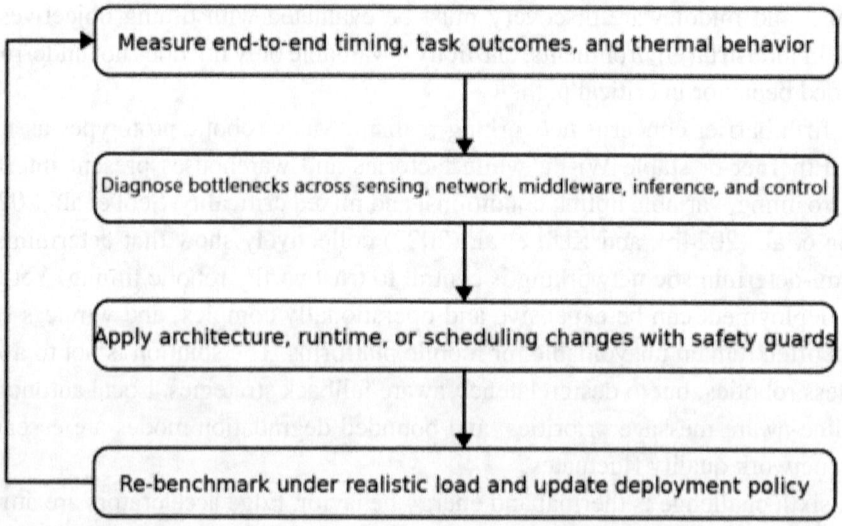

Table 6 complements this figure with a deployment maturity roadmap. The roadmap is not meant as a rigid sequence, but as a practical guide for organizations progressing from isolated prototypes to repeatable industrial edge AI operations.

Table 6. Deployment maturity roadmap for industrial edge AI robotic control.

Phase	Primary objective	Representative activities	Evidence of readiness
1. Visibility baseline	Measure the existing loop	Trace instrumentation, clock audit, baseline task metrics	Trusted latency decomposition for the current system
2. Local optimization	Stabilize single-robot or single-cell timing	Model compression, callback isolation, network cleanup	Lower tail latency and fewer missed deadlines
3. Architectural hardening	Separate critical and noncritical services	Fallback policies, safety gating, deterministic transport	Predictable degradation under induced faults
4. Shared edge integration	Support multiple services or robots safely	Admission control, orchestration, resource quotas	Stable timing under mixed workloads
5. Continuous governance	Institutionalize benchmarking and change control	Regression benchmarks, staged rollouts, thermal audits	Repeatable production performance across updates

7. PRACTICAL DESIGN GUIDELINES

The chapter's analysis suggests a set of practical design guidelines. First, start from timing budgets, not platforms. Teams should define the fastest loop that matters, the maximum tolerable data age, and the acceptable tail-latency envelope before evaluating hardware. This forces the design to remain task-driven and prevents the common error of purchasing an impressive accelerator that does not fit the control problem.

Second, separate hard real-time, soft real-time, and supervisory functions explicitly. Fast servo loops, safety monitors, and emergency behavior should remain in the most deterministic environment available. AI-driven perception, quality estimation, or local planning can sit at the cell edge if their timing is bounded and their outputs are mediated by safe control primitives. Cloud or central services should remain outside hard control paths unless the communication guarantees are exceptionally strong.

Third, instrument the full chain from the beginning. Tracing, causal message-flow analysis, synchronized timestamps, and thermal monitoring should be built into the development workflow. Bédard et al. (2022, 2023) and Puck et al. (2021) show why chain-level visibility matters; without it, teams cannot optimize responsibly.

Fourth, benchmark in the mode in which the system will actually operate. If the robot will process one frame at a time, then single-sample latency is the primary benchmark mode. If the system will run in a hot enclosure, then long-duration thermal benchmarking is mandatory. If the deployment will use Wi-Fi or mesh networks, benchmark those networks rather than only Ethernet. Synthetic laboratory benchmarks are useful, but only when paired with application-realistic scenarios.

Fifth, choose middleware and communication mechanisms according to work-load structure, not convention. DDS, Zenoh, MQTT, TSN-enabled Ethernet, and 5G-enabled paths each bring different trade-offs. Studies by Zhang et al. (2024a), Chovet et al. (2025), Corsaro et al. (2024), and Kehl et al. (2022) show that there is no universal winner across all scenarios.

Sixth, treat graceful degradation as part of latency optimization. A system that occasionally misses deadlines but degrades predictably can be more valuable than one that is faster on average but fails opaquely. Fallback policies, safe local autonomy, and model confidence gating should therefore be designed alongside performance tuning.

Seventh, align optimization with business outcomes. In some settings, the main benefit of lower latency is higher throughput; in others it is safer coordination, fewer quality escapes, or less downtime. Benchmarking programs should therefore track task-level effects such as cycle time, reject rate, tracking error, or fleet continuity rather than only infrastructure statistics.

8. FUTURE RESEARCH DIRECTIONS

Future research should push edge AI robotics in three directions. The first is benchmark standardization around control-age, percentile latency, and long-duration stability rather than average inference time alone. The second is cross-layer co-design linking model architecture, middleware scheduling, network determinism, and fallback control more explicitly; today these are too often optimized separately. The third is broader industrial evidence. More comparative studies are needed across fixed workcells, mobile robotics, and digital-twin-heavy manufacturing environments so that architectural recommendations can be tied to repeatable deployment outcomes rather than to isolated prototypes. Research on orchestration, thermal behavior, and secure real-time middleware will be especially important as multi-robot and edge-cloud deployments scale.

9. CONCLUSION

Latency optimization of edge AI-driven robotics can be viewed as a problem of full-stack control architecture. The key takeaway of this chapter is that there is no accelerator, or middleware, that can generate low latency, but rather the interaction of sensing, synchronization, communication, inference, control, and governance. When the layers are aligned, edge AI has the capability to enhance responsiveness, durability, and fleet industrial utility across robotic cells and fleets. With the absence of this, AI is another element of randomness within the loop. The chapter provides a practical approach to the evaluation and enhancement of real-time robotic systems by integrating a layered reference architecture, end-to-end timing relevance-based benchmarking framework, and literature-based industrial case studies. Where Edge AI will have the most impact will be where it is not used as an isolated feature, but as a disciplined timing-aware operating capability of industrial robotics.

REFERENCES

Bédard, C., Lajoie, P.-Y., Beltrame, G., & Dagenais, M. (2023). Message flow analysis with complex causal links for distributed ROS 2 systems. *Robotics and Autonomous Systems, 161*, 104361. DOI: 10.1016/j.robot.2022.104361

Bédard, C., Lütkebohle, I., & Dagenais, M. (2022). ros2_tracing: Multipurpose low-overhead framework for real-time tracing of ROS 2. *IEEE Robotics and Automation Letters, 7*(3), 6511–6518. DOI: 10.1109/LRA.2022.3174346

Bergs, L., Tréheux, G., Schäper, L., Pinheiro de Souza, L. M., Münker, S., Göppert, A., & Schmitt, R. H. (2024). Edge computing framework for enhanced robotic adaptivity in line-less mobile assembly systems. *Procedia CIRP, 129*, 26–31. DOI: 10.1016/j.procir.2024.07.006

Cantero, D., Esnaola-Gonzalez, I., Miguel-Alonso, J., & Jauregi, E. (2022). Benchmarking object detection deep learning models in embedded devices. *Sensors (Basel), 22*(11), 4205. DOI: 10.3390/s22114205

Chen, J., & Ran, X. (2019). Deep learning with edge computing: A review. *Proceedings of the IEEE, 107*(8), 1655–1674. DOI: 10.1109/JPROC.2019.2921977

Choi, H., Xiang, Y., & Kim, H. (2021). PiCAS: New design of priority-driven chain-aware scheduling for ROS2. In 2021 IEEE 27th Real-Time and Embedded Technology and Applications Symposium (RTAS) (pp. 251-263). DOI: 10.1109/RTAS52030.2021.00028

Chovet, L. P., Garcia, G. M., Bera, A., Richard, A., Yoshida, K., & Olivares-Mendez, M. A. (2025). Performance comparison of ROS2 middlewares for multi-robot mesh networks in planetary exploration. *Journal of Intelligent & Robotic Systems, 111*(1), 18. DOI: 10.1007/s10846-024-02211-2

Corsaro, A., Cominardi, L., Hecart, O., Baldoni, G., Avital, J. E. P., Loudet, J., Guimares, C., Ilyin, M., & Bannov, D. (2024). Zenoh: Unifying communication, storage and computation from the cloud to the microcontroller. In 2023 26th Euromicro Conference on Digital System Design (pp. 422-428). DOI: 10.1109/DSD60849.2023.00065

Dechouniotis, D., Spatharakis, D., & Papavassiliou, S. (2022). Edge robotics experimentation over next generation IIoT testbeds. In *2022 IEEE/IFIP Network Operations and Management Symposium* (pp. 1-3). DOI: 10.1109/NOMS54207.2022.9789889

Galliera, R., & Suri, N. (2022). Object detection at the edge: Off-the-shelf deep learning capable devices and accelerators. *Procedia Computer Science, 205*, 239–248. DOI: 10.1016/j.procs.2022.09.025

Gautam, M. K., Pati, A., Mishra, S. K., Appasani, B., Kabalci, E., Bizon, N., & Thounthong, P. (2021). A comprehensive review of the evolution of networked control system technology and its future potentials. *Sustainability (Basel), 13*(5), 2962. DOI: 10.3390/su13052962

Jayanth, R., Gupta, N., & Prasanna, V. (2024). Benchmarking edge AI platforms for high-performance ML inference. In *2024 IEEE High Performance Extreme Computing Conference*. DOI: 10.1109/HPEC62836.2024.10938499

Kang, P., & Somtham, A. (2022). An evaluation of modern accelerator-based edge devices for object detection applications. *Mathematics, 10*(22), 4299. DOI: 10.3390/math10224299

Kapusi, T. P., Erdei, T. I., Husi, G., & Hajdu, A. (2022). Application of deep learning in the deployment of an industrial SCARA machine for real-time object detection. *Robotics (Basel, Switzerland), 11*(4), 69. DOI: 10.3390/robotics11040069

Kehl, P., Ansari, J., Jafari, M. H., Becker, P., Sachs, J., König, N., Göppert, A., & Schmitt, R. H. (2022). Prototype of 5G integrated with TSN for edge-controlled mobile robotics. *Electronics (Basel), 11*(11), 1666. DOI: 10.3390/electronics11111666

Kronauer, T., Pohlmann, J., Matthé, M., Smejkal, T., & Fettweis, G. (2021). Latency analysis of ROS2 multi-node systems. In *2021 IEEE International Conference on Multisensor Fusion and Integration for Intelligent Systems (MFI)* (pp. 1-7). DOI: 10.1109/MFI52462.2021.9591166

Laurenzi, A., Antonucci, D., Tsagarakis, N. G., & Muratore, L. (2023). The XBot2 real-time middleware for robotics. *Robotics and Autonomous Systems, 163*, 104379. DOI: 10.1016/j.robot.2023.104379

Lema, D. G., Usamentiaga, R., & García, D. F. (2024). Quantitative comparison and performance evaluation of deep learning-based object detection models on edge computing devices. *Integration (Amsterdam), 95*, 102127. DOI: 10.1016/j.vlsi.2023.102127

Li, Z., Hasegawa, A., & Azumi, T. (2022). Autoware.perf: A tracing and performance analysis framework for ROS 2 applications. *Journal of Systems Architecture, 123*, 102341. DOI: 10.1016/j.sysarc.2021.102341

Macenski, S., Foote, T., Gerkey, B., Lalancette, C., & Woodall, W. (2022). Robot Operating System 2: Design, architecture, and uses in the wild. *Science Robotics, 7*(66), eabm6074. Advance online publication. DOI: 10.1126/scirobotics.abm6074

Maruyama, Y., Kato, S., & Azumi, T. (2016). Exploring the performance of ROS2. In *Proceedings of the 13th International Conference on Embedded Software* (pp. 1-10). DOI: 10.1145/2968478.2968502

Mtowe, D. P., Long, L., & Kim, D. M. (2025). Low-latency edge-enabled digital twin system for multi-robot collision avoidance and remote control. *Sensors (Basel), 25*(15), 4666. DOI: 10.3390/s25154666

Murshed, M. G. S., Murphy, C., Hou, D., Khan, N., Ananthanarayanan, G., & Hussain, F. (2021). Machine learning at the network edge: A survey. *ACM Computing Surveys, 54*(8), 170. Advance online publication. DOI: 10.1145/3469029

Ozer, A. S., & Cinar, A. (2025). Real-time and fully automated robotic stacking system with deep learning-based visual perception. *Sensors (Basel), 25*(22), 6960. DOI: 10.3390/s25226960

Park, J., Delgado, R., & Choi, B. W. (2020). Real-time characteristics of ROS 2.0 in multiagent robot systems: An empirical study. *IEEE Access : Practical Innovations, Open Solutions, 8*, 154637–154651. DOI: 10.1109/ACCESS.2020.3018122

Puck, L., Keller, P., Schnell, T., Plasberg, C., Tanev, A., Heppner, G., Rönnau, A., & Dillmann, R. (2021). Performance evaluation of real-time ROS2 robotic control in a time-synchronized distributed network. In 2021 IEEE 17th International Conference on Automation Science and Engineering (CASE) (pp. 1670-1676). DOI: 10.1109/CASE49439.2021.9551447

Satyanarayanan, M. (2017). The emergence of edge computing. *Computer, 50*(1), 30–39. DOI: 10.1109/MC.2017.9

Seol, Y., Hyeon, D., Min, J., Kim, Y., Kim, J., Lee, S., & Kim, H. (2021). Timely survey of time-sensitive networking: Past and future directions. *IEEE Access : Practical Innovations, Open Solutions, 9*, 142506–142527. DOI: 10.1109/ACCESS.2021.3120769

Shata, E., Chen, B., Hu, L., Seskar, I., Guo, Y., Mahmoudi, C., Shekhar, S., & Zou, Q. (2024). 5G-cloud-based real-time robotic part repairing for advanced manufacturing via computer vision. *Manufacturing Letters, 41*, 1398–1404. DOI: 10.1016/j.mfglet.2024.09.166

Shi, W., Cao, J., Zhang, Q., Li, Y., & Xu, L. (2016). Edge computing: Vision and challenges. *IEEE Internet of Things Journal, 3*(5), 637–646. DOI: 10.1109/JIOT.2016.2579198

Sze, V., Chen, Y.-H., Yang, T.-J., & Emer, J. S. (2017). Efficient processing of deep neural networks: A tutorial and survey. *Proceedings of the IEEE, 105*(12), 2295–2329. DOI: 10.1109/JPROC.2017.2761740

Tahir, N.. (2025). Edge computing and its application in robotics: A survey. *Robotics (Basel, Switzerland), 14*(4), 65. DOI: 10.3390/robotics14040065

Tipsuwan, Y., & Chow, M.-Y. (2003). Control methodologies in networked control systems. *Control Engineering Practice, 11*(10), 1099–1111. DOI: 10.1016/S0967-0661(03)00036-4

Wang, T., Guo, J., Zhang, B., Yang, G., & Li, D. (2025). Deploying AI on edge: Advancement and challenges in edge intelligence. *Mathematics, 13*(11), 1878. DOI: 10.3390/math13111878

Yang, Y., & Azumi, T. (2020). Exploring real-time executor on ROS 2. In *2020 IEEE International Conference on Embedded Software and Systems (ICESS)* (pp. 1-8). DOI: 10.1109/ICESS49830.2020.9301530

Zhang, C., Zhou, G., Li, J., Chang, F., Ding, K., & Ma, D. (2023a). A multi-access edge computing enabled framework for the construction of a knowledge-sharing intelligent machine tool swarm in Industry 4.0. *Journal of Manufacturing Systems, 66*, 56–70. DOI: 10.1016/j.jmsy.2022.11.015

Zhang, J., Yu, X., Ha, S., Queralta, J. P., & Westerlund, T. (2024a). Comparison of middlewares in edge-to-edge and edge-to-cloud communication for distributed ROS 2 systems. *Journal of Intelligent & Robotic Systems, 110*(4), 162. DOI: 10.1007/s10846-024-02187-z

Zhang, T., Wang, G., Xue, C., Wang, J., Nixon, M., & Han, S. (2024b). Time-sensitive networking (TSN) for industrial automation: Current advances and future directions. *ACM Computing Surveys*. Advance online publication. DOI: 10.1145/3695248

Zhang, Y., Wurll, C., & Hein, B. (2023b). KubeROS: A unified platform for automated and scalable deployment of ROS2-based multi-robot applications. In *2023 IEEE International Conference on Robotics and Automation (ICRA)* (pp. 9097-9103). DOI: 10.1109/ICRA48891.2023.10160632

Chapter 2
Edge AI Swarm Robotics for Real-Time Supply Chain Intelligence

Mohammad Al Khaldy
http://orcid.org/0009-0009-7502-4668
University of Petra, Amman, Jordan

Aquil Mirza Mohammed
http://orcid.org/0000-0001-7756-4363
The Hong Kong Polytechnic University, Hong Kong

Tsz Kin Wong
http://orcid.org/0009-0002-6862-834X
The Hong Kong Polytechnic University, Hong Kong

ABSTRACT

Edge AI swarm robotics is reshaping supply chain business intelligence by moving perception, coordination, and decision support closer to operational events. This chapter examines how decentralized robot collectives, embedded sensing, and edge analytics can create real-time visibility for warehouses, fulfillment centers, ports, and cold-chain networks. It synthesizes literature on edge computing, swarm robotics, digital twins, and supply chain analytics to develop an architectural and managerial framework for low-latency intelligence. The discussion integrates benchmarking metrics, governance requirements, and literature-grounded industrial case analyses. Attention is given to latency budgets, model deployment patterns, task allocation, resilience, cybersecurity, and human oversight. The chapter concludes with imple-

DOI: 10.4018/979-8-2600-0701-3.ch002

mentation recommendations and a future research agenda for trustworthy, scalable, and economically viable swarm-enabled intelligence in contemporary global multi-echelon supply networks under uncertainty.

1. INTRODUCTION

Supply chains today increasingly compete on how quickly they can sense, interpret, and act on operational events. Traditional business intelligence systems were built for periodic reporting, managerial dashboards, and historical trend analysis. But the reality of modern logistics has become far more immediate. Whether it's e-commerce fulfillment, same-day replenishment, cold-chain monitoring, cross-docking, or autonomous warehouse operations, decisions quickly lose their value if they're separated from the live event stream by even a few minutes or hours. Although data still accumulates in central ERP platforms, warehouse management systems, and cloud analytics stacks, real action is now often required right at the aisle, dock, pallet, tote, or robot level. This growing tension between centralized analytical depth and the need for instant local decisions is one of the main reasons edge computing and edge AI have gained such strategic importance in operations and supply chain research (Shi et al., 2016; Satyanarayanan, 2017; Waller & Fawcett, 2013).

At the same time, robotics in logistics has moved far beyond isolated, fixed automation. We're now seeing coordinated fleets of mobile robots that can sense their environment, communicate with one another, and adapt together in shared spaces. Fulfillment centers rely on autonomous mobile robots for transport, picking assistance, and inventory movement. Ports and yards are experimenting with synchronized vehicle systems, while cold-chain facilities use mobile sensing platforms that monitor conditions while goods are in motion. In these settings, the real value no longer comes only from what a single robot can do, but from how groups of robots interact intelligently with shelves, workers, inventory, and digital control systems. Swarm robotics provides a useful way to think about this shift, emphasizing decentralized coordination, scalable collective action, simple local rules, and robustness achieved through redundancy rather than dependence on a single central decision-maker (Brambilla et al., 2013; Chung et al., 2018; Kolling et al., 2016).

The intersection of these developments creates the subject of this chapter: Edge AI swarm robotics as an infrastructure for real-time supply chain business intelligence. In this context, edge AI refers to machine learning and related inferential capabilities executed on robots, gateways, access points, or nearby micro-data centers rather than only in distant cloud environments. Swarm robotics refers to groups of robots whose task allocation, navigation, sensing, and adaptation depend substantially on distributed coordination. Real-time supply chain business intelligence refers to the production of operationally actionable insights with latency low enough to support immediate interventions as well as near-term planning adjustments. The combination is important because raw robotic activity does not automatically produce intelligence. Intelligence emerges when perception, coordination, analytics, and managerial interpretation are integrated into a closed loop that connects local events with network-level visibility and decision support.

The strategic argument is straightforward. First, supply chain digitalization has expanded the volume and variety of data available for operational analysis. Internet of Things infrastructures provide streams from scanners, cameras, wearables, shelves, vehicles, environmental sensors, and robotic platforms, thereby widening the empirical base for monitoring and control and creating new demand for streaming analytics at the edge (Atzori et al., 2010; Gubbi et al., 2013; Mohammadi et al., 2018; Ben-Daya et al., 2019). Second, artificial intelligence methods have become sufficiently efficient to run under tighter memory, power, and latency budgets than earlier generations of models, especially when combined with compression, quantization, and model architecture optimization (Sze et al., 2017; He et al., 2016; Sandler et al., 2018; Jacob et al., 2018). Third, digital supply chain thinking has moved from simple visibility toward resilience, reconfiguration, and digital twin enabled experimentation, making local event detection more valuable because it improves the fidelity of planning models and control logic (Christopher & Peck, 2004; Ivanov & Dolgui, 2021; Ivanov, 2022; Tao et al., 2019). Fourth, analytics scholars have shown that data value depends not only on possession of data but on the organizational capabilities that convert data into process improvements and better decisions (Wamba et al., 2017; Gunasekaran et al., 2017; Queiroz et al., 2021).

Yet deploying intelligence through robotic swarms is not merely a technical extension of cloud analytics. It changes the spatial and temporal structure of decision making. In centralized business intelligence, event data are often transmitted upward, consolidated, cleansed, and modeled before they are reinterpreted into recommendations for managers. In a swarm-enabled edge architecture, parts of the analytic cycle occur on the move and in place. A robot may classify an obstacle, estimate queue congestion, infer shelf scarcity, detect a thermal excursion, or renegotiate a route locally, while the broader system simultaneously learns from aggregated fleet behavior. This redistribution of analytical work can reduce communication burden, protect sensitive operational data, and support graceful degradation during connectivity problems, but it also creates new challenges in governance, benchmarking, interoperability, cybersecurity, and managerial accountability (Mao et al., 2017; Zhou et al., 2019; Chen & Ran, 2019; Baryannis et al., 2019).

For public and private supply networks alike, the managerial implication is that robotics should no longer be evaluated solely as a labor-saving or throughput-enhancing technology. Robot fleets are also mobile sensors and distributed analytical actors.

They generate a fine-grained picture of asset location, congestion, compliance, and exception events that can materially improve situational awareness. Conversely, if their data remain trapped in local control loops or are transmitted without semantic normalization, much of that intelligence potential is lost. The business intelligence question is therefore not only whether robots can execute tasks faster, but whether their collective perception can support better decisions about replenishment, task reprioritization, exception handling, risk mitigation, and service performance.

This chapter develops a literature-grounded framework for understanding that question. It synthesizes work on edge computing and edge intelligence (Shi et al., 2016; Satyanarayanan, 2017; Zhou et al., 2019), swarm and multi-robot systems (Cao et al., 1997; Arai et al., 2002; Brambilla et al., 2013; Khamis et al., 2015), digital manufacturing and digital twins (Lee et al., 2015; Lu et al., 2020; Tao et al., 2019), and supply chain analytics and digitalization (Waller & Fawcett, 2013; Wamba et al., 2017; Brintrup et al., 2020; Huang et al., 2023). The chapter is intentionally integrative rather than narrowly algorithmic. It treats technical design decisions as inseparable from operational metrics, governance choices, and managerial use cases.

Three specific contributions organize the discussion. First, the chapter clarifies the architectural role of edge AI swarms in real-time business intelligence by separating sensing, actuation, coordination, and analytical functions. Second, it proposes a benchmarking lens that joins robotics metrics such as fleet utilization, path efficiency, and coordination latency with business intelligence outcomes such as forecast responsiveness, service-level recovery, and exception resolution speed. Third, it presents literature-grounded industrial case analyses showing how the same architectural logic applies differently in e-commerce fulfillment, cold-chain warehousing, and port or cross-dock operations. Across these settings, the objective is not to romanticize autonomy but to identify the conditions under which swarm-enabled intelligence is operationally credible and economically justified. Table 1 introduces the distinction between edge-level, swarm-level, facility-level, and network-level intelligence so that later sections can connect technical mechanisms to managerial horizons in a consistent way.

Table 1. Decision layers in edge AI swarm robotics for supply chain business intelligence

Decision layer	Primary scope	Typical latency window	Representative decisions	Business intelligence output
Edge level	Single robot, sensor, or local gateway	Milliseconds to seconds	Obstacle avoidance, object recognition, local anomaly detection, battery-aware motion adaptation	Immediate state estimates and event flags
Swarm level	Fleet or zone of interacting robots	Seconds to tens of seconds	Task allocation, routing negotiation, congestion balancing, collaborative scanning	Zone-level workload, queue, and coordination insights
Facility level	Warehouse, cold room, dock, or yard	Minutes to hours	Wave release, labor balancing, replenishment priorities, maintenance dispatch	Operational dashboards and near-real-time KPI shifts
Network level	Multi-site supply network	Hours to days	Capacity reallocation, contingency planning, inventory repositioning, resilience analysis	Strategic visibility, scenario evaluation, and risk intelligence

2. BACKGROUND

Edge computing emerged as a response to applications whose performance degrades when all computation is deferred to remote data centers. The central insight in foundational work is that bandwidth, latency, context sensitivity, privacy, and intermittent connectivity create a need for computational capability near the data source rather than only in centralized clouds (Satyanarayanan et al., 2009; Shi et al., 2016; Satyanarayanan, 2017). In logistics environments, this need is especially acute. A cloud platform may be suitable for enterprise-scale optimization, historical modeling, and cross-site reporting, but a robot navigating a busy aisle or coordinating a pallet transfer cannot afford the jitter associated with distant round trips for every inference or control action.

Mobile edge computing research deepened this argument by showing that communication and computation must be jointly managed when devices are resource constrained and applications are latency sensitive (Mao et al., 2017). Later work on edge intelligence extended the discussion from task offloading toward the design of end-to-end AI pipelines in which training, inference, adaptation, and data curation are distributed across device, edge, and cloud layers (Zhou et al., 2019). The literature now treats the edge not merely as a cache or relay but as an analytical layer where models can be deployed, updated, and monitored in relation to local context. For swarm robotics, this is crucial because the collective must often coordinate under changing spatial conditions. If every perception or allocation decision requires a centralized service call, the architecture becomes brittle and congested precisely when throughput pressure is highest.

The technical feasibility of edge AI has been strengthened by progress in model and hardware efficiency. Surveys on deep learning at the edge and efficient DNN processing show that compression, hardware-aware design, pruning, memory hierarchy optimization, and accelerator support make low-latency inference increasingly practical on constrained platforms (Chen & Ran, 2019; Sze et al., 2017). At the model level, residual networks, mobile architectures, and quantization techniques have broadened the design space for robotics applications that require visual detection, pose estimation, or scene classification under tight timing budgets (He et al., 2016; Sandler et al., 2018; Jacob et al., 2018). The implication for business intelligence is significant: when perceptual analysis can happen close to the operational event, robot fleets can transform raw sensory observations into structured data products before transmission, reducing both bandwidth demand and analytic delay.

Resource constraints nonetheless remain important. Edge nodes in logistics facilities differ in compute power, energy budget, network quality, and environmental robustness. As a result, training and updating models across fleets raises issues of heterogeneity and synchronization. Adaptive federated learning has become attractive in such environments because it can coordinate distributed model improvement without requiring continuous central access to raw local data (Wang et al., 2019). While federated learning is not a complete solution to swarm intelligence, it offers an intermediate governance model in which local adaptation and centralized oversight can coexist.

Swarm robotics draws inspiration from distributed biological systems while adapting those ideas to engineered collectives. The field emphasizes scalability, redundancy, self-organization, and robustness emerging from local interactions rather than solely from top-down planning (Brambilla et al., 2013). Earlier multi-robot research had already established many of the core questions: how to divide tasks, coordinate movement, share information, and preserve collective performance when communication is limited or uncertainty is high (Cao et al., 1997; Arai et al., 2002). Later reviews systematized the state of the art in task allocation, aerial swarms, human-swarm interaction, and path planning, showing that different application settings require different balances between decentralization and coordination discipline (Khamis et al., 2015; Kolling et al., 2016; Chung et al., 2018; Abujabal et al., 2023).

For supply chain operations, the relevance of swarm logic lies in the combination of density, dynamism, and repetitive micro-decisions. Consider a fulfillment center in which hundreds of robots share aisles, storage zones, charging areas, and picking interfaces. The system must allocate jobs, prevent deadlock, adapt to worker availability, avoid congestion, and recover from faults, all while preserving throughput and service quality. Swarm methods are attractive in such conditions because they can reduce dependence on a single global scheduler and improve resilience to local disruptions. However, practical deployments typically blend swarm principles with centralized constraints. Safety rules, task priorities, geofencing, and enterprise KPIs impose boundary conditions that pure self-organization cannot ignore. In other words, industrial swarm robotics is usually hybrid rather than fully decentralized.

Another important theme is the relationship between autonomy and human oversight. Surveys on human interaction with robot swarms show that as collectives scale, direct teleoperation becomes less feasible and supervisory control becomes more important (Kolling et al., 2016). In supply chains, this means managers and floor supervisors need not command every robot, but they do need intelligible summaries of what the swarm is doing, why priorities shift, and where exceptions are emerging. Business intelligence thus acts as a translation layer between swarm behavior and managerial cognition. Without this layer, operational trust in robot collectives is difficult to sustain, especially when failures or delays occur.

Supply chain business intelligence has expanded from static scorecards to predictive and prescriptive analytics, supported by richer data ecosystems and stronger analytical methods. Foundational discussions in logistics and operations emphasized that big data and predictive analytics could reshape supply chain design and performance measurement, but only when organizations built complementary capabilities for interpretation and action (Waller & Fawcett, 2013; Wamba et al., 2017; Gunasekaran et al., 2017). More recent reviews connect IoT, machine learning, and digitalization to risk management, resilience, and dynamic capability development, arguing that the value of analytics lies in timely intervention rather than mere visibility (Ben-Daya et al., 2019; Baryannis et al., 2019; Brintrup et al., 2020).

Digital twin research strengthens this managerial perspective. In manufacturing and logistics, cyber-physical architectures and digital twins provide a way to maintain synchronized representations of physical assets, processes, and environmental states (Lee et al., 2015; Tao et al., 2019; Lu et al., 2020). For supply chains, digital twins have been framed as tools for disruption management, scenario testing, and resilience analysis, particularly when coupled with event-driven data and simulation (Ivanov & Dolgui, 2021; Ivanov, 2022). Robot swarms enrich digital twins because they generate moving, context-rich observations rather than only fixed sensor feeds. A fleet can report not simply that inventory exists but whether access routes are congested, handling conditions are deteriorating, or specific storage zones are accumulating exception risk.

Supply chain digitalization studies further demonstrate that successful transformation depends on capabilities rather than technology procurement alone. Frameworks on digital supply chain capabilities, food supply chain digitalization, and Industry 4.0 adoption show that governance, integration, data semantics, interoperability, and workforce adaptation shape outcomes as strongly as hardware or software selection (Queiroz et al., 2021; Kittipanya-ngam & Tan, 2020; Huang et al., 2023; Ricárdez-Estrada et al., 2024). This observation is particularly relevant to edge AI swarm robotics. A technically sophisticated robot fleet can still fail as a business intelligence platform if its data do not align with enterprise master data, if decision rights are unclear, or if local optimizations distort network-level objectives.

The synthesis of these research streams suggests that edge AI swarms should be understood as cyber-physical intelligence systems. They are not only automation assets and not only data generators. They are intermediate entities that collect, interpret, and act on operational signals while also feeding managerial intelligence. Figure 1 presents the chapter's reference architecture for that integration. After the architecture is introduced, Table 2 summarizes the main deployment patterns through which AI functions may be distributed across robot, edge, and cloud layers.

Figure 1. Architecture for edge AI swarm robotics in supply chain business intelligence.

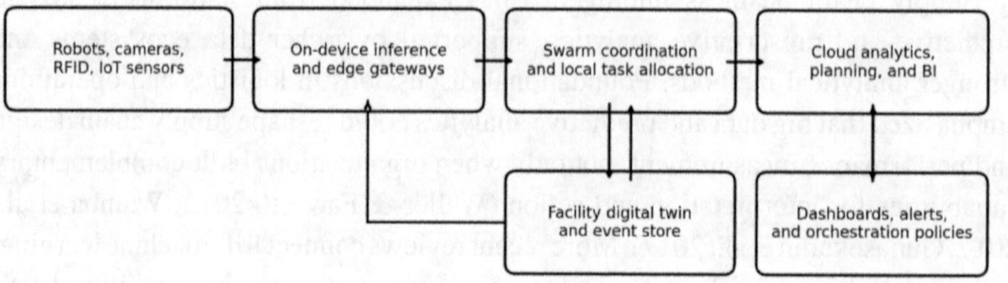

Table 2. AI deployment patterns in swarm-enabled logistics systems

Deployment pattern	Where inference runs	Where learning runs	Advantages	Limitations	Suitable use cases
On-device dominant	Robot processors	Cloud or scheduled edge retraining	Lowest control latency, reduced bandwidth, graceful degradation	Tight memory and energy budgets, limited model size	Collision avoidance, barcode reading, shelf localization
Edge-assisted swarm	Robots plus zone gateways	Edge aggregation with periodic cloud synchronization	Balances local autonomy with stronger coordination	Requires reliable local networking and orchestration	Task allocation, congestion prediction, local digital twin updates
Cloud-supervised hybrid	Robots for immediate control; cloud for richer inference	Centralized training with edge deployment	Stronger model governance, easier enterprise integration	Higher dependence on backhaul connectivity	Demand surges, cross-zone optimization, policy analytics
Federated adaptive	Robots or gateways locally	Distributed local training with federated aggregation	Privacy preservation, context adaptation, reduced raw data transfer	Complex validation, heterogeneity management, model drift risk	Multi-site fleets, changing layouts, varying local demand patterns

3. MAIN FOCUS OF THE CHAPTER

3.1 Architectural Foundations for Real-Time Intelligence

The central design question in edge AI swarm robotics is where intelligence should reside at each moment of the operational cycle. A useful answer begins by separating five layers: sensing, perception, coordination, synchronization, and interpretation. Sensing captures environmental signals through cameras, LiDAR, RFID, inertial units, environmental probes, and transactional events. Perception converts those signals into structured observations such as item identity, route obstruction, occupancy, or thermal deviation. Coordination assigns or renegotiates work among robots. Synchronization updates the facility or network model so that enterprise systems retain a coherent view of state. Interpretation turns these event streams into managerial insights, alerts, and decision options. When all five layers are forced into the cloud, the system becomes slow and communication heavy; when all five are trapped at the edge, enterprise visibility and governance deteriorate. Real-time business intelligence therefore depends on a principled distribution of these functions rather than a simple preference for decentralization.

In warehouse and intralogistics settings, the most latency-sensitive functions are usually perception and motion adaptation. A robot that detects an obstacle, a damaged tote, or an unexpectedly occupied picking zone should not wait for cloud confirmation before making an immediate operational adjustment. Here, edge AI on the robot or nearby gateway is appropriate because it minimizes decision delay and preserves continuity during short network failures (Shi et al., 2016; Chen & Ran, 2019). By contrast, functions such as replenishment prioritization across zones, labor rebalance recommendations, or site-level congestion forecasting often benefit from broader context and can tolerate slightly longer latency. These may be executed on facility edge servers or cloud analytics platforms depending on the required time horizon and data dependencies.

The architectural challenge becomes more complex when many robots interact. Swarm coordination requires each robot to act locally while still contributing to collective efficiency. In practice, logistics swarms often use hybrid control: decentralized local rules for navigation and collision avoidance, bounded negotiation for task exchange, and supervisory orchestration to enforce priorities, safety, and service objectives. Task allocation reviews suggest that no single method dominates across all contexts because environment structure, task heterogeneity, and communication constraints all matter (Khamis et al., 2015). For supply chain business intelligence, this means the architecture must record not only what task was executed but also why certain allocation decisions emerged, how coordination overhead changed, and whether local adaptations improved or harmed global KPIs.

A further issue concerns data semantics. Robot fleets create large volumes of event data, but raw telemetry has limited managerial value unless it is mapped to business entities. The same path deviation may imply very different risks depending on whether the robot is carrying a high-value order, a perishable item, or maintenance equipment. Consequently, edge data products should be semantically enriched near the source with order, SKU, zone, exception type, and timestamp metadata. This enrichment supports later aggregation into business intelligence outputs such as service-level risk, cycle-time instability, dwell-time anomalies, and cold-chain compliance. Without such semantic alignment, swarm data remain technically rich but managerially opaque.

Figure 2 depicts this closed-loop relationship between edge analytics and business intelligence. The figure emphasizes that the intelligence loop does not end with local robot action. Instead, every actuation event can become a new business signal when it is normalized, logged, and incorporated into the facility or network model. This is the step that converts robot fleets from operational tools into intelligence infrastructures.

Figure 2. Closed-loop relationship between swarm operations and business intelligence.

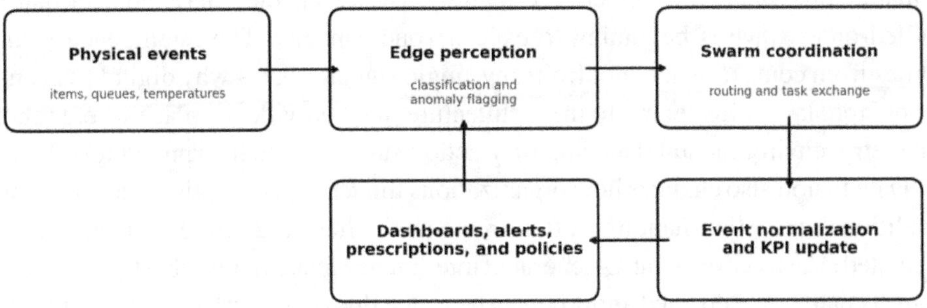

3.2 From Edge Events to Managerial Intelligence Products

A recurring weakness in automation programs is the assumption that more operational data automatically produce better managerial insight. In practice, the path from local edge events to usable business intelligence products requires deliberate transformation. Robot streams are high frequency, heterogeneous, and highly contextual. They include positional traces, battery states, route choices, sensor confidence scores, image features, environmental readings, and machine-status messages. Most of these data elements are not directly meaningful to planners or supervisors. The business intelligence challenge is to convert them into information objects that answer operational questions: Which zones are accumulating service risk? Which orders are likely to miss promise windows? Where is a handling anomaly likely to become a quality issue? Which route patterns indicate layout redesign opportunity?

This conversion process typically involves four analytic stages. The first is event filtration. Not every robot message should become an enterprise event. Systems need thresholds that distinguish normal variance from meaningful exceptions. The second is semantic mapping. Selected events are associated with orders, inventory units, storage locations, time windows, and responsibility centers. The third is aggregation. Events are combined into interpretable indicators such as congestion indices, heatmaps, utilization bands, exception clusters, or compliance trajectories. The fourth is prioritization. Indicators are ranked according to their likely business consequence so that human attention is focused on the highest-value interventions. Organizations that neglect the fourth stage often overwhelm supervisors with technically accurate but managerially unusable alerts.

This logic is especially important when multiple data modalities are present. For example, a robot in a grocery micro-fulfillment center may simultaneously report slightly elevated battery draw, a temporary route diversion, and a short wait at a handoff point. Individually these signals may be trivial. Combined with local demand spike data and temperature readings, however, they may indicate that a chilled-order wave is beginning to exceed zone capacity. The intelligence value comes from context fusion, not from any single signal. That is why digital twin and event-store layers are central to the architecture: they provide the place where robot telemetry, environmental data, and transaction states can be interpreted together.

Data fusion also changes how organizations think about exception management. Traditional exception handling often treats anomalies as discrete incidents to be escalated case by case. Swarm-generated intelligence makes it possible to see exceptions as patterns. Repeated short delays near one docking interface, repeated route negotiation failures around one aisle turn, or repeated brief thermal spikes during one transfer sequence may each seem operationally minor. Yet together they can reveal process fragility. Edge AI is valuable not merely because it triggers immediate responses but because it preserves the texture of these micro-events long enough for facilities to learn from them.

Another design principle follows from this observation: business intelligence products should be layered in the same way as the decision architecture. At the operational edge, products should be terse and prescriptive, for example a reroute command or a local anomaly classification. At the swarm or zone level, products should summarize collective conditions, such as congestion maps, task-imbalance indicators, and charging forecasts. At the facility level, intelligence should support supervisors and planners with dashboards on throughput risk, service-level vulnerability, and exception backlog. At the network level, the focus shifts toward scenario evaluation, resilience, and comparative learning across sites. Attempting to present all levels to all users usually reduces rather than improves decision quality.

The design of alerting deserves particular scrutiny. In many industrial deployments, alert systems fail because thresholds are static while operations are dynamic. A queue length that is tolerable at one time of day may be intolerable during a departure cutoff window. Likewise, a temporary thermal excursion may be acceptable for one product class but critical for another. Edge AI swarm systems should therefore support context-aware thresholds that combine physical state, service promise, product sensitivity, and network importance. This capability pushes business intelligence beyond generic monitoring toward situational prioritization.

An additional opportunity lies in retrospective intelligence. Because swarms generate granular spatiotemporal traces, they create a basis for process mining and root-cause analysis that is richer than conventional system logs. Facilities can reconstruct how local conditions evolved before service failures, how congestion propagated, and which interventions had the greatest stabilizing effect. Such retrospective intelligence can improve layout design, preventive maintenance, slotting logic, and staffing models. The point is that real-time intelligence and continuous improvement are not separate use cases; the same data infrastructure can support both when designed appropriately.

From a methodological standpoint, future implementations should treat intelligence product design as seriously as model design. Precision, recall, and latency remain important, but they do not by themselves guarantee that an alert, heatmap, or dashboard contributes to better action. Human-centered evaluation, task analysis, and intervention studies are needed to determine which formats actually improve operational performance. This is where supply chain analytics and HCI scholarship should meet more directly in future research and practice.

3.3 Benchmarking Metrics and Latency Logic

Real-time supply chain business intelligence requires a benchmarking language that can compare architectures without reducing performance to a single robotics metric. Robotics studies often emphasize throughput, collision avoidance, localization accuracy, or path efficiency. Supply chain analytics studies often emphasize service level, order cycle time, risk exposure, or forecast responsiveness. Edge AI swarm robotics links these domains, so the evaluation framework must do the same.

Table 3 summarizes a practical metric set organized around six families: latency, perception quality, coordination efficiency, resource efficiency, resilience, and managerial impact. Latency metrics include end-to-end event-to-insight time, local inference time, network round-trip time, and policy update propagation time. Perception quality includes precision, recall, false alarm rates, and localization error. Coordination efficiency covers task completion balance, route conflict rates, negotiation overhead, and idle time. Resource efficiency addresses compute utilization, bandwidth consumption, and battery draw. Resilience metrics include recovery time after node failure, degradation under packet loss, and continuity under cloud disconnection. Managerial impact metrics translate technical performance into business outcomes such as reduction in exception resolution time, order lead-time variance, spoilage risk, or queue-related service failures.

The importance of latency deserves special attention because low latency is often invoked imprecisely. For business intelligence, latency is not only a control engineering parameter; it is the delay between an operationally meaningful event and a useful decision. A system may have fast robot response but still poor intelligence latency if data normalization, alert ranking, or dashboard propagation are slow. Conversely, enterprise dashboards may update quickly while local robot control lags. A useful benchmarking approach therefore decomposes latency into sensing delay, inference delay, coordination delay, data integration delay, and decision presentation delay. Figure 3 provides this latency-budget view so that design teams can identify where milliseconds matter for actuation and were seconds or minutes matter for managerial interpretation.

This decomposition also clarifies the difference between operational and managerial real time. In a dense robotic aisle, sub-second reaction may be necessary for safety and navigation. In replenishment control, a delay of thirty seconds may be acceptable, while a delay of thirty minutes may already destroy decision value. In cold-chain monitoring, a local thermal anomaly might require immediate robot or gateway intervention, but cross-site reallocation decisions can tolerate slower aggregation. Benchmarking should therefore define thresholds relative to decision class rather than relying on an abstract universal latency target.

Another benchmarking issue concerns data quality drift. Edge AI systems learn from changing environments: packaging changes, lighting conditions shift, layouts evolve, and demand patterns reconfigure work zones. As a result, the same model may produce different error profiles over time or across facilities. Federated and hybrid training strategies can address some of this drift, but they also introduce version control, validation, and comparability challenges (Wang et al., 2019). For business intelligence teams, this means model governance must be part of performance evaluation. A perception model that improves local accuracy but fragments KPI consistency across sites may weaken managerial intelligence overall.

Table 3. Benchmarking metrics for edge AI swarm robotics in real-time supply chain business intelligence

Metric family	Illustrative indicators	Why it matters	Typical data source
Latency	Inference time, event-to-insight time, policy propagation delay	Determines whether intelligence is actionable before conditions change	Robot logs, gateway telemetry, BI event timestamps
Perception quality	Precision, recall, false alarm rate, localization error	Shapes trust in alerts and exception detection	Model validation sets, operator confirmation, video audits
Coordination efficiency	Task balance, congestion rate, negotiation overhead, route conflict count	Indicates whether the swarm scales without hidden inefficiency	Fleet manager, scheduler traces, robot-to-robot messages
Resource efficiency	Bandwidth use, CPU/GPU load, battery drain, memory occupancy	Reveals sustainability of deployment under operational pressure	Device monitors, gateway dashboards, energy telemetry
Resilience	Recovery time, packet-loss tolerance, fallback performance, fault isolation	Shows robustness during disruptions or degraded connectivity	Failure drills, simulation, incident records
Managerial impact	Cycle-time variance reduction, exception resolution speed, spoilage prevention, service-level recovery	Links technical gains to business value	WMS/TMS/ERP data, quality systems, customer service records

Figure 3. Decomposition of the latency budget in swarm-enabled supply chain intelligence.

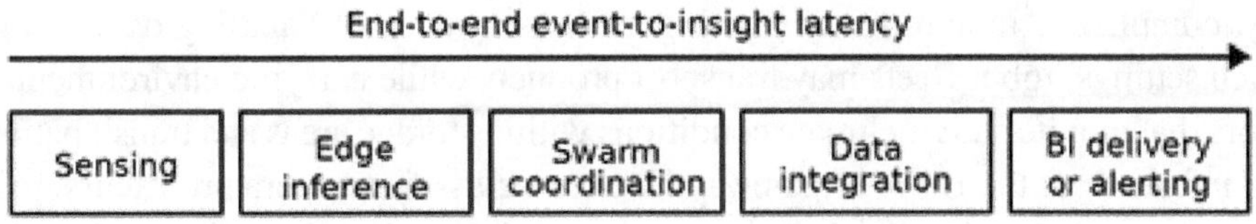

3.4 Industrial Case Analyses

To show how the architecture behaves in context, this section offers three literature-grounded industrial case analyses. These are not presented as primary field studies generated by the author. Instead, they synthesize recurring design patterns and operational constraints documented across the referenced literature on edge computing, digital twins, logistics analytics, and robotic coordination.

Case 1: Goods-to-Person E-Commerce Fulfillment

In a high-throughput e-commerce fulfillment center, autonomous mobile robots transport inventory pods or totes between storage and picking stations. The central managerial problem is not merely moving items faster; it is sustaining throughput during demand volatility while preventing congestion, stock access delays, and labor imbalance. Edge AI contributes by enabling robots and local gateways to detect aisle blockage, shelf occupancy anomalies, dropped items, or irregular queue buildup without waiting for cloud inference. Swarm logic contributes by allowing fleets to redistribute movement paths and exchange tasks when zones become congested.

From a business intelligence perspective, the critical innovation is that local routing and queue signals become early indicators of fulfillment risk. A spike in robot waiting time near a subset of SKUs may indicate demand concentration, replenishment error, station undercapacity, or poor slotting. If those patterns are normalized and streamed into facility dashboards, supervisors can adjust wave release, picker assignment, and replenishment priorities before service levels deteriorate. This linkage between local robot behavior and facility-level decision making is where edge AI swarm robotics surpasses conventional automation. The robots are not only moving inventory; they are detecting process stress in motion.

Case 2: Cold-Chain Warehousing and Pharmaceutical Handling

Cold-chain environments introduce additional constraints: temperature sensitivity, compliance requirements, and higher consequence of handling deviations. In such settings, robot fleets may transport products while carrying environmental sensors that monitor microclimate conditions within storage areas and transit paths. Edge inference at the robot or gateway level can classify temperature excursions, door-open anomalies, or dwell-time deviations quickly enough to trigger immediate containment actions. For example, the system may reroute a robot away from an affected zone, prioritize unloading of temperature-sensitive items, or escalate a targeted alert to quality control.

The business intelligence value here lies in preserving not only operational visibility but evidentiary traceability. Managers need to know which item lots were exposed, for how long, under what conditions, and with what likely quality implications. Swarm-generated data can feed a digital twin that supports forensic analysis and predictive prevention. Over time, repeated micro-anomalies around particular dock doors, staging points, or route intersections can reveal systemic design flaws that would be invisible in aggregate temperature averages. Thus, edge AI swarms support both immediate intervention and structured learning from near misses.

Case 3: Port, Yard, or Cross-Dock Coordination

In ports, container yards, or large cross-dock environments, coordination complexity expands across outdoor mobility, variable connectivity, safety-critical geofencing, and heterogeneous assets. Robot or vehicle collectives may interact with cranes, trailers, and manual operators in partially structured spaces. Here, the edge layer is important because connectivity can be uneven and environmental uncertainty is higher. On-device or nearby edge inference can detect route obstruction, container mismatch, unsafe proximity, or queue spillback. Swarm coordination can then smooth local flows while a higher-level orchestration layer aligns movements with vessel schedules, departure windows, and dock priorities.

For business intelligence teams, the key issue is how to transform distributed movement intelligence into network-relevant insights. Local path adaptation may be beneficial from an engineering standpoint yet detrimental if it hides accumulating delay against departure commitments. Effective architectures therefore tie local performance to schedule adherence, dwell time, handoff reliability, and disruption response. In this context, digital supply chain twin concepts are especially useful because they allow managers to evaluate what observed swarm behavior implies for the broader logistics network (Ivanov & Dolgui, 2021).

Table 4 consolidates the three case analyses into a comparative format. The comparison shows that the architectural pattern is similar across domains, but the dominant business intelligence outputs vary: throughput balance in fulfillment, compliance and spoilage prevention in cold chains, and synchronized handoff visibility in yard or cross-dock operations.

Table 4. Literature-grounded industrial case analyses

Context	Swarm role	Dominant edge AI tasks	Critical business intelligence outputs	Primary managerial decisions
E-commerce fulfillment	Fleet movement of totes/pods and collaborative scanning	Queue detection, obstacle classification, local demand hotspot recognition	Congestion heatmaps, cycle-time risk, slotting and replenishment signals	Wave release, labor balancing, replenishment priorities, route policy tuning
Cold-chain warehouse	Transport and environmental sensing across controlled zones	Thermal anomaly detection, dwell-time monitoring, route adaptation under compliance risk	Exposure traceability, spoilage-risk alerts, quality deviation patterns	Exception containment, QA escalation, dock prioritization, maintenance intervention
Port, yard, or cross-dock	Vehicle coordination under schedule and safety constraints	Queue spillback detection, mismatch recognition, geofence compliance checks	Dwell-time variance, handoff reliability, departure-risk visibility	Dock sequencing, asset allocation, contingency planning, disruption recovery

3.5 Issues, Controversies, and Persistent Problems

Despite its promise, edge AI swarm robotics raises substantive controversies that should temper simplistic narratives of full autonomy. The first concerns the trade-off between decentralization and control. Swarm principles encourage local autonomy and adaptation, but supply chains are accountable systems with service contracts, compliance rules, and explicit priority structures. Excessive local optimization can create emergent behavior that is efficient in one zone but misaligned with network goals. This is especially problematic when local task exchanges or routing changes alter workload visibility or distort KPI interpretation. Managers therefore need architectures that preserve local agility while maintaining auditable policy boundaries.

The second controversy concerns explainability and trust. Edge AI systems often make rapid perceptual judgments or coordination recommendations that are difficult for supervisors to inspect in real time. Yet operations personnel need usable reasons when the fleet repeatedly avoids a route, escalates a quality alert, or reprioritizes a set of orders. Research on explainability and human-AI interaction suggests that trust depends less on technical opacity in the abstract than on whether systems provide timely, context-appropriate, and actionable explanations for stakeholders (Miller, 2019; Amershi et al., 2019). In supply chain settings, explanations must be tied to business objects and consequences, not merely model confidence values.

A third issue is cybersecurity and data integrity. Robot fleets expand the attack surface of digital operations because they combine mobile endpoints, local wireless communication, sensor data, and control actions. Surveys on IoT security show that heterogeneous device environments are vulnerable not only to confidentiality breaches but also to integrity attacks that could corrupt situational awareness (Al-Garadi et al., 2020). In a swarm-enabled intelligence architecture, corrupt data can propagate into both actuation and business intelligence layers, creating double harm: unsafe operations and misleading managerial insight. Security controls therefore need to cover device identity, message authentication, edge model provenance, and anomaly detection for both cyber and physical events.

The fourth issue is data governance and cross-site comparability. As more inference happens at the edge, data products may be preprocessed differently across zones or facilities. This improves local relevance but can complicate enterprise benchmarking. A KPI derived from one site's local object taxonomy or alert threshold may not be directly comparable with another's. Organizations must therefore define common semantic models and governance rules for when local adaptation is allowed and how it is documented. The problem is not unique to robot swarms, but it becomes more acute because physical layouts and operating conditions differ sharply across sites.

The fifth issue is workforce adaptation. Industrial case evidence repeatedly suggests that automation projects fail when they underinvest in role redesign, exception governance, and supervisory competence. Robot swarms do not eliminate managerial work; they redistribute it. New tasks emerge around intervention thresholds, model validation, fleet health interpretation, and digital twin stewardship. If workforce systems continue to expect either manual control or passive dashboard monitoring, the hybrid human-swarm system underperforms. Human oversight must be designed as an active capability rather than a residual backup.

Finally, there is an economic controversy. Not every logistics context requires swarm-enabled edge intelligence. In some facilities, simpler automation and periodic analytics may be sufficient. The business case depends on process volatility, service sensitivity, disruption cost, and the marginal value of earlier intervention. Table 5 summarizes these issues as a managerial risk matrix. The lesson is not that the technology is immature in principle, but that deployment success depends on explicit design responses to identifiable governance and operational problems.

Table 5. Key issues, controversies, and design implications

Issue	Core concern	Operational consequence	Recommended design response
Decentralization versus control	Local optimization can drift from enterprise priorities	Zone gains may reduce network service performance	Use bounded autonomy and auditable rules
Explainability and trust	Supervisors may not understand why fleets adapt	Low adoption, slow intervention, weak accountability	Provide role-specific explanations tied to orders
Cybersecurity and data integrity	Compromised devices or messages distort insight and action	Unsafe motion, false alerts, misleading dashboards	Harden identity, signed updates, edge anomaly detection
Data governance	Local model drift and semantic inconsistency reduce comparability	Weak cross-site benchmarking and executive confidence	Standardize ontologies, metadata, and release controls
Workforce redesign	Roles and skills lag behind hybrid operations	Exception overload and underuse of intelligence outputs	Train supervisors in intervention logic and fleet analytics
Economic fit	Not every process benefits from very low-latency swarm intelligence	Overinvestment or unclear ROI	Link architecture to volatility, criticality, and disruption cost

3.6 Organizational and Economic Design Choices

When it comes to deploying edge AI in swarm robotics, there are these organizational decisions that people often don't think about enough. Like, who really owns all this stuff. In a lot of companies, the robotics side reports to operations engineering, and then business intelligence teams are off in IT or digital transformation or even finance. Edge AI kind of bridges that gap between them, it acts like this shared thing. But if the data pipelines for the robots are just set up for controlling the engineering part, they might skip over the details needed for reports that managers use. It seems like that could cause problems down the line. On the other hand, if the BI people set the data needs without getting the constraints of the control loops, they end up asking for way too much transmission or super low latencies that just aren't realistic in practice.

I think governance works better when there is some joint setup for overseeing it all. You know, where robotics, operations, quality control, and analytics teams all share the load on things like defining events, mapping out KPIs, and rules for when to escalate issues. That part feels a bit messy to figure out, but it probably helps avoid those mismatches. Sometimes firms overlook how intertwined it gets. Another organizational choice concerns where experimentation is permitted. Logistics leaders are often tempted to scale quickly once early productivity gains are observed, but swarm-enabled intelligence systems are highly sensitive to local layout, workforce routines, and exception patterns. A more robust approach is to treat early deployments as bounded learning environments. Facilities can compare alternative routing policies, alert thresholds, edge-cloud partitioning choices, and explanation interfaces under controlled conditions. The goal is not only to tune the technology but to refine the operating model around it. This mirrors the dynamic capability argument in the supply chain analytics literature: value arises when organizations build routines for sensing, seizing, and reconfiguring around data, not merely when they acquire tools (Wamba et al., 2017).

Economic evaluation should likewise move beyond simplistic labor substitution logic. Edge AI swarms generate value through several mechanisms: throughput improvement, error prevention, service recovery, reduced spoilage, resilience under disruption, and better planning information. Some of these benefits are direct and measurable, while others are contingent and become visible mainly during volatility. For example, the ability of a cold-chain facility to trace exposure and intervene rapidly during a brief equipment fault may rarely be exercised, yet the avoided quality loss or compliance penalty can justify substantial investment. Similarly, better congestion visibility in an e-commerce facility may not reduce average cycle time dramatically during ordinary days but can materially protect service performance during promotions and peaks.

This implies that the business case should use scenario-based analysis rather than static averages. Managers should ask how the architecture performs under normal demand, high-demand stress, temporary connectivity degradation, and local equipment failure. A system that looks expensive under average conditions may prove economical when disruption costs are considered. The resilience literature in supply chain management strongly supports this broader view of value, especially after repeated episodes of operational instability in global networks (Christopher & Peck, 2004; Ivanov, 2022).

There is also a pacing decision regarding infrastructure investment. Some facilities may benefit from lightweight edge augmentation of existing automation before moving toward fuller swarm intelligence. Others, particularly greenfield fulfillment centers or specialized cold-chain nodes, may justify more integrated designs from the outset. The right pace depends on process volatility, regulatory pressure, site complexity, and the maturity of enterprise data governance. What should be avoided is a mismatch in which highly capable robot fleets are deployed into poorly integrated information environments or, conversely, where ambitious dashboard strategies are built on top of weak local sensing and coordination.

A final organizational design issue is inter-site learning. One of the long-term advantages of swarm-enabled intelligence is that it can reveal how similar process patterns manifest differently across facilities. However, this only happens if organizations preserve both local contextual richness and enterprise-level comparability. Shared taxonomies, controlled experiments, and digital twin templates can help transform isolated pilots into a genuine learning network. Without that discipline, each site becomes a bespoke system and the broader strategic advantage of the technology is diluted.

4. SOLUTIONS AND RECOMMENDATIONS

The preceding analysis suggests that successful deployment of edge AI swarm robotics for real-time supply chain business intelligence requires a layered implementation strategy. The first recommendation is to design around decision classes rather than around technology categories. Organizations should begin by cataloging which decisions must occur in milliseconds, seconds, minutes, and hours, and then map those classes to the appropriate compute layer. Safety, collision avoidance, and immediate anomaly containment belong at the device or local edge. Zone balancing, local workload redistribution, and queue management often belong at the edge or facility layer. Cross-site optimization, long-horizon forecasting, and strategic resilience analysis belong in cloud and enterprise analytics layers. This decision-centered approach prevents both overcentralization and indiscriminate edge deployment.

The second recommendation is to engineer semantic continuity between robot events and business intelligence objects. Every edge event should be linked to business identifiers such as SKU, order, shipment, lot, asset, customer promise window, or compliance category. Doing so allows local signals to accumulate into decision-relevant intelligence rather than isolated technical telemetry. A digital twin or event-store layer is particularly useful here because it can synchronize physical states, robot actions, and enterprise entities in a common representation (Lee et al., 2015; Tao et al., 2019; Lu et al., 2020). In practical terms, this means robot software teams and BI teams should co-design event schemas instead of working sequentially.

Third, organizations should adopt progressive autonomy. Early deployments should use robot swarms to augment visibility and recommend actions before granting wider autonomous discretion. For example, edge AI can first detect congestion and propose zone rebalancing, while supervisors retain confirmation authority. Over time, as accuracy and governance confidence improve, bounded forms of autonomous task exchange or route adaptation can be enabled within predefined guardrails. This staged model is more consistent with the human-AI interaction literature than sudden full automation because it allows trust to build through observable performance and comprehensible system behavior (Amershi et al., 2019; Kolling et al., 2016).

Fourth, benchmarking must be institutionalized, not improvised. Pilot projects should establish baseline measures for cycle time, order lead-time variance, congestion rates, exception resolution speed, and recovery under connectivity loss before full deployment. Technical metrics alone are insufficient. Management teams should ask whether faster local inference produced measurable reductions in spoilage, missed dispatches, order backlog growth, or service penalties. The combined metric architecture in Table 3 is intended to support precisely this kind of cross-functional evaluation.

Fifth, model lifecycle governance must be treated as an operational discipline. Edge AI introduces risks of silent model drift, inconsistent updates, and site-specific behavior that may be invisible to executives until performance erodes. A governance loop should therefore include data quality monitoring, model validation, signed deployment packages, rollback mechanisms, and periodic review of threshold settings. Federated learning can be useful when local context differs across facilities, but it should be coupled with centralized validation and comparability checks so that local learning does not undermine enterprise coherence (Wang et al., 2019).

Sixth, security and resilience must be built into the architecture rather than appended later. Because swarm-enabled systems couple mobile cyber-physical assets with business dashboards, defense should span endpoint hardening, network segmentation, encrypted communications, provenance tracking for models and configuration files, and anomaly detection that considers both digital and physical inconsistencies. Resilience engineering should also include degraded-mode operations: what the fleet does if the cloud is unreachable, if one zone gateway fails, or if a subset of robots becomes unavailable. Systems that cannot degrade gracefully are poor candidates for mission-critical intelligence roles.

Seventh, implementation should be organized around high-value use cases instead of general automation ambition. The literature points repeatedly to use cases where earlier intervention has disproportionate value: congestion avoidance in high-throughput fulfillment, spoilage prevention in cold chains, and disruption visibility in networked logistics (Ben-Daya et al., 2019; Baryannis et al., 2019; Ivanov & Dolgui, 2021). Choosing such contexts improves the business case and clarifies what success looks like. Conversely, using swarm-enabled edge intelligence in low-volatility environments without clear intervention value may create complexity without commensurate return.

Finally, leadership teams should establish a governance loop that links fleet operations, digital twins, model management, and business policy revision. Figure 4 presents such a loop. Its purpose is to show that deployment is not a one-time engineering event. It is a continuing cycle in which local behavior generates data, data update the twin, the twin informs policy and model revision, and revised policies return to the operational edge.

Figure 4. Governance loop for continuous improvement of swarm-enabled supply chain intelligence.

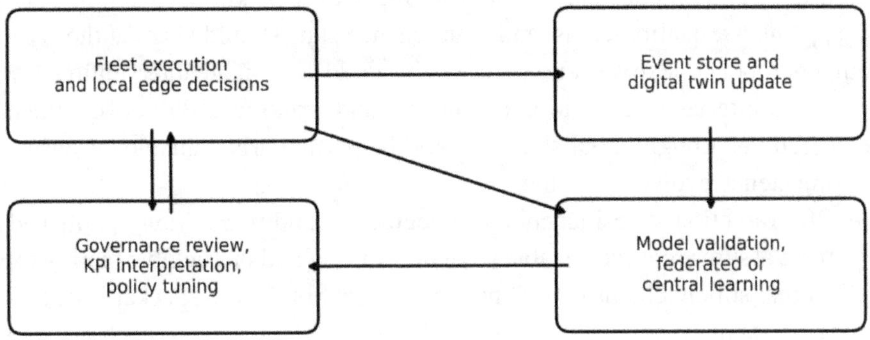

5. FUTURE RESEARCH DIRECTIONS

The first research direction concerns benchmark standardization. Although edge AI and swarm robotics both have mature technical literatures, there is still no widely adopted benchmark that jointly evaluates robotic coordination, edge resource use, and supply chain intelligence value. Future research should develop cross-layer benchmark suites in which event streams, layouts, demand shocks, and disruption scenarios can be varied systematically. Such benchmarks would help compare architectures on the dimensions that matter to managers rather than only to algorithm designers.

A second direction is trustworthy adaptive learning. Supply chain environments change frequently because of assortment shifts, seasonal peaks, revised layouts, and new compliance requirements. Future work should examine how federated, continual, and transfer learning approaches can support adaptation without undermining model comparability or auditability. This research should include formal mechanisms for explaining when local adaptation is allowed, how drift is detected, and how enterprise dashboards remain semantically stable.

A third direction is digital twin fidelity. Current digital twin discussions often assume that the more data available, the better the twin. Yet swarm-generated data can be noisy, partial, or strategically distorted by local filtering choices. Research is needed on how to maintain twins that are simultaneously real-time, economically maintainable, and decision relevant. This includes questions about which robot-generated events deserve persistence, how to fuse mobile and fixed sensing, and how to align simulation with observed swarm behavior under disruption.

A fourth direction is human-swarm organizational design. Much of the technical literature focuses on coordination algorithms, but less is known about how supervisors, planners, quality teams, and maintenance staff should share authority with intelligent robot collectives. Future research should investigate intervention design, explanation interfaces, escalation thresholds, and workforce skill development in live operations. Longitudinal studies would be particularly valuable because trust and competence evolve over time.

A fifth direction is resilience under degraded conditions. Many published architectures assume relatively stable local networking and predictable facility topology. Yet real supply chains experience temporary blind spots, packet loss, battery constraints, and asynchronous updates. Research should examine how swarms preserve both safe operation and meaningful business intelligence when part of the infrastructure fails. This is especially important for ports, humanitarian logistics, and public-sector or remote operations where conditions may be harsher than in highly structured fulfillment centers.

A sixth direction concerns sustainability and economic efficiency. Edge AI is often justified by latency and privacy, but future studies should measure its energy footprint, lifecycle cost, and hardware obsolescence profile relative to centralized alternatives. The best architecture may not be the one with the lowest technical latency if its operating cost or environmental burden is disproportionate to its decision value.

A seventh direction is multi-echelon intelligence integration. Most research and industrial experimentation still focus on single facilities. Yet the full promise of edge AI swarm robotics may lie in how local intelligence propagates upward into network-level decisions on inventory positioning, risk pooling, and disruption recovery. This requires data models and orchestration frameworks that connect local swarm events with enterprise planning systems without overwhelming them. Table 6 summarizes the resulting research agenda.

Table 6. Future research agenda

Research area	Illustrative question	Why it matters	Likely methods
Cross-layer benchmarking	How can robot, edge, and BI metrics be assessed together?	Supports credible comparison across architectures	Digital twins, simulation, controlled pilots
Trustworthy adaptive learning	How can local learning improve context fit without harming comparability?	Balances adaptation with governance	Federated learning experiments, model audits
Digital twin fidelity	Which swarm-generated events most improve decision quality?	Reduces data overload and improves model usefulness	Event mining, causal analysis, simulation
Human-swarm organization	What supervisory interfaces and explanations improve intervention quality?	Strengthens adoption and accountability	Field studies, experiments, design science
Resilience under degradation	How should swarms behave during partial connectivity or asset failure?	Preserves operations during disruptions	Stress testing, resilience engineering, scenario analysis
Sustainability and economics	What is the energy and cost trade-off of lower-latency edge intelligence?	Improves responsible investment decisions	Techno-economic analysis, lifecycle assessment
Multi-echelon integration	How should local swarm insights feed network planning and resilience models?	Moves from facility optimization to supply network intelligence	Integration architecture studies, digital twin research

6. CONCLUSION

Edge AI swarm robotics, I think, fits into supply chain business intelligence in a bigger way than just some automation tool. It moves things like selective analytics and coordination right up close to where operations are happening. That setup with swarm-enabled edge architecture cuts down on decision latency a lot. It also gives better visibility into those small disruptions that pop up. Plus, it adds more context to digital twins through observations that are rich.

The strategic side shines in spots where local events shift fast, and if you wait to step in, it hits costs or quality or service hard. But the chapter points out that being technically advanced doesn't automatically mean you get real intelligence value from it. Success in deploying this stuff relies on mixing it semantically with business data. You need bounded autonomy too, along with solid benchmarking and model governance. Cybersecurity matters and redesigning the workforce thoughtfully is key. Hybrid architecture seems to work best. They keep quick local responses intact but still hold onto enterprise goals, making sure everything stays comparable and accountable.

For researchers, this interesting overlap between robotics, edge systems, analytics, and operations management. It offers a ton to explore. Practitioners should see robot fleets as assets for distributed intelligence. But govern them like socio-technical systems, where local actions stay legible, measurable, and lined up with strategy. The recommendation comes down to a balanced approach. Push intelligence to the edge when timing or context calls for it. Still, hold onto enterprise integration and human accountability with shared data models and ongoing governance. It feels like that middle path keeps things practical without losing control.

REFERENCES

Abujabal, N. A., Rabie, T., & Kamel, I. (2023). Path planning techniques for multi-robot systems: A systematic review. In 2023 15th International Conference on Innovations in Information Technology (IIT) (pp. 1-6). DOI: 10.1109/IIT59782.2023.10366472

Al-Garadi, M. A., Mohamed, A., Al-Ali, A. K., Du, X., Ali, I., & Guizani, M. (2020). A survey of machine and deep learning methods for Internet of Things (IoT) security. *IEEE Communications Surveys and Tutorials*, *22*(3), 1646–1685. DOI: 10.1109/COMST.2020.2988293

Amershi, S., Weld, D., Vorvoreanu, M., Fourney, A., Nushi, B., Collisson, P., & Horvitz, E. (2019). Guidelines for human-AI interaction. In *Proceedings of the 2019 CHI Conference on Human Factors in Computing Systems* (pp. 1-13). DOI: 10.1145/3290605.3300233

Arai, T., Pagello, E., & Parker, L. E. (2002). Editorial: Advances in multi-robot systems. *IEEE Transactions on Robotics and Automation*, *18*(5), 655–661. DOI: 10.1109/TRA.2002.806024

Atzori, L., Iera, A., & Morabito, G. (2010). The Internet of Things: A survey. *Computer Networks*, *54*(15), 2787–2805. DOI: 10.1016/j.comnet.2010.05.010

Baryannis, G., Validi, S., Dani, S., & Antoniou, G. (2019). Supply chain risk management and artificial intelligence: State of the art and future research directions. *International Journal of Production Research*, *57*(7), 2179–2202. DOI: 10.1080/00207543.2018.1530476

Ben-Daya, M., Hassini, E., & Bahroun, Z. (2019). Internet of things and supply chain management: A literature review. *International Journal of Production Research*, *57*(15-16), 4719–4742. DOI: 10.1080/00207543.2017.1402140

Brambilla, M., Ferrante, E., Birattari, M., & Dorigo, M. (2013). Swarm robotics: A review from the swarm engineering perspective. *Swarm Intelligence*, *7*(1), 1–41. DOI: 10.1007/s11721-012-0075-2

Brintrup, A., Kito, T., & Wakolbinger, T. (2020). A review of machine learning in supply chain management: Applications, challenges and opportunities. *Computers & Industrial Engineering*, *139*, 105774. DOI: 10.1016/j.cie.2019.106764

Cao, Y. U., Fukunaga, A. S., & Kahng, A. B. (1997). Cooperative mobile robotics: Antecedents and directions. *Autonomous Robots*, *4*(1), 7–27. DOI: 10.1023/A:1008855018923

Chen, J., & Ran, X. (2019). Deep learning with edge computing: A review. *Proceedings of the IEEE, 107*(8), 1655–1674. DOI: 10.1109/JPROC.2019.2921977

Christopher, M., & Peck, H. (2004). Building the resilient supply chain. *International Journal of Logistics Management, 15*(2), 1–14. DOI: 10.1108/09574090410700275

Chung, S. J., Paranjape, A. A., Dames, P., Shen, S., & Kumar, V. (2018). A survey on aerial swarm robotics. *IEEE Transactions on Robotics, 34*(4), 837–855. DOI: 10.1109/TRO.2018.2857475

Gubbi, J., Buyya, R., Marusic, S., & Palaniswami, M. (2013). Internet of Things (IoT): A vision, architectural elements, and future directions. *Future Generation Computer Systems, 29*(7), 1645–1660. DOI: 10.1016/j.future.2013.01.010

Gunasekaran, A., Papadopoulos, T., Dubey, R., Wamba, S. F., Childe, S. J., Hazen, B., & Akter, S. (2017). Big data and predictive analytics for supply chain and organizational performance. *Journal of Business Research, 70*, 308–317. DOI: 10.1016/j.jbusres.2016.08.004

He, K., Zhang, X., Ren, S., & Sun, J. (2016). Deep residual learning for image recognition. In *Proceedings of the IEEE Conference on Computer Vision and Pattern Recognition* (pp. 770-778). DOI: 10.1109/CVPR.2016.90

Huang, K., Wang, K., Lee, P. K. C., & Yeung, A. C. L. (2023). The impact of Industry 4.0 on supply chain capability and supply chain resilience: A dynamic resource-based view. *International Journal of Production Economics, 262*, 108913. DOI: 10.1016/j.ijpe.2023.108913

Ivanov, D. (2022). Viable supply chain model: Integrating agility, resilience and sustainability perspectives-lessons from and thinking beyond the COVID-19 pandemic. *Annals of Operations Research, 319*(1), 1411–1431. DOI: 10.1007/s10479-020-03640-6 PMID: 32836614

Ivanov, D., & Dolgui, A. (2021). A digital supply chain twin for managing the disruption risks and resilience in the era of Industry 4.0. *Production Planning and Control, 32*(9), 775–788. DOI: 10.1080/09537287.2020.1768450

Jacob, B., Kligys, S., Chen, B., Zhu, M., Tang, M., Howard, A., Adam, H., & Kalenichenko, D. (2018). Quantization and training of neural networks for efficient integer-arithmetic-only inference. In *Proceedings of the IEEE/CVF Conference on Computer Vision and Pattern Recognition* (pp. 2704-2713). DOI: 10.1109/CVPR.2018.00286

Khamis, A., Hussein, A., & Elmogy, A. (2015). Multi-robot task allocation: A review of the state-of-the-art. *Studies in Computational Intelligence*, *604*, 31–51. DOI: 10.1007/978-3-319-18299-5_2

Kittipanya-ngam, P., & Tan, K. H. (2020). A framework for food supply chain digitalization: Lessons from Thailand. *Production Planning and Control*, *31*(2-3), 158–172. DOI: 10.1080/09537287.2019.1631462

Kolling, A., Walker, P., Chakraborty, N., Sycara, K., & Lewis, M. (2016). Human interaction with robot swarms: A survey. *IEEE Transactions on Human-Machine Systems*, *46*(1), 9–26. DOI: 10.1109/THMS.2015.2480801

Lee, J., Bagheri, B., & Kao, H.-A. (2015). A cyber-physical systems architecture for Industry 4.0-based manufacturing systems. *Manufacturing Letters*, *3*, 18–23. DOI: 10.1016/j.mfglet.2014.12.001

Lu, Y., Liu, C., Kevin, I., Wang, K., Huang, H., & Xu, X. (2020). Digital twin-driven smart manufacturing: Connotation, reference model, applications and research issues. *Robotics and Computer-integrated Manufacturing*, *61*, 101837. DOI: 10.1016/j.rcim.2019.101837

Mao, Y., You, C., Zhang, J., Huang, K., & Letaief, K. B. (2017). A survey on mobile edge computing: The communication perspective. *IEEE Communications Surveys and Tutorials*, *19*(4), 2322–2358. DOI: 10.1109/COMST.2017.2745201

Miller, T. (2019). Explanation in artificial intelligence: Insights from the social sciences. *Artificial Intelligence*, *267*, 1–38. DOI: 10.1016/j.artint.2018.07.007

Mohammadi, M., Al-Fuqaha, A., Sorour, S., & Guizani, M. (2018). Deep learning for IoT big data and streaming analytics: A survey. *IEEE Communications Surveys and Tutorials*, *20*(4), 2923–2960. DOI: 10.1109/COMST.2018.2844341

Queiroz, M. M., Pereira, S. C. F., Telles, R., & Machado, M. C. (2021). Industry 4.0 and digital supply chain capabilities: A framework for understanding digitalisation challenges and opportunities. *Benchmarking*, *28*(5), 1761–1782. DOI: 10.1108/BIJ-12-2018-0435

Ricárdez-Estrada, J., Garay-Rondero, C. L., Romero, D., Wuest, T., & Pinto, R. (2024). Identifying digital supply chain capabilities. *Procedia Computer Science*, *232*, 1182–1191. DOI: 10.1016/j.procs.2024.01.116

Sandler, M., Howard, A., Zhu, M., Zhmoginov, A., & Chen, L.-C. (2018). MobileNetV2: Inverted residuals and linear bottlenecks. In *Proceedings of the IEEE/CVF Conference on Computer Vision and Pattern Recognition* (pp. 4510-4520). DOI: 10.1109/CVPR.2018.00474

Satyanarayanan, M. (2017). The emergence of edge computing. *Computer, 50*(1), 30–39. DOI: 10.1109/MC.2017.9

Satyanarayanan, M., Bahl, P., Caceres, R., & Davies, N. (2009). The case for VM-based cloudlets in mobile computing. *IEEE Pervasive Computing, 8*(4), 14–23. DOI: 10.1109/MPRV.2009.82

Shi, W., Cao, J., Zhang, Q., Li, Y., & Xu, L. (2016). Edge computing: Vision and challenges. *IEEE Internet of Things Journal, 3*(5), 637–646. DOI: 10.1109/JIOT.2016.2579198

Sze, V., Chen, Y.-H., Yang, T.-J., & Emer, J. S. (2017). Efficient processing of deep neural networks: A tutorial and survey. *Proceedings of the IEEE, 105*(12), 2295–2329. DOI: 10.1109/JPROC.2017.2761740

Tao, F., Zhang, H., Liu, A., & Nee, A. Y. C. (2019). Digital twin in industry: State-of-the-art. *IEEE Transactions on Industrial Informatics, 15*(4), 2405–2415. DOI: 10.1109/TII.2018.2873186

Waller, M. A., & Fawcett, S. E. (2013). Data science, predictive analytics, and big data: A revolution that will transform supply chain design and management. *Journal of Business Logistics, 34*(2), 77–84. DOI: 10.1111/jbl.12010

Wamba, S. F., Gunasekaran, A., Akter, S., Ren, S. J.-F., Dubey, R., & Childe, S. J. (2017). Big data analytics and firm performance: Effects of dynamic capabilities. *Journal of Business Research, 70*, 356–365. DOI: 10.1016/j.jbusres.2016.08.009

Wang, S., Tuor, T., Salonidis, T., Leung, K. K., Makaya, C., He, T., & Chan, K. (2019). Adaptive federated learning in resource constrained edge computing systems. *IEEE Journal on Selected Areas in Communications, 37*(6), 1205–1221. DOI: 10.1109/JSAC.2019.2904348

Zhou, Z., Chen, X., Li, E., Zeng, L., Luo, K., & Zhang, J. (2019). Edge intelligence: Paving the last mile of artificial intelligence with edge computing. *Proceedings of the IEEE, 107*(8), 1738–1762. DOI: 10.1109/JPROC.2019.2918951

Chapter 3
ROS-Based Mobile Robot for Environmental Monitoring and Data Analysis

Ka Chung Lai
http://orcid.org/0009-0001-6890-4140
The Hong Kong Polytechnic University, Hong Kong

Aquil Mirza Mohammed
http://orcid.org/0000-0001-7756-4363
The Hong Kong Polytechnic University, Hong Kong

Hewa Majeed Zangana
http://orcid.org/0000-0001-7909-254X
Duhok Polytechnic University, Iraq

ABSTRACT

Autonomous mobile robots face critical perception failures due to severe motion blur during rapid rotation. Standard deblurring models introduce geometric artifacts when applied to planar yaw motion. To resolve this, this chapter details a real-time, hardware-accelerated perception pipeline developed in ROS 1 for the memory-constrained NVIDIA Jetson Orin Nano. The authors propose a Cartesian Cascaded Refinement Network (CAR-Net) that eliminates polar coordinate biases, corrects linear yaw blur, and achieves a 29.52 dB PSNR. Additionally, an optimized YOLOv8 obstacle detection model was generated using feature extraction trained on unmodified data to prevent knowledge loss. Freezing specific layers preserved robust object tracking. Physical stress tests utilizing high-intensity steering (up to

DOI: 10.4018/979-8-2600-0701-3.ch003

1.2 rad/s) demonstrated that the TensorRT-optimized pipeline reduced frame latency from 32 to 23 frames. This edge-based solution maintains a stable 20Hz throughput, significantly enhancing autonomous obstacle avoidance capabilities.

1. INTRODUCTION

1.1 Background

There is great interest surrounding autonomous mobile robots for environmental monitoring and data gathering as this area continues to grow within the field of robotics (Dunbabin & Marques, 2012). Autonomous Mobile Robots have the advantage of providing safer, more scalable and productive options for the collection of data as compared to manual acquisition methods of the past. To be able to operate independently within the context of continually changing and unpredictable surroundings, Autonomous Mobile Robots use advanced sensing capabilities in order to determine both the location of objects in their environment as well as their actual position within the environment. According to (Cadena et al., 2017), Simultaneous Localization and Mapping (SLAM) continues to experience the beginnings of a transition from pure research into practical applications with current systems being made to operate under a common standardized formulation (that allows for both robust perception) when performing one or more of these functions. Autonomous Mobile Robot Operations are often limited due to constraints imposed by the relatively high price, complex design, and large power consumption of Laser-based (LiDAR) systems (Macario Barros et al., 2022); this makes them unsuitable for use on compact/environmentally monitoring types of platforms that are limited in available resources.

Vision-based navigation techniques, such as Visual Simultaneous Localization and Mapping (V-SLAM), have emerged as an affordable choice for robots, which can use either monocular or stereo camera systems to create detailed geometric spatial maps of their trajectories and to determine their locations in complex environments (Campos et al., 2021). Although this can be advantageous, the sensitivity of these systems to image degradation is inherent to camera-based navigation and can be a significant point of failure when rapid angular motion is present (Saputra et al., 2018) (i.e., a robot is moving and scanning its environment or making sharp angular changes). In such cases, there is a very large relative motion between a camera system and an observed scene within a short amount of time that the camera sensor integrates, leading to a high degree of motion blur. Therefore, the standard computer vision algorithms used for both feature based and modern object detection applications are reliant on visual features that have high-frequency, sharp edge, and gradient textures. When these features are degraded by motion blur, robots suffer critical localization drift and fail to detect critical obstacles, creating serious safety hazards.

Understanding the specific kinematics of robotic motion will help in understanding how to reduce camera motion blur caused by the dynamics of rotating (yaw) and turning (rotation around the Z axis for example, and resulting pattern of camera motion/blurring being viewed as round circle) and where the primary motion of ground/ground mobile (robot) is through a completely different set of motions from those of other forms of motion captured by typical cameras. Therefore, by using the current state-of-the-art models in this area as well as previously developed models we can take advantage of both new methods of reducing blur (non-blind methods) which incorporate joint kinematic (angular velocity, for example) and digital image data to perform better than older blind methods. The main element of this research effort is the CAR-Net (Lai & Cetinkaya, 2026) which represents the most successful method to date used in minimizing motion blur, as demonstrated by the CAR-Net having won both the AAIML 2026 Best Paper award and having been published in the IEEE Conference Proceedings. Although angular angle embedding has proven to be an effective mechanism for guiding CAR-Net image restoration, the system itself is originally designed around polar coordinates for correcting roll blurring. Therefore, applying this technique directly to planar motion will produce extreme geometric distortion and severe "ringing" artifacts, thereby resulting in much lower performance for both restoration and downstream object detection.

To eliminate the algorithmic limitations of this system while enabling it to run in real-time on small, resource-limited platforms, a complete awareness pipeline for the system was developed. Although this drove substantial changes to the system architecture throughout the design phase. Initially, the system was developed based on ROS 2 (Humble) (Macenski et al., 2022) and an NVIDIA Jetson Orin Nano 8GB, but encountered severe hardware-software incompatibility issues: the Wheeltec robot base controller lacked stable support for ROS 2, and attempting to run ROS 1 in a Docker container (Merkel, 2014) severely broke compatibility with TensorRT hardware acceleration. Hence, the system architecture was transferred to the native Robot Operating System 1 (ROS 1 Noetic) (Quigley et al., 2009) environment on a Jetson Orin Nano 4GB. This environment resolved driver and compatibility issues. To reduce 4GB of memory resulted in severe memory and processing bottlenecks which require aggressive optimization to maintain real-time performance.

1.2 Problem Statement and Objectives

This project address the critical failure of vision-based navigation systems on ground robots caused by linear yaw motion blur. This problem further affected by the strict memory and computational constraints of edge computing devices (e.g., 4GB RAM limit).

To solve this, this project establishes three primary objectives:

1. **Algorithmic Adaptation:** To identify and eliminate the polar-coordinate inductive bias in existing non-blind deblurring models (CAR-Net) by adapting the architecture to natively process linear horizontal motion in a Cartesian coordinate space.
2. **Edge Hardware Optimization:** To engineer a highly optimized, native ROS 1 processing pipeline utilizing TensorRT FP16 serialization and layer fusion to achieve a stable, real-time inference rate (>15Hz) on a severely memory-constrained NVIDIA Jetson Orin Nano.
3. **Robust Domain Adaptation:** To prevent "Catastrophic Forgetting" in downstream object detectors (YOLO) by utilizing a frozen-layer training approach, maximizing tracking stability across varying degrees of motion blur.

1.3 Contributions

To achieve real-time and robust sensing capabilities, this study introduces three main technical contributions:

First, we fundamentally redesigned the deblurring module to address limitations in the kinematic domain. The point spread function (PSF) has been updated to accurately model linear yaw blur; additionally, the CAR-Net architecture has been shifted from a polar coordinate system to a Cartesian coordinate system to eliminate rotational geometric artifacts (Lai & Mohammed, 2026).

Second, we implemented a highly optimized object detection node. We evaluated multiple versions of the YOLO architecture—including YOLOv8, YOLOv11, and the newly released YOLOv26 (Jocher et al., 2023; Sapkota et al., 2025). To avoid overfitting to specific degradation patterns, these models were trained on clear, high-fidelity keyframes from the nuScenes dataset. We evaluated both full-network retraining and frozen-layer feature extraction to maximize the stability of the detection.

Third, to address the 4GB memory limit, we performed extensive edge optimization. By leveraging the serialization capabilities of the NVIDIA TensorRT engine and strategic input resolution scaling (e.g., 128×128 and 256×256), this processing pipeline overcomes stringent hardware constraints and achieves a stable 30Hz throughput across the integrated camera, deblurring, and detection nodes.

The initial results of this study, particularly the identification of the kinematic domain gap between roll and yaw ambiguities, and the development of a physical information-based synthetic data generation pipeline, have been peer-reviewed and published in the proceedings of the ICWSNUCA 2026 International Conference on Wireless Sensor Networks, Ubiquitous Computing and Applications (Springer) (Lai & Mohammed, 2026). Building on this, this final report extends this approach to fully optimized, hardware-accelerated physical deployments.

1.4 Paper Organization

The following section of this report are structured as follows: Section 2- *Related Works*, reviews the literature on visual recognition, motion deblurring, and edge deployment. Section 3- *Design and Methodology*, details the implementation of the ROS 1 pipeline, the adaptive adjustment of the Cartesian coordinate system CAR-Net, and the YOLO optimization strategy. Section 4- *Experimentation and Benchmark Results*, presents the experimental setup and benchmarks the system under physical robotic deployment. Section 5- *Discussion*, discusses the trade-offs of efficiency, and Section 6- *Conclusion and Future Work*, summarizes this report and proposes directions for future research.

2. RELATED WORKS

2.1 Visual Perception and the Motion Blur Problem in Robotics

Visual perception is the foundation for achieving low-cost autonomous navigation. Algorithms such as Direct Sparse Odometry (DSO) (Engel et al., 2017) and ORB-SLAM3 (Campos et al., 2021) have demonstrated that monocular and stereo cameras can provide high-precision spatial mapping and localization. However, these systems fundamentally rely on two assumptions: constant brightness and the existence of high-frequency spatial gradients (e.g., sharp corners and edges).

In the field of ground-based mobile robots, rapid dynamic movements—particularly yaw rotations during turns or active scanning—violate these assumptions. When the relative motion between the camera and the scene exceeds the sensor's integration time, pixel brightness is averaged over the exposure window, which acts as a low-pass filter which eliminates high-frequency features (Luan et al., 2024). As a result, feature trackers lose its reference points, leading to a significant increase in positioning errors. Although traditional solutions involve the use of global-shutter cameras or flash illumination, these hardware interventions significantly increase cost and power consumption, thereby negating the primary advantages of visual systems over LiDAR. Therefore, recovering degraded visual data through algorithms remains a critical area of research.

2.2 Deep Learning for Motion Deblurring: Blind Vs. Non-Blind

Algorithmic motion deblurring is essentially a corrupted inverse problem; countless combinations of sharp original images and blur kernels can produce the same blurred image. Deep learning solutions to this problem can generally be categorized into two paradigms: "blind" and "non-blind."

Blind Deblurring Approaches: Blind methods attempt to reconstruct images using only degraded pixel data, implicitly estimating the blur kernel within the network layers. Architectures such as DeblurGAN-v2 (Kupyn et al., 2019) and Scale-Recurrent Networks (SRN) (Tao et al., 2018) have achieved state-of-the-art visual fidelity in the field of general photography. More recently, Vision Transformer (ViT) architectures have established new baselines for image restoration and motion deblurring (Liang et al., 2021; Zamir et al., 2022). However, a deeper analysis reveals two key limitations in deploying these methods for robotics: first, they rely on large and computationally expensive backbone networks (e.g., Inception-ResNet), which cannot meet the strict latency constraints (e.g., >15Hz) of edge devices such as the Jetson Orin Nano. Second, in the absence of specific motion prior information, these networks are easily confused by texture ambiguity, often generating structural illusions or failing to distinguish natural repetitive textures (e.g., brick walls) and motion blur.

Non-Blind Deblurring and Kinematic Priors: Non-blind methods impose constraints on the solution space by explicitly computing the blur kernel using physical sensor data (typically from an inertial measurement unit (IMU)). A major advancement in this field is the Cascaded Refinement Network (CAR-Net) (Lai & Cetinkaya, 2026), a non-blind model that directly embeds the camera's rotation angle into the feature extraction layer. By using the physical rotation angle as prior information, CAR-Net not only achieves higher reconstruction accuracy than blind methods but also significantly reduces computational overhead.

However, the existing literature reveals a common gap: most sensor fusion deblurring datasets and frameworks are optimized for camera *roll* (Z-axis rotation)- which mostly produced by handheld devices and drones is concentric circular blur. In contrast, ground robots experience planar *yaw* (rotation about the Y-axis), which produces linear horizontal blur. We applied the standard CAR-Net, which process in Polar coordinate, to the linear blur caused by yaw. This resulted in 'ring-shaped' artifacts and geometric distortion (Lai & Mohammed, 2026). This project addresses this domain gap by directly modifying the point spread function (PSF) and transferring CAR-Net into a Cartesian coordinate processing framework for the elimination of polar coordinate bias.

2.3 Evolution of Real-Time Object Detection in Dynamic Environments

Real-time object detection is essential for turning reconstructed images into useful navigation data. For boundary-based detection tasks, researchers typically default to the YOLO (You Only Look Once) architecture (Jocher et al., 2023). The release of YOLOv8 shifted the framework toward anchor-free detection, significantly cutting down on post-processing overhead. Base on it, newer iterations like YOLOv11 and YOLOv26 have focused on adjusting the depth-to-width ratio of the network. This allows the models to push up their mean average precision (mAP) without increasing the parameter count (Sapkota et al., 2025).

Although YOLO provides an excellent starting point, standard fine-tuning for autonomous driving tasks usually leads to "Catastrophic Forgetting". One common solution is synaptic consolidation, where updates to weights handling general feature extraction are penalized. This helps the network learn new domain-specific driving hazards without losing its foundational knowledge (Kirkpatrick et al., 2017). However, our project tests it more directly. To rely on clear and high-fidelity keyframes from the nuScenes dataset, we freeze the first 10 feature extraction layers of the YOLO model and retrain the deeper semantic layers and regression head only. If we trained the detector directly on blurry data, the network would inevitably learn degraded representations. By avoiding this, our approach sets up a clean detection baseline for isolate and evaluate exactly how well the upstream Cartesian CAR-Net module improves the final detection result.

2.4 Constraints of Edge Computing and Robotic Middleware

While pushing artificial intelligence to the network edge is critical for real-time robotic autonomy (Zhou et al., 2019), deploying large neural networks on mobile robots usually presents strict hardware and middleware bottlenecks. For example, the 4GB version of the NVIDIA Jetson Orin Nano operates under tight power and memory constraints. When PyTorch (Paszke et al., 2019) models run unoptimized in their default FP32 precision, it quickly consume all the unified memory. This inevitably leads to out-of-memory (OOM) crashes and frame rates dropping below <7Hz—far too slow for real-time autonomous navigation.

Therefore, edge deployment relies heavily on model compression, graph optimization, and quantization techniques (Cheng et al., 2017). Using tools such as NVIDIA TensorRT (NVIDIA Corporation, 2023), we can fuse network layers and reduce weight precision down to FP16 or INT8. This cuts down memory bandwidth usage and improve throughput.

The choice of middleware is also important for system stability. While ROS 2 offers the modern DDS protocol, containerizing it often causes overhead and hardware driver conflicts. Our initial setup on ROS 2 Humble triggered communication failures with the Wheeltec chassis controller and stopped Docker containers from properly mapping to TensorRT hardware. To bypass these issue, we reverted to a native ROS 1 (Noetic) architecture. Even without DDS, ROS 1 ultimately provided a much more reliable and deterministic bridge between the chassis kinematics and the GPU-accelerated perception stack.

3. DESIGN AND METHODOLOGY

Our methodology focuses on resolving the algorithmic and hardware bottlenecks that come with running deep learning vision systems on constrained mobile robots. We structured the system pipeline to prioritize low-latency data streams, physics-based model adaptation, and heavy hardware optimization.

3.1. System Architecture and Middleware Design

In edge robotics, getting the middleware right is often one of the biggest hurdles. The system requires a stable bridge between hardware sensors, GPU perception nodes, and low-level motor controllers. We initially built the perception stack using ROS 2 Humble, running it inside a Docker container on an 8GB NVIDIA Jetson Orin Nano. Under normal circumstances, ROS 2 is ideal because of its asynchronous DDS protocol. During field tests, however, we ran into major hardware incompatibilities.

Specifically, the serial driver for the Wheeltec chassis controller was extremely unstable in the ROS 2 environment, resulting in lost commands and abnormal vehicle behavior. On top of that, the containerized environment blocked clean access to the low-level NVIDIA CUDA and TensorRT APIs necessary for hardware acceleration. Because of these obstacles, we abandoned the Docker approach and reverted to a native ROS 1 (Noetic) on the 4GB version of the Jetson Orin Nano. While this choice means losing the benefits of DDS, running ROS 1 natively guarantees a much more predictable connection to the chassis controller and allows direct, unhindered access to the Jetson's GPU.

The overall system architecture and asynchronous data flow are illustrated in Figure 1. The final software logic is implemented as a directed acyclic graph (DAG) composed of asynchronous ROS 1 nodes:

1. **Camera Node (Source):** Connects to the V4L2 interface to capture 640×480 frames in Motion-JPEG (MJPG) format, avoiding USB 3.0 bandwidth saturation issues, and publishes to the /camera/raw_image topic.
2. **Deblurring Node (Processing):** Subscribes to the raw image stream, applies a modified CAR-Net inference algorithm to restore the image, and publishes it to /camera/deblurred_image.
3. **Detection Node (Processing):** Subscribe to the restored image frames, run the YOLO inference engine, and publish the bounding box arrays and confidence scores for use by downstream navigation logic.

Figure 1. System architecture and asynchronous data flow between the ROS 1 nodes on the Jetson Orin Nano.

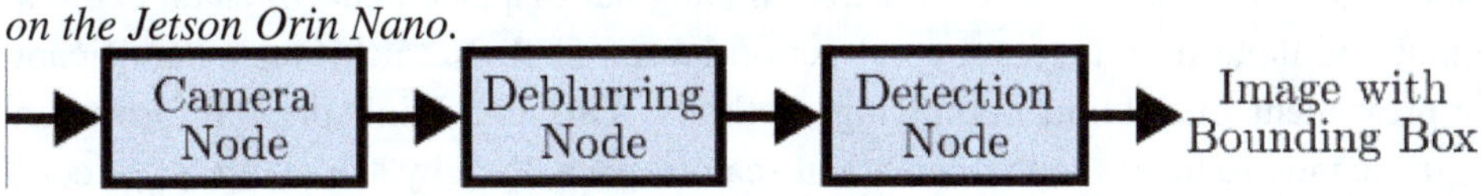

3.2. Physics-Informed Data Synthesis

Since deep learning networks are highly sensitive to the distribution of training data, bridging the domain gap between theoretical models and real-world robots is crucial. Collecting perfectly aligned pairs of blurred and clear images from moving physical robots is mechanically challenging. Therefore, we developed a physics-based data synthesis workflow.

We extracted high-fidelity, clear keyframes from the nuScenes-mini autonomous driving dataset (Caesar et al., 2020). To simulate specific image distortions that occur when a robot turns, we implemented a linear motion kernel generator driven by rigid-body dynamics. Unlike standard rolling blur (rotation around a central axis), the yaw of a ground robot produces horizontal linear motion blur. The angular velocity ω is uniformly sampled within the physically realistic range $[-60^{\circ}/s, +60^{\circ}/s]$. Assuming a standard robot's low-light exposure time is t=0.05s, the total angular displacement θ is calculated accordingly. A normalized 1D horizontal box filter (point spread function, or PSF) is constructed and convolved with a sharp nuScenes image. This ensures that the "pseudo-IMU" label (θ) is mathematically coupled with visual blur, thereby providing the precise non-blind prior information required to train the CAR-Net model.

Using the physical synthesis pipeline, we generated a custom dataset of 2,425 images based on nuScenes keyframes. We reserved 10% of the data (243 images) as a test set. The remaining 90% was split 80/20 ratio, with 1,745 images for training and 437 for validation.

Architecturally, CAR-Net are not using the standard encoder-decoder setup. It uses a cascaded refinement architecture. During training, we optimized the network by using a combination of $L1$ reconstruction, physical, angle, and SSIM losses (Wang et al., 2004). The combination of these specific loss functions ensures the model preserves the high-frequency spatial gradients that the downstream detector needs to function properly.

3.3. Algorithmic Adaptation: Cartesian CAR-Net

The core deblurring engine is based on the Cascade Refinement Network (CAR-Net) (Lai & Cetinkaya, 2026). By treating angular displacement (θ) as an explicit input and input into the feature extraction layers, CAR-Net achieves a significant improvement over blind deblurring methods. This non-blind prior information significantly reduces the mathematical search space, thereby achieving faster convergence and higher fidelity.

However, applying the original model directly to ground robots requires fundamental architectural adjustments. The original CAR-Net is designed to correct rotational and translational motion (such as camera shake on a drone) and relies on two specific components: a point spread function (PSF) generator and a polar coordinate transformation layer, which is designed to unfold concentric circular blurring into straight lines for processing.

Step 1: Point Spread Function (PSF) Adaptation. First, we replaced the original rotational PSF generator with a custom linear yaw version to match the physical movement of ground robots and solving the domain gap bewteen yaw blur and rotational blur. This new architecture helps the model can accurately simulate horizontal motion blur. However, applying this updated PSF within the existing polar coordinate space caused major algorithmic issues. Specifically, the polar transformation forcedly to warp our new, straight yaw trails into unnatural curves inside the network's latent space. When the model performed the inverse transformation, these distortions resulted in heavy "ringing" artifacts and false structural details (see Figure 2).

Step 2: The Cartesian Shift. To fix this geometric mismatch, we redesigned the architecture a second time by completely removed the polar coordinate transformation layer. We shifted the model to process both the images and the new point spread function (PSF) in Cartesian coordinates. This change aligned the network's focus with the actual physical movement of the ground robot. By operating natively in Cartesian space, CAR-Net handles horizontal stripes much better. As shown in

Figure 2c, it successfully eliminates ring artifacts while keeping the sharp structural edges—like lane lines and obstacles—intact for downstream detection nodes.

Figure 2. Visual comparison illustrating the geometric distortion caused by Polar CAR-Net versus the structurally accurate restoration of the adapted Cartesian CAR-Net.

(a) Original blurred input (b) Polar CAR-Net (c) Cartesian CAR-Net (d) Ground truth

3.4. Object Detection Strategy and Domain Adaptation

To evaluate the practical value of the reconstructed images and enable autonomous obstacle avoidance, we integrated a robust object detection module into our processing workflow. Given its industry-leading inference speed, we selected the YOLO architecture. For this evaluation, we specified "pedestrians" as the detection target to simulate critical real-world safety hazards in dynamic environments.

Although mean average precision (mAP) is a standard metric in computer vision literature, it is essentially a theoretical metric calculated during validation on static datasets. Real-world robot navigation relies heavily on temporal continuity; therefore, mAP is insufficient for evaluating real-time video streams on edge devices. To address this, we have established custom time-stability metrics: *Total Processed Frames*, *Frames with Detected Targets*, *Longest Blind Spot* (the maximum number of consecutive missed frames), and *Average Confidence Score*. In the context of autonomous navigation, minimizing the *longest blind spot* is critical, as prolonged perceptual failures directly correlate with an increased probability of collision.

To avoid overfitting to specific degraded features, we trained the model using only the raw, clear keyframes from the nuScenes dataset. We employed an empirical step-by-step elimination method to determine the optimal model configuration:

Phase 1: Evaluation of training methods. We compared the performance of retraining the entire network (updating all weights) against a layer-freezing approach (freezing the first 10 backbone layers and training only the semantic/regression head). Using YOLOv8 as a baseline, both models were tested on real-world *rosbag* recordings captured from a Wheeltec robot (rotating at 0.6 rad/s). We hypothesize that the layer-freezing method acts as a structural constraint to prevent "Catastrophic Forgetting" (Kirkpatrick et al., 2017), specifically by preserving the low-level geometric filters from the basis of general feature extraction.

Phase 2: Architecture Evaluation. After establishing a superior training method in Phase 1, we applied the "freeze-layer" method to three different architecture variants: the established YOLOv8, the optimized YOLOv11, and the experimental YOLOv26 (Jocher et al., 2023; Sapkota et al., 2025). These variants were evaluated against a *rosbag* reference at 0.6 rad/s, with the aim of identifying the architecture that provides the highest temporal stability (lowest blind spots) and the highest detection rate for the pedestrian class.

The optimal configuration resulting from this rigorous elimination process was subsequently designated as the final downstream detection node. This "winning" model was then applied to the final system evaluation to test the actual performance of the upstream Cartesian coordinate CAR-Net de-blurring module, which processed raw physical *rosbags* under a series of progressively increasing rotational yaw rates (0.6, 1.2, and 1.8 rad/s).

For the YOLO training and domain adaptation, models were trained using the Ultralytics framework for 50 epochs on a single GPU. To ensure memory efficiency on edge hardware, the image size was scaled to 256×256 pixels (imgsz = 256) with a batch size of 16. To evaluate the layer-freezing hypothesis, the first 10 layers (backbone layers 0 through 9) were frozen during training using the freeze = 10 parameter, while the control group was trained with all weights updated.

3.5 Edge Hardware Optimization and Latency Mitigation

Transitioning to a native ROS 1 environment on the NVIDIA Jetson Orin Nano 4GB resolved hardware driver instability issues but introduced significant computational and memory bottlenecks. The entire perception stack (operating system, ROS middleware, camera driver, Cartesian CAR-Net, and YOLOv8) must operate within a strictly enforced 4GB memory limit.

Initial benchmarking used an unoptimized PyTorch model and ran at the camera's default capture resolution, resulting in system failures. The camera outputs images at a rate of 30Hz, but the PyTorch-based CAR-Net could only process frames at 7Hz, while the YOLOv8 node failed to initialize entirely due to an out-of-memory (OOM) error. To achieve real-time operation, we implemented a multi-level optimization strategy.

1. Engine Serialization via TensorRT: Running native PyTorch (.pt) models on edge devices incurs significant interpreter overhead, and the default 32-bit floating-point (FP32) precision wastes memory bandwidth. Both the CAR-Net and YOLOv8 models were exported and compiled into the NVIDIA TensorRT (.trt) engine format. TensorRT applies layer fusion (merging convolution, bias, and activation layers into a single operation) and quantizes the model weights from FP32 to 16-bit half-precision (FP16) (Krishnamoorthi, 2018). This strategy was hypothesized to significantly reduce the memory footprint, allowing both models to fit within the 4GB RAM limit without triggering Linux's OOM-killer.

2. Resolution Scaling and Hardware Scheduling: To further mitigate the computational bottleneck of processing high-resolution images on the Nano's GPU, strategic input resolution scaling was planned. We designed a testing matrix to evaluate the system at 128×128 and 256×256 resolutions to find the optimal trade-off between spatial semantic fidelity (required by YOLO) and temporal latency (required for robot obstacle avoidance). The quantitative results of these optimization strategies are detailed in Section 4- *Experimentation and Benchmark Results.*

4. EXPERIMENTATION AND BENCHMARK RESULTS

The validation of the proposed robot perception workflow consists of two main phases. First, quantitative benchmarking of the individual modules (Cartesian coordinate system CAR-Net and YOLO) is conducted to verify the effectiveness of the architectural and training adjustments. Second, the fully integrated workflow is applied to real-world *rosbag* data captured by the Wheeltec robot to assess its practical navigational feasibility on Jetson Orin Nano (4GB) hardware.

4.1. Deblurring Module Validation: Polar Vs. Cartesian Alignment

To validate the systematic architectural adjustments made to CAR-Net, we evaluated these variants on a custom synthetic yaw dataset using Peak Signal-to-Noise Ratio (PSNR) and the Structural Similarity Index (SSIM) (Wang et al., 2004).

As shown in Table 1, the original baseline CAR-Net (using a rotational point spread function and polar coordinates) achieved a baseline PSNR of 25.94dB. Upgrading the model architecture by integrating the correct yaw PSF while still processing in polar coordinates yields measurable improvements, but the gains are limited, reaching only 27.69dB. These results confirm the hypothesis proposed in the methodology: if the network's coordinate processing space physically distorts the shape, simply providing the network with the correct blur shape (yaw) is insufficient.

Table 1. Quantitative Comparison of CAR-Net Architectural Adaptations

Model Configuration	PSNR (dB) ↑	SSIM ↑
Baseline (Original Polar Architecture)	25.94	0.7773
New Architecture in Polar	27.69	0.8184
Cartesian CAR-Net (No physical loss)	28.89	0.8492
Cartesian CAR-Net (With physical loss)	**29.52**	**0.8476**
Cartesian CAR-Net (With physical parameter noise)	29.27	0.8413

Following the transition to a Cartesian processing workflow, a key engineering breakthrough was achieved. By completely bypassing polar coordinate-induced bias and evaluating images in the native Cartesian coordinate system, the PSNR value achieved a significant quantitative leap to 29.52dB when using precise physical priors. Even when simulating real-world sensor errors by introducing parametric noise into the priors, the Cartesian model still maintains a highly robust performance of 29.27dB. Synchronizing point spread function (PSF) generation and geometric processing with the robot's physical kinematics has proven to be the key to achieving optimal structural reconstruction.

4.2. Object Detection Optimization: The Elimination Process

After validating the deblurring module, we conducted performance tests on the downstream YOLO detection node. The target class was set to "pedestrians," which represent a critical dynamic safety hazard.

Phase 1: Training Methodology. The first phase evaluated whether to perform a comprehensive retraining of the network, or adopt feature extraction using frozen layer. Using YOLOv8 as the test architecture, both models were trained on the high-resolution nuScenes dataset and then tested on a physical *rosbag* recording (containing a total of 105 frames) of a robot turning at 0.6 rad/s.

Table 2. YOLOv8 Training Methodology Evaluation (0.6 rad/s Yaw Blur)

Method	Detected Frames	Detection Rate	Max Consecutive Misses	Avg Confidence
Full Retrain	4 / 105	3.81%	50 frames	0.3404
Frozen Layer	44 / 105	41.90%	23 frames	0.4961

As shown in Table 2, we observed a major difference in performance between the two approaches. The fully retrained YOLOv8 model struggled heavily, detecting pedestrians in only 4 out of 105 frames (3.81%). At one point, it failed to detect anything for 50 consecutive frames. In a real-world autonomous driving scenario, losing sight of the road for that long would almost certainly lead to a collision.

In contrast, the layer-freezing method successfully detected targets in 41.90% of the frames, raising the average confidence score to approximately 0.50 and reducing the longest blind spot to 23 frames—a reduction of more than half. This provides strong evidence for the "Catastrophic Forgetting" hypothesis. If the network were fully retrained, the model would overfit the nuScenes data and lose its robust pre-trained COCO baseline filter (Lin et al., 2014) (which performs exceptionally well in recognizing general human silhouettes). By freezing the first 10 layers, the network successfully retained its fundamental edge detection capabilities, enabling it to more effectively "penetrate" moderate blurring.

Phase 2: Architectural Evaluation. After establishing the "freeze-layer" approach as the superior training method, we evaluated three YOLO variants (YOLOv8, YOLOv11, and YOLOv26). Before analyzing real-world temporal stability, we first established a theoretical baseline by evaluating these models on a static validation set containing motion-blurred frames. Table 3 presents the standard computer vision metrics, including Precision, Recall, F1-score, and Mean Average Precision (mAP@50 and mAP@50-95).

Table 3. Baseline Evaluation of YOLO Variants on Static Validation Set

Model	Precision	Recall	F1-Score	mAP@50	mAP@50-95
YOLOv11	**0.686**	0.432	**0.530**	0.496	**0.314**
YOLOv8	0.667	**0.433**	0.525	**0.497**	0.313
YOLOv26	0.600	0.405	0.484	0.463	0.284

While standard metrics such as mAP provide a snapshot of spatial accuracy
While standard metrics such as mAP provide a snapshot of spatial accuracy on individual frames, they do not account for temporal continuity. To determine which architecture prevents critical navigation failures, we tested these models against the continuous 0.6 rad/s *rosbag* video stream using our custom time-stability metrics.

Table 4. YOLO Architectural Comparison (Frozen Layers, 0.6 rad/s Yaw Blur)

Model Variant	Detected Frames	Detection Rate	Max Consecutive Misses	Avg Confidence
YOLOv11 (Frozen)	32 / 105	30.48%	42 frames	0.3425
YOLOv8 (Frozen)	**44 / 105**	**41.90%**	**23 frames**	**0.4961**
YOLOv26 (Frozen)	3 / 105	2.86%	77 frames	0.4856

As shown in Table 4, YOLOv8 outperforms its successors by a wide margin. While YOLOv11 still maintains a usable baseline performance (30.48% detection rate), the highly experimental YOLOv26 suffered a near-total collapse, failing to meet the target for 77 consecutive frames. These results provide a key academic insight: although newer models in the YOLO family have optimized the depth-to-width ratio for ideal, static benchmark datasets (such as COCO), they often sacrifice the generalization robustness present in earlier, more mature achitectures (such as YOLOv8). The core architecture of YOLOv8 demonstrates significantly higher robustness against interference caused in high-frequency information loss due to motion blur. Therefore, YOLOv8 which adopts a frozen-layer architecture, was selected as the detection node for this processing workflow.

4.3. Hardware Optimization and Throughput Benchmarks

To quantitatively validate the edge optimization strategy outlined in Section 3.5, we benchmarked the system on a Jetson Orin Nano (4GB). Initial attempts is running unoptimized native PyTorch models and failed completely; the PyTorch-based CAR-Net model was severely limited to approximately ~7Hz, and the YOLOv8 node triggered an out-of-memory (OOM) crash during initialization.

As shown in Table 5, translate to a serialized TensorRT FP16 engine successfully bypassed the 4GB memory bottleneck.

Table 5. Hardware Optimization Impact on Jetson Orin Nano (4GB Unified Memory)

Framework / Engine	Precision	Memory	CAR-Net	YOLO	Pipeline Status
Native PyTorch (.pt)	FP 32	>4 GB (OOM)	~7 Hz	0 Hz (failed)	Crash / Unusable
TensorRT (128×128)	FP 16	>4 GB (Stable)	30 Hz	30 Hz	Real-time
TensorRT (256×256)	FP 16	<4 GB (Stable)	20 Hz	20 Hz	**Optimal Trade-off**

While the 128×128 resolution achieved a perfect 30 Hz synchronization with the camera, excessive downsampling destroyed the spatial fidelity required by the YOLO regression head to draw bounding boxes around distant pedestrians. Consequently, the 256×256 TensorRT configuration was selected. Although this increased computational load reduced the inference update rate to 20 Hz, it provided the optimal balance: retaining sufficient semantic detail for accurate object detection while comfortably exceeding the standard 15 Hz threshold required for dynamic obstacle avoidance.

4.4. Full Pipeline Integration and Speed Stress Test

The final evaluation of the system measured the end-to-end perception processing workflow running on a Jetson Orin Nano (4GB). The physical robot recorded *rosbags* at progressively increasing yaw rates: 0.6 rad/s, 1.2 rad/s and 1.8 rad/s. A frozen YOLOv8 node processed these streams under two conditions: blurred (raw camera images) and deblurred (CAR-Net processing in Cartesian coordinates optimized via TensorRT).

Table 6. End-to-End Pipeline Stress Test across Escalating Yaw Velocities.

Rotational Speed (Total Frames)	**Detection Rate (%)**		**Max Consecutive Misses**		**Avg. Confidence**	
	No Deblur	**With Deblur**	**No Deblur**	**With Deblur**	**No Deblur**	**With Deblur**
0.6 rad/s (105)	41.90%	**43.81%**	23 frames	**21 frames**	0.4961	0.4949
1.2 rad/s (65)	12.31%	12.31%	32 frames	**23 frames**	0.4146	**0.5031**
1.8 rad/s (35)	0.00%	0.00%	35 frames	35 frames	N/A	N/A

The data in Table 6 illustrates the real-world value of the deblurring module:

- **Low Velocity (0.6 rad/s):** Image blurring is relatively minor. YOLOv8 is robust enough to handle the raw image data effectively. Cartesian CAR-Net offers a slight improvement, not only slightly increasing the overall detection rate but also reducing the longest blind spot by 2 frames.

Figure 3. Comparison of maximum consecutive missed frames. At (1.2 rad/s), the deblurring module significantly reduces the perceptual blind spot by 9 frames.

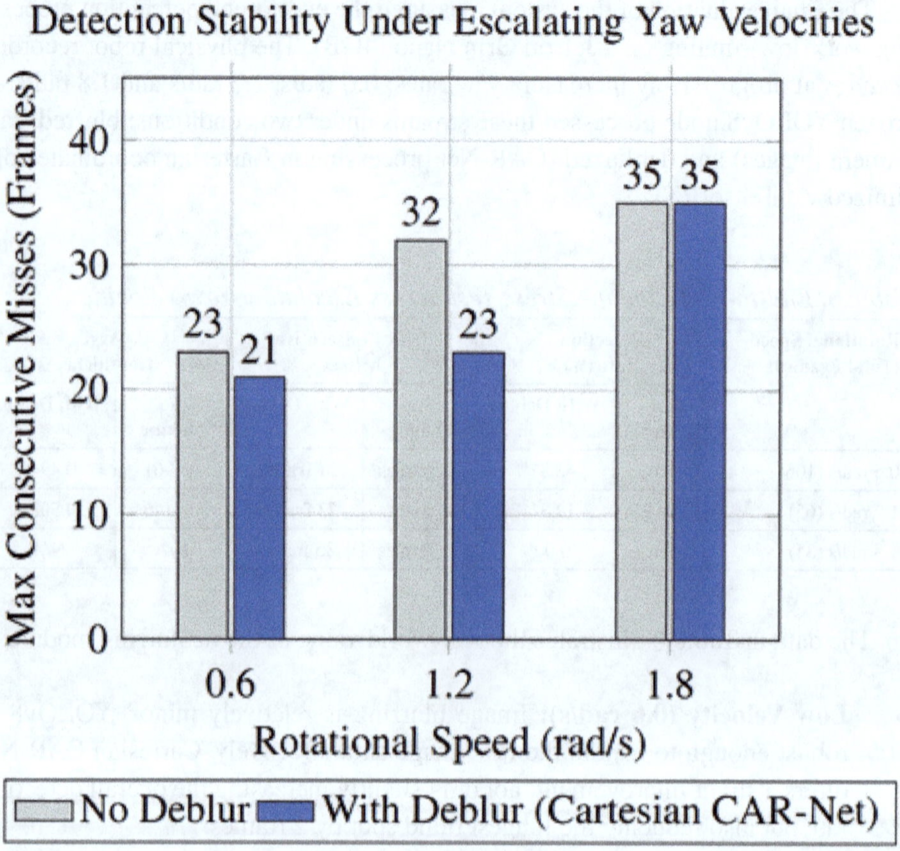

- **Medium Velocity (1.2 rad/s):** This represents the "optimal point" for demonstrating the effectiveness of the processing workflow, visualized in Figure 3. At this speed, motion blur significantly degrades the quality of semantic features, resulting in a total of 8 detected frames in both scenarios. However, the distribution of these detection results changes dramatically. Without deblurring, the robot would face a critical continuous blind spot lasting up to 32 frames. After enabling the Cartesian coordinate system CAR-Net, the maximum consecutive missed frames dropped significantly to 23 frames, and the average confidence score of detections increase by nearly 10%. In practical robotic applications, recovering these 9 frames of visual information during sharp turns can provide the navigation system with the critical milliseconds needed to trigger an emergency stop and avoid collisions.

- **High Velocity (1.8 rad/s):** At such extreme rotational speeds, the system encounters severe physical limitations. The speed of motion far exceeds the camera's integration time, resulting in the complete loss of all semantic geometric information. Regardless of the physical prior information used, no algorithm can reconstruct information that the sensor has completely failed to capture. Both scenarios result in a detection rate of 0%, highlighting the limitations of current camera hardware in high velocity scenarios.

In conclusion, the experimental results confirm that the TensorRT-optimized Cartesian-coordinate CAR-Net effectively reduce yaw blur at medium ranges, stabilizes downstream YOLOv8 tracking, reduces critical blind spots, and runs smoothly under the strict computational constraints of a 4GB edge device.

4.5. Qualitative Evaluation of Pipeline Integration

To visually validate the quantitative metrics presented in Table 6, a qualitative analysis was conducted on the integrated perception pipeline during physical deployment. Figure 4 presents a side-by-side comparison of a specific frame captured during the 0.6 rad/s rotational stress test.

Figure 4. Qualitative evaluation of the perception pipeline under a yaw rotation of 0.6 rad/s. (Left) The raw camera image is affected by horizontal motion blur, resulting in the loss of gradient information and subsequent detection errors. (Right) The same frame processed by Cartesian CAR-Net at a resolution of 256×256. The deblurring module successfully recovers enough structural edges for the YOLOv8 model to register a positive detection (confidence: 0.26), effectively preventing a tracking failure.

(a) Raw Input (No Detection) (b) Deblurred Output (Pedestrian 0.26)

As shown in the figure 4a, the original camera footage exhibits horizontal motion blur due to the robot's yaw rotation. This linear blurring acts like a low-pass filter, smearing the high-frequency structural edges of pedestrians. Consequently, the YOLOv8 detector in the freeze layer is completely unable to extract the necessary geometric features, resulting in false negatives (blind spots).

In contrast, the figure 4b shows the result of processing the same frame using a Cartesian coordinate CAR-Net module optimized with TensorRT. Despite employing aggressive downsampling limited by hardware (reducing the resolution to 256×256, resulting in noticeable pixelation), the algorithm successfully recovered the core structural boundaries of the target object. The recovered gradient information was sufficient for the downstream YOLOv8 regression head to successfully draw bounding boxes and register pedestrians into the system with a confidence level of 0.26. Although the confidence level is relatively low, this qualitative result perfectly demonstrates the system's main objectives: recovering edge visual data, preventing continuous tracking failures, and providing actionable obstacle data for navigation stacking.

5. DISCUSSION

Building and deploying this ROS-based pipeline highlighted the massive gap between deep learning theory and actual edge robotics. Today, most computer vision research focuses on chasing the highest possible accuracy on static datasets using nearly unlimited cloud computing power. However, our work shows that deploying a robot in the real world is a completely different challenge. Success in the field depends almost completely on overcoming strict hardware limits, maintaining stable middleware, and handling actual physical motion.

5.1. Middleware Pragmatism: The ROS 2 Abstraction Penalty

A major turning point in our project was the decision to drop our containerized ROS 2 (Humble) setup and move to a native ROS 1 (Noetic) deployment. In theory, ROS 2 is the standard of modern robotics by utilizing the DDS protocol for robust network communication. However, in practice, we found that running it on resource-constrained edge device like the Jetson Orin Nano resulted in excessive abstraction overhead.

Specifically, we encountered severe stability issues with the Wheeltec robot's low-level controller (responsible for processing motor encoder data and serial commands).The combination of DDS overhead and Docker's network bridging caused random latency spikes that cause the serial driver miss critical drive frames. This aligns with existing literature, which demonstrates that the ROS 2 abstraction layer must convert data for DDS (Maruyama et al., 2016). This process incurs additional transport latency and overhead—up to a 50% increase compared to native DDS (Kronauer et al., 2021)—which acts as a severe bottleneck for distributed, embedded systems. Additionally, placing our deep learning stack inside a container made unnecessarily complicated memory mapping. Docker's abstraction layer repeatedly interfered with the Jetson's native CUDA path and our TensorRT initialization, causing the system to crash repeatedly.

Reverting to native ROS 1 solved these issues. Even though it relies on the older roscore master node and standard TCP/UDP protocols, it allowed us to connect to the serial hardware directly. By completely removing Docker and running natively on Ubuntu 20.04, we regained uninterrupted access to the Jetson's GPU memory. This ensured reliable communication between our vision stack and the drive wheels. In the end, this experience demonstrated that middleware selection must be based on strict hardware realities, rather than simply adopting the latest software framework.

5.2. Edge Resource Scarcity and the 4GB Memory Crisis

Running two heavy neural networks (CAR-Net and YOLO) simultaneously on a 4GB Jetson Orin Nano was one of our biggest challenges. Unlike a standard desktop with a dedicated GPU, the Jetson relies on a unified memory architecture. This means the CPU (handling the operating system, ROS, and camera drivers) and the GPU (handling CUDA tensors) must share that strict 4GB limit.

When we initially attempted to run standard PyTorch (.pt) models, the system immediately crashed due to an out-of-memory (OOM) error. Since PyTorch uses 32-bit floating-point (FP32) precision by default and employs dynamic computation graphs, each pixel and network weight occupies 4 bytes of RAM, making model compression and acceleration techniques a necessity for edge robotics (Cheng et al., 2017). Additionally, running PyTorch natively incurs significant overhead due to the Python interpreter.

To resolve this issue, we switched to NVIDIA TensorRT. This was not merely a minor performance tweak but the only way to get the system running. By exporting the model as a .trt engine, we leveraged layer fusion—combining convolution, bias, and activation layers into a single operation—thereby reducing the number of memory accesses. We also quantized the weights from FP32 to FP16. As demonstrated in recent quantization (Krishnamoorthi, 2018). This successfully halved memory usage without significantly reducing the network's accuracy.

5.3. The Spatial Vs. Temporal Trade-off in Resolution Scaling

Despite our TensorRT optimizations, input resolution remained a major bottleneck in the workflow. The camera hardware naturally captures and outputs frames at 30Hz. If we set the processing pipeline to a low 128×128 resolution, the Jetson easily runs both the deblurring and detection nodes at that full 30Hz rate. However, doing this reveals a major limitation in how YOLO handles spatial data.

Object detectors need enough spatial detail to calculate accurate bounding boxes. At a 128×128 resolution, the image is simply too compressed. Distant or smaller objects—like a pedestrian a few meters away—shrink down to an unrecognizable cluster of three or four pixels. Without sufficient gradient information, the YOLO network cannot draw clear bounding boxes, which completely negates the benefit of running at a fast 30Hz.

To fix this, we increased the resolution to 256×256. Quadrupling the pixel count restored the spatial gradients that the YOLO regression head needs to work correctly. However, this heavier computational load caused a synchronization mismatch: the camera still output frames at 30Hz, but the inference nodes dropped to a 20Hz update rate. Ultimately, we found this 20Hz rate to be the best compromise. It retains enough semantic detail for accurate detection while staying fast enough to support dynamic obstacle avoidance, which generally requires at least 15Hz to run stably. In the end, these results show that deploying edge AI is fundamentally an exercise in balancing temporal speed against spatial accuracy.

5.4. Algorithmic Induction Bias: Polar Vs. Cartesian Coordinates

A significant scientific contribution of this research lies in identifying and addressing a gap in the kinematic domain present in standard deblurring architectures. The initial failure of the baseline CAR-Net model when applied to camera footage from a Wheeltec robot highlighted a critical deficiency in deep learning models regarding encoding inductive bias.

The initial CAR-Net model was highly optimized for drone and handheld photography, scenarios where rotation (Z-axis) produces concentric circle blur. To handle this blur, the network employed a polar coordinate transformation, essentially unfolding the circle into a straight line, applying deconvolution, and then unfolding it again. However, ground robots primarily operate in a two-dimensional plane; their turning motion produces purely horizontal linear yaw (Y-axis) blur.

When the Polar CAR-Net receives this linear yaw blur, the mathematical transformation excessively distorts the straight-line motion trajectory in the network's latent space, turning it into complex and unnatural curves. This forces the deconvolutional layers to attempt to solve a non-existent geometric problem, resulting in severe "ringing" artifacts and illusory geometry observed during intermediate testing.

By fundamentally refactoring the network to handle point spread function (PSF) and pixel data strictly in Cartesian coordinates, the architecture's inductive bias is consistent with the physical characteristics of ground robots. Empirical results showing a PSNR jump to 29.52 dB confirm that neural network architectures cannot be blindly copied from one robotic domain (aerial) to another (ground). The geometric priors embedded in the network layers must be explicitly matched to the kinematic degrees of freedom of the physical hardware.

5.5. Mitigating Catastrophic Forgetting in YOLO

Results from YOLO detection nodes provide empirical evidence for domain adaptation in autonomous driving systems. Regarding motion blur at 0.6 rad/s, the performance of the entirely retrained YOLOv8 model is quite low at 3.81%, as compared to a 41.90% performance rate for the frozen-layer method.

Because the frozen weights have already been optimized for color, the general low-level edges and shape filters, which are already assigned their own general purpose classifications using the general low-level features learned during pre-training on the large-scale COCO (Common Objects in Context) dataset- therefore, the new network would cause "catastrophic forgetfulness" since it was trained upon a different dataset than what was previously optimized. The result is that the new YOLO has "lost" its original capabilities of identifying general geometric shapes via the previously stored filter weights by utilizing the real world, slightly motion blurred images that have been captured by Wheeltec's robotic camera systems. Thus, the new model will now not be able to provide sufficient low-level filter capabilities to reliably detect general human geometric objects within those types of images.

By freezing the first 10 layers, the network is forced to retain its generalized COCO filters. Only the deeper semantic layers and the bounding box regression head are allowed to adapt to autonomous driving scenarios. This strategy results in a highly robust detector capable of "penetrating" moderate blurring and distortion. Which lead to the failure of the newer YOLOv26 architecture further validates this concept. Newer models are highly optimized to chase benchmark scores on the original dataset, often pruning redundant parameters to improve speed. However, in noisy and dynamic robotic environments, it is precisely this parameter redundancy that often provides the robustness needed to withstand visual degradation.

5.6. The Physical Ceiling of Hardware Integration

Although the algorithm successfully reduced mild (0.6 rad/s) and moderate (1.2 rad/s) motion blur, at rotation speeds as high as 1.8rad/s, the system's detection rate plummeted to zero even with the CAR-Net Cartesian coordinate system enabled.

It must be clearly understood that algorithmic deblurring is essentially a data recovery process, not data generation. At a rotational speed of 1.8 rad/s, the Wheeltec robot rotates fast enough for the relative motion of the environment to completely overwhelm the standard exposure time (approximately 0.05s) of a low-cost CMOS rolling-shutter camera. The resulting images are not merely blurred; high-frequency semantic information (edges, texture, color) has been completely erased, resulting in blurring and diffusion across hundreds of pixels.

Regardless of how sophisticated our deep neural networks or inertial measurement unit (IMU) information is, software cannot replicate the visual information received from the camera hardware that was unable to completely record that information. The fact that this is true shows that we cannot depend solely on vision when we build autonomous systems for use in mobile robotic systems. The design of future systems must include both algorithmic techniques and improvements to hardware; for mobile robots that will operate at high rates, it will be necessary to use global shutter cameras with microsecond-long exposure times, or to implement secondary sensors such as 2D LiDAR. All of these additional features are required in order to overcome the limits imposed by real-world lighting conditions, and the integration times of cameras in robotic systems.

5.7. Limitations of the Study

While our physical hardware stress tests involved relatively short rosbag recordings over the course of 35 to 105 frames per sequence, we agree that additional evaluation against a large multi-environment dataset is warranted to validate long-term reliability of the pipeline. Future iterations of this effort must also include substantially longer physical tests across multiple real-world environments and various lighting conditions.

6. CONCLUSION AND FUTURE WORK

6.1. Conclusion

One serious challenge for implementing vision based mobile robots is how fast visual data degrades when there is quick motion involved. Therefore, during this project, a Real Time, ROS based visual perception pipeline was created and implemented that actively decreases motion blur effects caused by rotation and yaw motions. Additionally, the overall system was designed so that it can operate using only low power edge hardware.

Furthermore, by redesigning the overall structure of the perception system and employing Domain Adaptation approaches, we have successfully closed the distance between the theoretical use of image restoration techniques, and their real-world application to autonomous mobile robot navigation.

When we started the system engineering phase of our project, it became apparent to us that choosing the right middleware and maximizing our hardware configuration would be very important. To accomplish this goal, we transitioned from a containerized ROS 2 environment to a native ROS 1 (Noetic) configuration. This allowed us to eliminate our issues with serial driver instability and create a robust link between the GPU on our Jetson Orin Nano and the base controller of the Wheeltec Robot. In addition, we worked very hard to optimize our computations such that we would remain below the strict limit of 4GB of unified memory on our edge device. By exporting our PyTorch model to an NVIDIA TensorRT engine and applying FP16 quantization and layer fusion, we were able to dramatically reduce the occurrence of out-of-memory (OOM) errors. As a result of these optimizations, our system is now able to maintain a stable inference rate of 20Hz with an input resolution of

256×256, thus achieving the requirement for low-latency performance required for real-time obstacle avoidance.

From an algorithmic perspective, we found a critical flaw in the basis of operation of many of the existing Non-blind deblurring models as a result of their design. Many of these models operate in polar coordinate systems with respect to roll motion of aerial cameras. This is a fundamental conflict with the flat linear yaw motion characteristic of ground robots. In order to solve this, we rewritten the PSF and moved the Cascaded Refined Network (CAR-Net) completely into Cartesian space for processing. This change resolved geometric "ringing" artifacts and resulted in a significantly clearer recovery result with an improvement to our baseline of PSNR to 29.52dB.

Our research on Downstream Detection has provided valuable insights into Transfer Learning. When we retrained modern Object Detectors (like YOLO) for the purpose of testing the robot on specific datasets using a full re-training approach, these Retrainings produced "Catastrophic Forgetting" of the object detection model's filters used for obtaining edge features. To provide maximum stability for Tracking, we froze the initial processing layers of the YOLOv8 Object Detector. The entire integrated system has been verified through end-to-end physical stress tests. Even when the robot was aggressively turned (1.2 rad/s), the Cartesian-based CAR-Net was able to reduce its maximum continuous detection blind spot from 32 frames to 23 frames. As a result, the Navigation Stack had significantly more time to activate its Collision Avoidance Protocols.

6.2. Future Work

Even if this project established an excellent initial basis for the edge device deblurring technology, it demonstrated practical limits when operating at extreme rotational speeds, which suggest some distinct routes to pursue in future care for both research and development.

1. Hardware Upgrades and Global Shutter Integration: The detection process fails completely at 1.8 rad/s, indicating a physical limit that is difficult to overcome. The low-cost CMOS cameras currently in use employ a rolling shutter mechanism, which inherently introduces local distortion during high-speed translation. More importantly, the sensor's integration time causes high-frequency structural information to become completely blurred, making it impossible to recover via algorithms. Future versions of this platform must incorporate global-shutter cameras capable of microsecond-level exposures. Although this will result in darker raw images due to the shorter light integration time, low-light enhancement algorithms have lower computational costs and are mathematically more stable than extreme motion deblurring.

2. Hardware-Synchronized IMU Fusion: Currently, the CAR-Net module employs a physics-based simulated angular angle prior (θ) derived from the robot's known kinematic commands. To apply this system in unpredictable and highly challenging environments, the network must be capable of receiving real-time physical data from an inertial measurement unit (IMU). Future research should focus on developing hardware-synchronized triggering loops that lock the camera shutter and IMU gyroscope timestamps together at the hardware level, and feed the fully synchronized ω vectors directly and in real time into the CAR-Net latent space in Cartesian coordinates.

3. Adaptive Resolution and Dynamic Routing: Processing each frame of video using deep learning for de-blurring consumes a significant amount of power and computational resources. However, when the robot is moving in a straight line, the degree of image blurring is negligible. Future software architectures should implement an adaptive dynamic routing node. By monitoring real-time odometry data from the hub encoder and yaw angular velocity, the system can conditionally bypass the CAR-Net node entirely during linear motion, routing the raw camera image directly to the YOLO detector at a frequency of 30Hz. The deblurring node will only be dynamically triggered when the angular velocity exceeds a specific threshold (e.g., >0.4 rad/s). This will significantly reduce thermal throttling on the Jetson Orin Nano and extend battery life, representing the next step toward building a truly intelligent and context-aware robot perception stack.

REFERENCES

Cadena, C., Carlone, L., Carrillo, H., Latif, Y., Scaramuzza, D., Neira, J., Reid, I., & Leonard, J. J. (2017). Past, present, and future of simultaneous localization and mapping: Toward the robust-perception age. *IEEE Transactions on Robotics*, *32*(6), 1309–1332. DOI: 10.1109/TRO.2016.2624754

Caesar, H., Bankiti, V., Lang, A. H., Vora, S., Liong, V. E., Xu, Q., Krishnan, A., Pan, Y., Baldan, G., & Beijbom, O. (2020). Nuscenes: A multimodal dataset for autonomous driving. *Proceedings of the IEEE/CVF Conference on Computer Vision and Pattern Recognition*, 11621–11631. DOI: 10.1109/CVPR42600.2020.01164

Campos, C., Elvira, R., Rodríguez, J. J. G., Montiel, J. M., & Tardós, J. D. (2021). Orb-slam3: An accurate open-source library for visual, visual–inertial, and multimap slam. *IEEE Transactions on Robotics*, *37*(6), 1874–1890. DOI: 10.1109/TRO.2021.3075644

Cheng, Y., Wang, D., Zhou, P., & Zhang, T. (2017). A survey of model compression and acceleration for deep neural networks. *arXiv Preprint arXiv:1710.09282*.

Dunbabin, M., & Marques, L. (2012). Robots for environmental monitoring: Significant advancements and applications. *IEEE Robotics & Automation Magazine*, *19*(1), 24–39. DOI: 10.1109/MRA.2011.2181683

Engel, J., Koltun, V., & Cremers, D. (2017). Direct sparse odometry. *IEEE Transactions on Pattern Analysis and Machine Intelligence*, *40*(3), 611–625. DOI: 10.1109/TPAMI.2017.2658577 PMID: 28422651

Jocher, G., Chaurasia, A., & Qiu, J. (2023). *Ultralytics YOLOv8* (Version 8.0.0). https://github.com/ultralytics/ultralytics; GitHub.

Kirkpatrick, J., Pascanu, R., Rabinowitz, N., Veness, J., Desjardins, G., Rusu, A. A., Milan, K., Quan, J., Ramalho, T., Grabska-Barwinska, A., Hassabis, D., Clopath, C., Kumaran, D., & Hadsell, R. (2017). Overcoming catastrophic forgetting in neural networks. *Proceedings of the National Academy of Sciences of the United States of America*, *114*(13), 3521–3526. DOI: 10.1073/pnas.1611835114 PMID: 28292907

Krishnamoorthi, R. (2018). Quantizing deep convolutional networks for efficient inference: A whitepaper. *arXiv Preprint arXiv:1806.08342*.

Kronauer, T., Pohlmann, J., Matthé, M., Smejkal, T., & Fettweis, G. (2021). Latency analysis of ros2 multi-node systems. *arXiv Preprint arXiv:2101.02074*. DOI: 10.1109/MFI52462.2021.9591166

Kupyn, O., Martyniuk, T., Wu, J., & Wang, Z. (2019). Deblurgan-v2: Deblurring (orders-of-magnitude) faster and better. *Proceedings of the IEEE/CVF International Conference on Computer Vision*, 8878–8887.

Lai, K. C., & Cetinkaya, A. (2026). Car-net: A cascade refinement network for rotational motion deblurring under angle information uncertainty. *Proceedings of the International Conference on Artificial Intelligence and Machine Learning (AAIML).* DOI: 10.1109/AAIML67890.2026.11498173

Lai, K. C., & Mohammed, A. M. (2026). Robust visual perception for mobile robots: Integrating physics-informed deblurring in ROS2. *International Conference on Wireless Sensor Networks, Ubiquitous Computing and Applications (ICWSNUCA).*

Liang, J., Cao, J., Sun, G., Zhang, K., Van Gool, L., & Timofte, R. (2021). Swinir: Image restoration using swin transformer. *Proceedings of the IEEE/CVF International Conference on Computer Vision*, 1833–1844.

Lin, T.-Y., Maire, M., Belongie, S., Hays, J., Perona, P., Ramanan, D., Dollár, P., & Zitnick, C. L. (2014). Microsoft coco: Common objects in context. *European Conference on Computer Vision*, 740–755.

Luan, S., Yang, C., Qin, X., Chen, D., & Sui, W. (2024). Towards robust visual odometry by motion blur recovery. *Frontiers in Signal Processing*, *4*, 1417363. DOI: 10.3389/frsip.2024.1417363

Macario Barros, A., Michel, M., Moline, Y., Corre, G., & Carrel, F. (2022). A comprehensive survey of visual slam algorithms. *Robotics (Basel, Switzerland)*, *11*(1), 24. DOI: 10.3390/robotics11010024

Macenski, S., Foote, T., Gerkey, B., Lalancette, C., & Woodall, W. (2022). Robot operating system 2: Design, architecture, and uses in the wild. *Science Robotics*, *7*(66), eabm6074. DOI: 10.1126/scirobotics.abm6074 PMID: 35544605

Maruyama, Y., Kato, S., & Azumi, T. (2016). Exploring the performance of ROS2. *Proceedings of the 13th International Conference on Embedded Software*, 1–10.

Merkel, D. (2014). Docker: Lightweight linux containers for consistent development and deployment. *Linux J., 2014*(239).

NVIDIA Corporation. (2023). *NVIDIA TensorRT documentation.* NVIDIA Corporation. https://developer.nvidia.com/tensorrt

Paszke, A., Gross, S., Massa, F., Lerer, A., Bradbury, J., Chanan, G., Killeen, T., Lin, Z., Gimelshein, N., Antiga, L., Desmaison, A., Kopf, A., Yang, E., DeVito, Z., Raison, M., Tejani, A., Chilamkurthy, S., Steiner, B., Fang, L., . . . Chintala, S. (2019). PyTorch: An imperative style, high-performance deep learning library. In H. Wallach, H. Larochelle, A. Beygelzimer, F. d'Alché-Buc, E. Fox, & R. Garnett (Eds.), *Advances in neural information processing systems* (Vol. 32). Curran Associates, Inc. https://proceedings.neurips.cc/paper_files/paper/2019/file/bdbca288fee7f92f2bfa9f7012727740-Paper.pdf

Quigley, M., Gerkey, B., Conley, K., Faust, J., Foote, T., Leibs, J., Berger, E., Wheeler, R., & Ng, A. Y. (2009). ROS: An open-source robot operating system. *ICRA Workshop on Open Source Software, 3*, 5.

Sapkota, R., Cheppally, R. H., Sharda, A., & Karkee, M. (2025). YOLO26: Key architectural enhancements and performance benchmarking for real-time object detection. *arXiv Preprint arXiv:2509.25164*. https://arxiv.org/abs/2509.25164

Saputra, M. R. U., Markham, A., & Trigoni, N. (2018). Visual SLAM and structure from motion in dynamic environments: A survey. *ACM Computing Surveys*, *51*(2), 1–36. DOI: 10.1145/3177853

Tao, X., Gao, H., Shen, X., Wang, J., & Jia, J. (2018). Scale-recurrent network for deep image deblurring. *Proceedings of the IEEE Conference on Computer Vision and Pattern Recognition*, 8174–8182.

Wang, Z., Bovik, A. C., Sheikh, H. R., & Simoncelli, E. P. (2004). Image quality assessment: From error visibility to structural similarity. *IEEE Transactions on Image Processing*, *13*(4), 600–612. DOI: 10.1109/TIP.2003.819861 PMID: 15376593

Zamir, S. W., Arora, A., Khan, S., Hayat, M., Khan, F. S., & Yang, M.-H. (2022). Restormer: Efficient transformer for high-resolution image restoration. *Proceedings of the IEEE/CVF Conference on Computer Vision and Pattern Recognition*, 5728–5739. DOI: 10.1109/CVPR52688.2022.00564

Zhou, Z., Chen, X., Li, E., Zeng, L., Luo, K., & Zhang, J. (2019). Edge intelligence: Paving the last mile of artificial intelligence with edge computing. *Proceedings of the IEEE*, *107*(8), 1738–1762. DOI: 10.1109/JPROC.2019.2918951

APPENDIX

A. Hardware Specifications

Figure 5 presents the hardware specification and system status of the NVIDIA Jetson Orin Nano (4GB) used in this project, as captured by the jtop system monitoring tool.

Figure 5. NVIDIA Jetson Orin Nano (4GB) hardware specifications as reported by jtop.

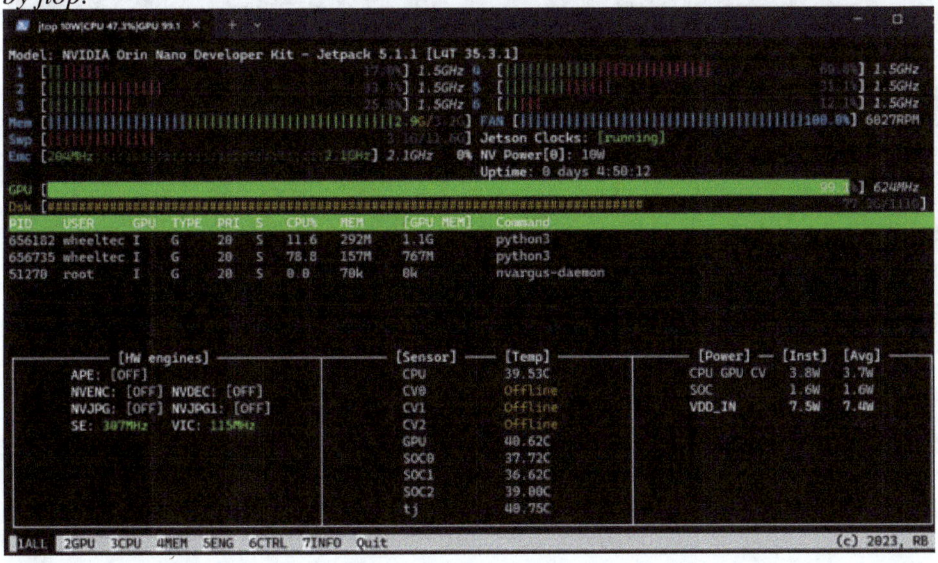

B. ROS Pipeline Throughput Verification

Figures 6, 7, and 8 present the real-time publishing frequency of each ROS node in the perception pipeline, captured during live deployment at 256×256 resolution on the Jetson Orin Nano (4GB).

Figure 6. Camera node /camera/raw_image

```
wheeltec@wheeltec:~$ rostopic hz /camera/raw_image
subscribed to [/camera/raw_image]
average rate: 30.289
        min: 0.032s max: 0.034s std dev: 0.00028s window: 22
average rate: 29.926
        min: 0.000s max: 0.046s std dev: 0.00937s window: 52
average rate: 30.308
        min: 0.000s max: 0.048s std dev: 0.01190s window: 82
average rate: 30.064
        min: 0.000s max: 0.048s std dev: 0.01255s window: 112
average rate: 30.305
        min: 0.000s max: 0.048s std dev: 0.01306s window: 143
average rate: 30.304
        min: 0.000s max: 0.048s std dev: 0.01347s window: 174
average rate: 30.243
        min: 0.000s max: 0.048s std dev: 0.01354s window: 204
average rate: 30.304
        min: 0.000s max: 0.048s std dev: 0.01378s window: 234
average rate: 30.304
        min: 0.000s max: 0.048s std dev: 0.01395s window: 265
average rate: 30.271
        min: 0.000s max: 0.048s std dev: 0.01395s window: 295
```

Figure 7. Deblurring node /camera/deblurred_image

```
wheeltec@wheeltec: ~
average rate: 20.938
        min: 0.009s max: 0.617s std dev: 0.02570s window: 656
average rate: 20.932
        min: 0.009s max: 0.617s std dev: 0.02536s window: 676
average rate: 20.930
        min: 0.009s max: 0.617s std dev: 0.02497s window: 697
average rate: 20.929
        min: 0.009s max: 0.617s std dev: 0.02460s window: 718
average rate: 20.898
        min: 0.009s max: 0.617s std dev: 0.02436s window: 738
average rate: 20.897
        min: 0.009s max: 0.617s std dev: 0.02403s window: 759
average rate: 20.896
        min: 0.009s max: 0.617s std dev: 0.02371s window: 780
average rate: 20.893
        min: 0.009s max: 0.617s std dev: 0.02340s window: 801
average rate: 20.890
        min: 0.009s max: 0.617s std dev: 0.02315s window: 821
```

Figure 8. Detection node /detection/result

```
wheeltec@wheeltec:~$ rostopic hz /detection/result
subscribed to [/detection/result]
average rate: 21.129
        min: 0.024s max: 0.065s std dev: 0.00711s window: 20
average rate: 20.588
        min: 0.024s max: 0.069s std dev: 0.00671s window: 41
average rate: 20.676
        min: 0.024s max: 0.071s std dev: 0.00770s window: 61
average rate: 20.694
        min: 0.024s max: 0.071s std dev: 0.00664s window: 82
average rate: 20.735
        min: 0.024s max: 0.071s std dev: 0.00593s window: 103
average rate: 20.748
        min: 0.024s max: 0.071s std dev: 0.00588s window: 124
average rate: 20.666
        min: 0.024s max: 0.071s std dev: 0.00658s window: 144
```

Chapter 4
SkyWasteTracker: An Edge-Cloud Architecture for Connectivity-Constrained Aviation Environments

Wui Sum Cheng
http://orcid.org/0009-0000-0928-1329
The Hong Kong Polytechnic University, Hong Kong

ABSTRACT

Airlines discard an estimated $4 billion in untouched food annually, yet biosecurity regulations mandate immediate incineration of cabin waste, precluding post-flight auditing. This chapter presents SkyWaste, a resilient edge-cloud IoT architecture that transforms crew tablets into waste sensors producing item-level telemetry without cloud connectivity during flight. The architecture comprises three planes: an Edge Plane producing structured telemetry (~300 bytes/tray) within 67 ms, a Cloud Plane with six modular serverless stacks, and an MLOps Plane for active learning. A store-and-forward protocol with crash-resilient session linkage ensures data integrity. Manifest integration uses embedding similarity to bridge catering vocabulary and detection taxonomy. Validated across 489 tests and backed by an airline partner agreement.

1. INTRODUCTION

The SkyWaste initiative originated as the 2nd Runner-Up entry at an industry hackathon in 2024, where the initial concept envisioned a robotic hardware solution for inflight waste sorting. The global aviation industry produces an estimated 5.7 million tonnes of cabin waste annually (Bailey, 2022; International Air Transport

DOI: 10.4018/979-8-2600-0701-3.ch004

Association [IATA], 2019). Approximately 20% of this waste consists of untouched food and beverages, representing roughly $4 billion in wasted catering value each year (IATA, 2019; Food Control Consultants, 2024). When disposal costs estimated at nearly $1 billion are included, the total economic burden exceeds $5 billion. Despite this scale, airlines have no systematic mechanism for determining what passengers actually consume. They track aggregate disposal weights (a 200-gram tray deficit could indicate one untouched dish or three partially consumed items) but lack the item-level granularity required to optimize catering procurement, substantiate Environmental, Social, and Governance (ESG) reporting claims, or close the feedback loop between passenger consumption patterns and provisioning decisions.

The root cause of this data gap is not technological indifference but regulatory constraint. International Catering Waste protocols, enforced by national biosecurity agencies worldwide, classify inbound cabin waste as a high-priority biosecurity risk (Food Control Consultants, 2024). All waste from international flights must be immediately incinerated or sterilized upon arrival. The post-flight waste segregation and audit procedures that are standard practice in terrestrial food service are legally prohibited in aviation (IATA, 2019). This regulatory firewall renders conventional waste auditing methods structurally infeasible. The only window for data collection is during the flight itself, in the galley, as cabin crew collect trays from passengers.

The proposed solution must resolve three distinct classes of challenge. The Operational Challenge: restrictive International Catering Waste protocols establish a regulatory firewall that precludes conventional physical auditing. Inbound waste is classified as a high-priority biosecurity risk mandated for immediate incineration or sterilization, legally prohibiting post-flight waste segregation methodologies standard in terrestrial hospitality. The Strategic Challenge: the existing feedback loop between inflight consumption and upstream catering provisioning is fractured by a lack of granular telemetry. Without item-level insight, the airline cannot optimize procurement at the origin, compromising the auditability of ESG reporting. The Technical Challenge: the aircraft galley imposes severe environmental constraints including intermittent connectivity and structural vibrations, necessitating an edge-native computing approach.

Edge-native IoT, processing sensor data entirely on the device and transmitting only lightweight structured telemetry when connectivity becomes available, is the only architecturally viable approach. This pattern is not unique to aviation. Maritime IoT systems face similar constraints: vessels at sea rely on expensive, bandwidth-limited satellite connectivity, necessitating on-device processing and store-and-forward data synchronization (Fall, 2003). Underground mining operations deploy edge sensors in environments where radio frequency propagation is unreliable and data must be buffered locally for hours before reaching surface infrastructure (Shi, Cao, Zhang, Li, & Xu, 2016). Remote agricultural monitoring systems operate in areas with no cellular coverage, accumulating sensor readings locally and uploading in batches when a connectivity window opens (Satyanarayanan, 2017). In each domain, the common architectural pattern is the same: perform computation at the edge, store results locally with crash resilience, and synchronize opportunistically when network access is detected. The Delay Tolerant Networking (DTN) architecture formalized by Fall (2003) provides the theoretical foundation for this approach, treating intermittent connectivity as a first-class architectural concern rather than an exceptional failure mode. The SkyWaste architecture applies this pattern to aviation waste telemetry, with domain-specific adaptations for the regulatory, privacy, and operational constraints of the airline galley.

This chapter presents the systems engineering of the SkyWaste architecture. The contributions are fourfold. First, a resilient three-plane edge-cloud architecture that produces audit-grade waste telemetry with zero cloud dependency during active operation. Second, a store-and-forward synchronization protocol with crash-resilient session-tray linkage, presigned URL-based image upload, and privacy-preserving data handling. Third, a manifest integration subsystem that uses text embedding similarity to bridge the vocabulary mismatch between free-text catering menu items and the fixed detection taxonomy, supporting seasonal menu changes through date-range validity. Fourth, a modular serverless backend comprising six independent infrastructure-as-code stacks with write-time aggregation, append-only audit logging, and a nine-page operational analytics dashboard. The machine learning pipeline (the detection and waste estimation models) is treated throughout this chapter as a black box that processes camera frames and produces structured telemetry. The model architecture, training methodology, and evaluation metrics are the subject of a separate publication focused on computer vision methodology.

2. RELATED WORK

2.1 Edge-Cloud IoT Architectures for Constrained Environments

The shift from centralized cloud computing to distributed edge architectures has been driven by applications where latency, bandwidth, or connectivity constraints make round-trip cloud processing impractical. Shi et al. (2016) articulated this vision in their foundational survey, identifying edge computing as a paradigm where "the applications, data, and services are pushed away from centralized nodes to the logical extremes of a network" (p. 637). Their framework identifies three primary drivers for edge placement: latency sensitivity (applications requiring sub-second response), bandwidth efficiency (applications generating data volumes that exceed available uplink capacity), and privacy preservation (applications where raw sensor data should not leave the device). Aviation waste telemetry satisfies all three criteria simultaneously, a rare convergence that makes edge-native architecture not merely preferable but structurally necessary.

Satyanarayanan (2017) extended the edge computing vision to what he termed "the emergence of edge computing," arguing that edge nodes should be capable of autonomous operation for extended periods, not merely serve as caches or proxies for cloud services. This autonomy requirement maps directly to the SkyWaste use case, where edge devices must operate through entire flight durations, potentially 12 or more hours on long-haul routes, without any expectation of cloud connectivity.

More recent work has examined specific architectural patterns for edge-cloud coordination. Chen and Ran (2019) proposed a tiered edge architecture for video analytics that partitions processing between edge devices and edge servers based on computational complexity, demonstrating that significant bandwidth savings can be achieved by performing initial feature extraction on-device before transmitting compressed representations. The SkyWaste architecture takes this principle further: the entire inference pipeline runs on-device, and only the final structured output (approximately 300 bytes per tray event) is transmitted, achieving a bandwidth reduction exceeding 99.9% compared to raw frame streaming.

2.2 Store-and-Forward Protocols and Offline-First Design

The concept of intermittent connectivity as a first-class architectural concern, rather than an exceptional failure mode, was formalized by Fall (2003) in the Delay Tolerant Networking (DTN) architecture. Fall's key insight was that networks characterized by long delays, frequent disconnections, and asymmetric data rates require fundamentally different protocol assumptions than the always-connected model underlying TCP/IP. The DTN architecture introduced the "bundle" as a store-and-forward unit, a self-contained data package that can be held in persistent storage at intermediate nodes until a forwarding opportunity arises. While the SkyWaste system does not implement the full DTN bundle protocol, it adopts the core architectural principle: all telemetry is treated as delay-tolerant, stored locally with ACID guarantees, and forwarded opportunistically when a trusted network is detected.

The offline-first design pattern, building applications that function fully without network connectivity and treat synchronization as an asynchronous background process, has gained prominence in mobile application engineering. Feyerke (2013) articulated the core principle: "We can't predict what the network will be like, if it's available at all. We must assume that the network is hostile and design accordingly." This philosophy inverts the traditional client-server assumption, treating offline operation as the default state and connectivity as an opportunistic bonus.

Several synchronization strategies exist for offline-first applications. Conflict-free Replicated Data Types (CRDTs) enable eventually consistent replication without coordination (Shapiro, Preguica, Baquero, & Zawirski, 2011), but their complexity is unnecessary when data flows unidirectionally from edge to cloud, as in telemetry collection. Simple append-only event logs with idempotent upload operations provide the same guarantees with far less implementation complexity. The exponential backoff pattern for retry scheduling, first formalized for Ethernet collision management (Metcalfe & Boggs, 1976) and subsequently adapted for distributed systems (Nygard, 2018), prevents devices from overwhelming a recovering network with simultaneous retry attempts. In fleet deployments where dozens of devices may detect network availability simultaneously after landing, coordinated backoff ensures that the backend is not subjected to a thundering herd of concurrent uploads.

2.3 Serverless Backend Patterns for IoT Telemetry

Serverless computing, where cloud providers dynamically allocate compute resources per request with billing based on actual execution time rather than provisioned capacity, offers natural advantages for IoT backends characterized by bursty, unpredictable traffic patterns (Jonas et al., 2019). An IoT fleet that generates zero traffic during flights and bursts of uploads after landing maps poorly to fixed-capacity server provisioning but maps naturally to the auto-scaling model of serverless functions.

However, serverless architectures introduce their own design constraints. The stateless execution model means that each function invocation starts from a cold state (or a cached warm state with no guaranteed persistence), requiring all persistent state to reside in external storage services. Cold start latency (the time to initialize a new function instance) can introduce multi-second delays for the first request after an idle period (Shahrad et al., 2020), though this is acceptable for asynchronous telemetry ingestion where sub-second response time is not required.

A critical design decision in serverless IoT backends is when to compute aggregate statistics. The naive approach, scanning all stored records at query time, scales linearly with data volume. The alternative, write-time aggregation, where each incoming record atomically updates a pre-computed summary, shifts the computational cost from reads to writes and makes dashboard queries constant-time regardless of data volume (Kleppmann, 2017). This pattern, sometimes called "materialized aggregation" in the event sourcing literature, is particularly well-suited to IoT telemetry where the write-to-read ratio is heavily skewed toward writes.

Infrastructure-as-code (IaC) tools such as AWS Cloud Development Kit (CDK), Terraform, and Pulumi enable cloud infrastructure to be defined, versioned, and deployed through the same software engineering practices applied to application code

(Hashimoto, 2016). For startup-stage deployments where cost control is paramount, IaC enables selective stack deployment and teardown. Individual infrastructure components can be provisioned when needed and destroyed when their cost exceeds their value, without affecting other system components.

2.4 Semantic Embeddings for Knowledge Integration

Recent advances in text embedding models map natural language strings into dense vector spaces where semantically related descriptions occupy proximate regions, enabling similarity computation between arbitrary text pairs through geometric operations such as cosine distance. The emergence of lightweight, high-quality embedding APIs, including both commercial offerings and open-source alternatives based on the Sentence-BERT architecture, (Reimers & Gurevych, 2019), has made semantic matching accessible as a composable building block for domain-specific applications.

In supply chain and provisioning contexts, vocabulary mismatch is a persistent challenge. Product descriptions, item codes, and categorical labels often share no lexical surface forms despite referring to the same physical item. Traditional approaches to bridging this gap, keyword matching, fuzzy string similarity, and manually curated lookup tables, degrade rapidly when the source and target vocabularies differ in language, granularity, or domain-specific jargon (Wang, Qu, Shen, & Araki, 2023). Embedding similarity provides a more robust alternative: by operating in semantic space rather than lexical space, it naturally handles synonymy, multilingual variation, and domain-specific paraphrase without explicit enumeration of equivalences.

The cosine similarity threshold approach, automatically accepting matches above a high-confidence threshold and flagging matches below it for human review, instantiates a human-in-the-loop pattern that balances automation efficiency with quality assurance (Monarch, 2021). This pattern is well-established in information retrieval and entity resolution, where the cost of false matches exceeds the cost of human review for ambiguous cases.

2.5 Dashboard Design for Operational Intelligence

Operational dashboards serve as the interface between IoT telemetry systems and human decision-makers. Effective dashboard design distinguishes between leading indicators (metrics that predict future outcomes and enable proactive intervention) and lagging indicators (metrics that summarize past performance) (Few, 2006). In waste management, the proportion of untouched items on a particular route is a leading indicator, it suggests that the catering mix for that route may need adjustment. Total waste tonnage is a lagging indicator, it confirms a problem but does not identify its cause.

Auto-refresh mechanisms in operational dashboards must balance data freshness against computational cost. Polling-based refresh (the client periodically requests updated data) is simpler to implement than push-based approaches (the server notifies clients of changes) but generates predictable load regardless of whether data has changed. For IoT telemetry dashboards where data arrives in bursts (after flight landings) with long idle periods in between, the cost of polling during idle periods is wasted but typically negligible compared to the infrastructure cost of maintaining persistent push connections (Grigorik, 2013).

The unified export pattern, ensuring that data exported from any dashboard page follows a consistent schema, addresses a practical challenge in operational analytics. When different stakeholders export data from different pages and attempt to combine or compare exports, schema inconsistencies create friction and errors. A standardized export schema (consistent column names, data types, and formats across all pages) eliminates this integration burden (Wickham, 2014).

3. SYSTEM ARCHITECTURE

3.1 Three-Plane Topology

The SkyWaste architecture is organized into three decoupled planes, each with distinct operational responsibilities, deployment targets, and failure domains.

The Edge Plane runs autonomously on crew tablets with no cloud dependency during active operation. It encompasses the complete inference pipeline, from camera frame capture through detection, tracking, and waste estimation, producing structured JSON telemetry as its output. Each tray event generates approximately 300 bytes of structured data: a session identifier, tray number, a list of detected items with class labels and waste percentages, confidence scores, and device telemetry (inference latency, thermal state, battery level). The on-device inference pipeline processes frames within 67 milliseconds producing structured JSON telemetry, providing cabin crew with immediate visual feedback. The Edge Plane also manages local data persistence, network detection, synchronization scheduling, privacy protection (face detection and blurring, metadata stripping), device integrity verification, and active learning data collection for model improvement.

The Cloud Plane handles data ingestion, persistent storage, query serving, and analytics. It comprises six modular infrastructure stacks deployed to AWS, each independently deployable, updatable, and removable. The Cloud Plane receives telemetry from edge devices after flights, validates and stores events, maintains pre-aggregated statistics for dashboard queries, and serves the operational analytics dashboard. It operates entirely passively during flights, no edge device depends on cloud availability for its core sensing function.

The MLOps Plane facilitates continuous model improvement through active learning. Low-confidence detections identified by the Edge Plane are queued for upload alongside regular telemetry. Each active learning sample includes the cropped image region, the full probability distribution across classes, the model version identifier, and detection metadata. These samples feed an annotation pipeline where human annotators review uncertain cases, producing labeled training data for subsequent model iterations. The MLOps Plane operates asynchronously and its availability has no impact on the Edge Plane's sensing or the Cloud Plane's analytics capabilities.

This three-plane decomposition ensures that the failure of any single plane does not cascade to the others. If the Cloud Plane is temporarily unavailable, edge devices continue sensing and accumulating data locally. If the MLOps Plane is offline, model improvement pauses but operational data collection continues uninterrupted. The only hard dependency is the Edge Plane's on-device inference pipeline, which requires no external services.

3.2 Network Traffic Design

The edge-native architecture reduces upstream bandwidth by more than 99.9%, transmitting only lightweight JSON telemetry (~300 bytes per event) instead of streaming raw video. Three strategies govern network traffic.

First, latency. Geostationary satellite links, the dominant inflight connectivity technology, introduce over 600 milliseconds of round-trip latency (Grous, n.d.). Adding server-side processing time and response serialization, the total round-trip for a single frame would exceed 800 ms, well beyond the 100 ms threshold for perceived instantaneous feedback (Nielsen, 1993; Card, Moran, & Newell, 1983). Cabin crew processing trays at a rate of one every two to three seconds require immediate visual confirmation that the system has captured the tray; an 800 ms delay would disrupt workflow and erode trust.

Second, bandwidth. A camera capturing frames at 15 FPS at typical inference resolution produces approximately 300 megabytes of raw pixel data per minute. Inflight satellite bandwidth is shared among hundreds of passengers and is typically throttled to single-digit megabits per second per device. Streaming even compressed video would saturate the available bandwidth within seconds, degrading connectivity for all passengers.

Third, availability. Inflight connectivity is not guaranteed, many flights lack any satellite link, and even equipped flights experience outages during route segments with poor satellite coverage. A cloud-dependent architecture would produce zero data during these outages, creating unpredictable gaps in coverage. The edge-native architecture produces data continuously regardless of connectivity state.

Extrapolating from measured per-device data volumes, a fleet-scale deployment of 100 or more devices would generate approximately 1.5 megabytes of structured telemetry and 500 megabytes of evidence images per day. The structured telemetry volume is negligible for commodity cloud infrastructure. The evidence image volume (500 MB/day, approximately 15 GB/month) is modest by cloud storage standards, and a Glacier lifecycle policy reduces long-term storage costs for images older than 30 days.

3.3 Infrastructure Composition: Six Modular CDK Stacks

The Cloud Plane is implemented as six independent AWS Cloud Development Kit (CDK) stacks, each encapsulating a distinct infrastructure concern: Table 1 summarizes: Six Modular CDK Stacks and Their Purposes.

Table 1. Six Modular CDK Stacks and Their Purposes

Stack	Purpose
Security Stack	Cognito User Pool (TOTP MFA), KMS key, Secrets Manager
Storage Stack	S3 data lake (KMS encrypted, Glacier lifecycle), DynamoDB
Database Stack	Aurora PostgreSQL Serverless v2, RDS Proxy, VPC
API Stack	API Gateway + 4 Lambda functions + AWS WAF
Hosting Stack	S3 static website + CloudFront CDN with OAC
Budgets Stack	AWS Budgets ($50/month threshold, email alerts)

The decision to use six stacks rather than a single monolithic deployment was driven by three requirements: independent lifecycle management (each stack can be updated without redeploying the others), selective cost control (expensive stacks can be torn down independently), and blast radius containment (a deployment failure in one stack does not affect the others).

The Database Stack is feature-flagged: it is deployed only when a configuration parameter is set. This flag proved critical for cost control, during development phases that did not require relational queries, the stack was torn down entirely, eliminating its idle costs while the remaining five stacks continued operating on the NoSQL layer alone. Cost analysis during initial deployment revealed that the Aurora PostgreSQL stack consumed disproportionate idle costs, even at minimum capacity, the combination of compute, storage, RDS Proxy, and NAT Gateway charges for the VPC significantly exceeded the cost of all other stacks combined. Rather than accepting the ongoing cost, the team tore down the stack entirely. Because the stacks are independent, this required no changes to the other five stacks, the parser function detected the absence of the database configuration and skipped its dual-write path, the query function continued serving dashboard data from DynamoDB, and the system operated normally with reduced capabilities and reduced cost. When manifest features were implemented months later and their data model genuinely required date-range overlap detection (a relational query pattern), the stack was redeployed with a single deployment command. This experience illustrates a broader principle for startup-stage IoT deployments: infrastructure-as-code modularity is not merely an engineering best practice, it is a cost management strategy.

The Hosting Stack is similarly feature-flagged to avoid CDN costs during local development. The Budgets Stack deploys to a separate region (the only region supporting the Budgets API), while all other stacks deploy to the region closest to the airline partner's operations for data residency proximity.

The parser Lambda validates and stores event data with comprehensive server-side validation: flight number must match a defined format, tray number must be within 1-999, items are capped at 20 per tray, class name must be in an allowlist of valid detection classes, waste percent must be 0-100, and confidence must be 0-1. A per-session rate limit of 200 trays prevents abuse. The Lambda atomically updates a pre-aggregated statistics record on each insert using atomic increment expressions, eliminating the need for expensive full-table scans to compute dashboard statistics. Daily snapshot records with 90-day TTL enable time-series analytics. The Lambda also writes append-only audit records with IAM deny policies on update and delete operations, ensuring tamper-evident logging.

The presigned URL Lambda generates time-limited S3 upload URLs for evidence images. Each URL is scoped to a UUID-based filename (preventing path traversal attacks), enforces a content-length range (1 byte to 10 megabytes, rejecting both empty uploads and oversized files), and restricts content types to an allowlist of image MIME types. The presigned URL expires after five minutes, limiting the window for misuse.

The query Lambda serves dashboard statistics, session listings, event details, analytics data, and three maintenance endpoints (system health, audit log, device metrics). Dashboard statistics are read from a single pre-aggregated record, making the most frequent query, the dashboard refresh, a constant-time, single-record read regardless of data volume.

The manifest Lambda manages the full lifecycle of flight meal manifests: upload with embedding-based class mapping, confirmation with date-range storage, retrieval by flight number and date, listing with active-only filtering, and deletion by identifier or date.

An AWS WAF (Web Application Firewall) applies rate limiting per client IP and managed rule sets for common web exploits, providing defense-in-depth protection at the API layer. The WAF rate limits also serve as a safety net against misconfigured or misbehaving edge devices in fleet deployments.

3.4 Write-Time Aggregation

The choice of when to compute aggregate statistics, at write time or at query time, was the single most impactful backend design decision, with direct implications for cost, latency, and scalability.

The default approach in most web applications is query-time aggregation: when a dashboard page loads, the backend scans all relevant records, computes sums and averages, and returns the result. However, its cost scales linearly with data volume. With nine dashboard pages auto-refreshing every 30 seconds and a data store that grows with every tray event, the number of read operations per hour becomes significant. For a fleet of 100 devices each uploading 50 tray events per day, the data store would accumulate 5,000 records daily. After one year, a single dashboard refresh would require scanning 1.8 million records.

Write-time aggregation inverts this cost profile. Each time the parser function inserts a new tray event, it atomically increments a single pre-aggregated statistics record containing running totals: total flights, total trays, total items, cumulative waste percentage. Dashboard reads then retrieve this single record, one read operation, regardless of whether the data store contains one thousand or one million events.

The write-time aggregation pattern ensures that dashboard query cost remains constant regardless of fleet size, whether the data store contains events from 10 devices or 1,000 devices, a dashboard refresh reads a single pre-aggregated record. The per-session rate limit (200 trays per session) prevents individual devices from generating disproportionate load.

Daily snapshot records, written alongside each event insertion, extend this pattern to time-series analytics. Each snapshot captures the day's cumulative statistics and carries a 90-day TTL, enabling trend charts over the past three months without historical scans and automatically expiring old snapshots to control storage costs.

3.5 Manifest System with Embedding Similarity

To integrate contextual knowledge, correlating what was served with what was detected, the manifest system uses a text embedding API to compute cosine similarity between uploaded item names and 8 class anchor descriptions. Anchors are rich semantic descriptions rather than raw class names (e.g., the main course anchor is described as “hot main meal, protein dish, rice or noodle entree”, a description that occupies the same region of embedding space as the diverse natural language descriptions that airlines use for their main course offerings). Cosine similarity above 0.75 triggers auto-mapping; below 0.75 flags items for operator review in the admin console upload preview. Confirmed manifests are stored with date-range validity (validFrom/validTo), supporting seasonal menu changes for the same flight number.

3.6 Operational Analytics Dashboard

The admin console provides 9 pages of operational intelligence, built as a single-page application with TypeScript, utility-first CSS, and charting libraries. Pages include: Dashboard (4 stat cards, waste trend line chart, waste-by-class bar chart, recent sessions table), Events (filterable table with expandable rows showing per-item details), Flights (cards grid with configurable sorting), Flight Detail (per-flight statistics, waste-by-tray chart, item distribution breakdown), Analytics (waste distribution histogram, trend area chart, class breakdown, top flights ranking), Models (inference pipeline specifications, class routing table, benchmarking placeholder), Menus (manifest upload with CSV/TXT parsing and embedding similarity preview), Settings (system information and API health check), and Maintenance (7 developer-only accordion sections covering health overview, device metrics, model performance, annotation platform integration, audit log, debug logs, and configuration). All data pages auto-refresh every 30 seconds with a "Last updated" indicator and support a unified Excel export schema (FlightNumber, Date, TrayNumber, ClassName, WastePercent, Confidence, ModelVersion) for consistent cross-page reporting.

4. OFFLINE-FIRST EDGE SYNCHRONIZATION

4.1 The Store-and-Forward Protocol

The SkyWaste synchronization protocol draws directly on the Delay Tolerant Networking principles articulated by Fall (2003), treating network connectivity as an opportunistic resource rather than a guaranteed service. All telemetry generated by the Edge Plane is first persisted to local storage with full ACID guarantees before any transmission is attempted. The network is never in the critical path of data capture.

An embedded SQLite database configured with Write-Ahead Logging (WAL) mode provides the local persistence layer. WAL mode enables concurrent reads and writes without blocking, which is critical because the inference pipeline produces new events at a higher rate than the synchronization subsystem can transmit them. The database transaction commits before the event is considered "captured", if the device loses power between inference completion and database commit, the event is lost, but this window is measured in microseconds. The database schema includes two tables: one for completed detection records and one for pending synchronization items.

A system-level network path monitor (NWPathMonitor) continuously evaluates connectivity state. The monitor distinguishes between network availability (an IP route exists) and network quality (the route leads to a functional backend). A network path becoming "satisfied" triggers the synchronization manager, but actual upload is gated on a successful lightweight health check to the backend API.

The synchronization manager iterates through unsynchronized events in the local database, ordered by creation timestamp. Each event is serialized to JSON and submitted to the parser API endpoint. Successful uploads are marked as synchronized; failures are retained for retry. Failed uploads are retried with an exponential backoff schedule that increases the delay between successive attempts, following the pattern first formalized for Ethernet collision management (Metcalfe & Boggs, 1976) and subsequently adapted for distributed systems (Nygard, 2018). This prevents a recovering network from being overwhelmed by retry storms, particularly in fleet scenarios where dozens of devices may detect network availability simultaneously after landing. Each retry attempt is idempotent, the backend either accepts the event (if new) or silently acknowledges it (if already stored), preventing duplication.

Evidence images are uploaded asynchronously and independently of telemetry. Each image upload uses a presigned S3 URL with a UUID-based filename, content-length enforcement (1 byte to 10 megabytes), and content-type validation. Image upload failures do not affect telemetry integrity, the structured tray event data (class labels, waste percentages, confidence scores) is complete and useful even without the accompanying image. Images provide supplementary evidence for audit and active learning purposes.

The entire upload process is opportunistic and non-blocking. The inference pipeline continues capturing new tray events regardless of synchronization state. A sync status badge in the camera view displays the number of queued items, offline state, or synchronization confirmation with the last timestamp, providing crew with immediate visibility into data integrity without requiring any crew action.

Compared to other offline-first architectures that rely on CRDTs for bidirectional synchronization, the SkyWaste protocol exploits the unidirectional nature of telemetry collection. Data flows only from edge to cloud, eliminating the need for conflict resolution entirely. Simple append-only event logs with idempotent upload operations provide the same guarantees as more complex synchronization frameworks with far less implementation overhead.

4.2 Crash-Resilient Session-Tray Linkage

A subtle but critical challenge in edge IoT systems is maintaining data integrity across application crashes. In the SkyWaste context, a scanning session may span 30 or more minutes and produce 50 or more tray events. If the application crashes mid-session, the tray events uploaded before the crash must be attributable to the correct session, orphaned tray records with no session context are analytically useless.

The system addresses this through a preliminary session record strategy. At the moment a cabin crew member initiates a new scanning session, the Edge Plane attempts to upload a lightweight session record to the cloud containing only the session identifier, flight number, and start timestamp. This upload is opportunistic, if no connectivity is available, the record queues locally alongside tray data. Each subsequent tray upload includes the session identifier as a foreign key.

If the application crashes mid-session, the preliminary record ensures that the cloud backend already knows the session exists. Any tray events uploaded before the crash are linked to it through the session identifier. When the application restarts and the synchronization manager processes the local queue, the remaining tray events are uploaded with the same session identifier, maintaining session coherence even across crashes.

When a session completes normally, a final upload enriches the preliminary record with summary statistics: total trays processed, per-class waste averages, session duration, and device health metrics. This enrichment is also idempotent, if the final upload is interrupted and retried, the summary statistics overwrite the previous values without duplication.

4.3 Privacy-Preserving Data Handling

The Edge Plane processes camera frames that may incidentally capture passenger faces, crew members, or other personally identifiable visual information. Compliance with privacy regulations, including GDPR Article 25 (Data Protection by Design and by Default), requires that privacy protection is embedded in the data processing pipeline, not added as a post-processing step.

Three privacy-preserving mechanisms operate at the edge. Face detection and blurring uses the platform's native computer vision framework to detect faces in evidence images and applies Gaussian blur before any image is stored locally or transmitted. This ensures that no identifiable face data exists in persistent storage at any point, the raw frame is processed in volatile memory, the blurred version is persisted, and the raw frame is discarded. EXIF metadata stripping removes GPS coordinates, device identifiers, timestamps, and other metadata from evidence images before storage or upload. Even if an image were intercepted during transmission, it would contain no location or device information. Device integrity verification performs six integrity checks at application launch, detecting indicators of device compromise. In debug builds, violations are logged; in production builds, violations block application operation entirely, preventing data collection on potentially compromised devices.

4.4 Active Learning Telemetry Collection

The Edge Plane collects additional telemetry specifically for model improvement, framed as IoT data collection for continuous system refinement. When the inference pipeline produces a detection with a confidence score falling within an empirically tuned uncertainty margin (high enough to be a plausible detection, low enough to indicate model uncertainty), the system captures a supplementary data package: the cropped image region, the full probability distribution across all classes (not just the top prediction), the model version identifier, and detection metadata including bounding box coordinates, frame timestamp, and device thermal state.

These active learning packages are queued for upload alongside regular telemetry but at lower priority, the synchronization manager transmits tray events first and active learning data second, ensuring that operational telemetry is never delayed by supplementary data collection. The collected data feeds the MLOps Plane's annotation pipeline, where human annotators review uncertain cases to produce labeled training examples for subsequent model versions. This creates a closed-loop feedback mechanism between operational deployment and model improvement, driven by the actual data distribution encountered in production rather than synthetic or laboratory data.

5. MANIFEST INTEGRATION VIA EMBEDDING SIMILARITY

5.1 The Vocabulary Mismatch Problem

Correlating waste telemetry with catering provisioning data requires mapping between two fundamentally different vocabularies. Flight manifests describe menu items in free-text natural language, “Hainanese Chicken Rice,” “Pollo con Arroz,” “Seasonal Vegetable Risotto”, while the detection pipeline operates on a fixed taxonomy of eight container-type classes. These vocabularies share virtually no lexical surface forms. “Hainanese Chicken Rice” contains no words in common with a class label like “main.”

A naive string-matching approach fails immediately. Even fuzzy string matching algorithms (Levenshtein distance, Jaccard similarity on character n-grams) cannot bridge this gap because the mismatch is semantic, not typographic, the menu item and the class label refer to the same physical object using entirely different descriptive frameworks.

A manually curated lookup table would work for a fixed menu but fails to scale. Airlines rotate menus seasonally, introduce regional variations for different routes, and describe the same dish differently across languages and catering providers. A lookup table would require constant maintenance, every new menu item, every seasonal rotation, every regional variation would need a manual mapping entry. For an airline operating hundreds of routes with quarterly menu changes, this maintenance burden is operationally unsustainable. This vocabulary mismatch challenge is not unique to aviation, it is a persistent problem in supply chain and provisioning contexts where product descriptions, item codes, and categorical labels often share no lexical surface forms despite referring to the same physical item (Wang et al., 2023).

5.2 The Embedding Similarity Approach

The manifest integration subsystem uses a text embedding API to compute cosine similarity between uploaded menu item descriptions and semantically rich anchor descriptions for each detection class. The key design decision is the construction of the anchor descriptions: rather than using the terse class labels themselves (which would provide poor embedding targets), each class is associated with a rich natural language description that captures its semantic scope. For example, the main course class anchor is described as "hot main meal, protein dish, rice or noodle entree", a description that occupies the same region of embedding space as the diverse natural language descriptions that airlines use for their main course offerings.

When an operator uploads a flight manifest (as a CSV file, plain text file, or direct text input), the system computes embeddings for each menu item description and each class anchor, then calculates pairwise cosine similarities. Items exceeding a similarity threshold of 0.75 are automatically mapped to their highest-similarity class. Items below the threshold are flagged as "needs review", the operator sees these items highlighted in the upload preview interface and can manually assign them to the correct class or reject them.

The 0.75 threshold was tuned empirically. A higher threshold (0.85+) would increase precision (fewer incorrect auto-mappings) but reduce recall (more items requiring manual review), creating operational friction. A lower threshold (0.60) would auto-map more items but risk incorrect mappings that corrupt downstream analytics. The 0.75 threshold was found to reliably auto-map straightforward items ("Chicken Rice" to the main course class, "Orange Juice" to the beverage class) while correctly flagging ambiguous items (such as "Assorted Dessert Platter") for human judgment. This threshold approach instantiates the human-in-the-loop pattern described by Monarch (2021), balancing automation efficiency with quality assurance.

5.3 Date-Range Validity and Overlap Detection

Airlines serve different menus on the same flight number across different time periods, a summer menu, a winter menu, a promotional seasonal offering. The manifest system supports this through date-range validity: each confirmed manifest is stored with validFrom and validTo dates, and manifest retrieval queries return only the manifest valid for the requested date (defaulting to the current date).

Overlap detection prevents conflicting manifests. If an operator attempts to confirm a manifest for a flight with a validity period that overlaps an existing manifest for the same flight, the system detects the conflict. When the validFrom date matches an existing manifest exactly, the new manifest upserts (replaces) the existing one, this supports corrections. When the date ranges overlap without an exact validFrom match, the system rejects the new manifest with an error describing the conflict. This logic is implemented as a relational query, which is why the manifest feature requires the database stack to be deployed.

5.4 The Manifest-Default Pattern

A critical question in waste analytics is: what should the system report when the camera does not detect an item that the manifest says should be on the tray? Three options exist. First, ignore the missing item, this understates waste by omitting items from the analysis. Second, flag it as unknown, this creates a data gap that reduces the completeness of analytics. Third, assume the item was fully consumed, the most likely explanation for a missing item is that the passenger ate it entirely and only residue or an empty container remains, too small or ambiguous for the detection pipeline to identify.

The system adopts the third option, termed the manifest-default pattern. At tray finalization, the system compares the set of detected classes against the loaded manifest. For every manifest item class not detected, a synthetic result is injected with a waste estimate of 0% (fully consumed). This default biases toward reporting higher consumption (lower waste), which is the conservative interpretation from an ESG reporting perspective, the system never overstates the airline's waste problem.

To maintain transparency, synthetic items are tagged with a distinct model version identifier ("manifest-default") that distinguishes them from camera-detected items in all downstream analytics and exports. Operations staff can monitor the manifest-default injection rate per flight, a high rate may indicate detection pipeline underperformance, while a low rate confirms comprehensive tray coverage.

5.5 Bias Risk and Validation Methodology

The manifest-default pattern introduces a known bias: if the detection pipeline misses an item due to poor lighting, an unusual angle, or a closed lid rather than actual consumption, the system incorrectly logs it as fully eaten. This bias systematically understates waste, which is problematic for both procurement optimization (the airline underestimates how much food is being wasted) and ESG reporting (waste figures appear more favorable than reality).

To quantify this bias, the planned ground pilot includes a manual sampling protocol. On a random subset of trays (approximately 10%), a human auditor will physically inspect tray contents and record ground-truth item presence and waste levels. Comparing auditor records against the system's manifest-default injections yields the false-consumption rate, the proportion of manifest-default items that were actually present but missed by the detection pipeline. This metric will determine whether the manifest-default heuristic's bias is operationally acceptable or whether detection recall must be improved before the pattern can be trusted for audit-grade reporting.

6. OPERATIONAL ANALYTICS DASHBOARD

6.1 Architecture and Navigation

The admin console is a single-page application that authenticates users through a Cognito-backed authentication flow including MFA challenges. The login page is branded with the system identity (custom logo, hero image background, branded color palette). The application provides nine pages organized into two navigation groups: operational pages accessible to all authenticated users, and developer pages (Maintenance) accessible to technical staff.

6.2 Dashboard and Core Pages

The landing page presents four summary statistic cards, total flights, total trays, total items, and average waste percentage, followed by two charts (a waste trend line chart showing waste percentage over time, and a waste-by-class bar chart showing average waste per detection class) and a table of recent scanning sessions. The statistic cards read from the single pre-aggregated statistics record described in Section 3.4, making this page's data retrieval cost-independent of total data volume.

The Events page presents a filterable table of all tray events with expandable rows that reveal per-item detail (class, waste percentage, confidence, model version).

Filtering groups events by session (date plus flight number combination) rather than flight number alone, enabling analysts to distinguish between morning and evening services on the same route. The Flights page displays flight sessions as a card grid with configurable sort controls (by date, flight number, tray count, average waste). Each card links to a Flight Detail page providing per-flight deep-dive analytics: waste-by-tray charts (sorted numerically by tray number), item distribution breakdowns, and a tray breakdown table with expandable rows.

The Analytics page provides four advanced visualization panels: a waste distribution histogram, a trend area chart over configurable time periods, a class breakdown panel with sortable columns, and a top-flights panel identifying routes with the highest waste rates, the leading indicators that drive procurement optimization.

6.3 Menus Page

The Menus page supports the manifest upload workflow described in Section 5. Operators upload a flight manifest (CSV file, text file, or direct text input), specify the flight number and date range, and submit for embedding-based class mapping. The preview interface displays each menu item alongside its auto-mapped class and similarity score, with items below the 0.75 threshold highlighted for manual review. After the operator reviews and confirms the mappings, the manifest is stored with date-range validity. A manifest table below the upload interface lists all stored manifests with delete capability.

6.4 Maintenance Page

The Maintenance page provides developer-oriented observability through seven collapsible accordion sections. Health Overview displays status cards for key infrastructure components with color-coded indicators. Device and Fleet Metrics aggregates device telemetry (inference latency distributions, thermal state trends, battery level averages) across recent sessions. Model Performance shows confidence histograms, waste distribution analysis, and inference latency tracking. Additional sections provide annotation platform integration controls, a filterable and paginated audit log, CloudWatch log stream pointers for backend troubleshooting, and current feature flag states and environment settings.

6.5 Cross-Cutting Features

All data pages implement auto-refresh with a 30-second polling interval and a "Last updated" timestamp indicator, giving operators confidence that displayed data is current without requiring manual refresh. The 30-second interval balances data freshness against computational cost, for IoT telemetry that arrives in post-flight bursts, more frequent polling would be wasteful during the long idle periods between flight landings.

A unified Excel export function is available on the Dashboard, Events, Flights, Flight Detail, and Analytics pages, producing consistently formatted spreadsheets with a standardized schema: FlightNumber, Date, TrayNumber, ClassName, WastePercent, Confidence, ModelVersion. This schema consistency ensures that exports from different pages can be combined or compared without transformation, a practical requirement when multiple stakeholders (catering operations, ESG reporting, route planning) consume data from different dashboard views.

7. EVALUATION

7.1 System-Level Testing

The system's non-probabilistic behavior, infrastructure configuration, input validation, API routing, data aggregation, UI rendering, is validated by 489 automated tests organized into 44 test suites across three system components.

The cloud backend comprises 194 tests across 12 suites. These tests validate the parser function's input sanitization across all accepted fields (flight number format, tray number range, items cap, class name allowlist, waste percentage bounds, confidence bounds), the presigned URL function's size enforcement and filename randomization, the query function's aggregation logic and response formatting, and the manifest function's embedding mapping, date-range validation, and overlap detection. Infrastructure tests verify that CDK stacks synthesize correctly, resource properties match specifications, and cross-stack references resolve properly.

The admin console comprises 263 tests across 32 suites. These tests validate component rendering (all nine pages render without errors with mock data), data transformation (API responses are correctly mapped to chart data structures), filtering logic (event grouping, flight sorting, date range selection), export formatting (Excel output matches the standardized schema), and authentication flow (login, MFA challenge, session expiry).

System tests exercise the full pipeline from simulated edge client through the API layer to database storage and back through the query endpoints to the admin console's data display. These end-to-end tests verify that data uploaded by an edge device appears correctly in the dashboard within the auto-refresh interval, that presigned URL uploads produce retrievable images, and that manifest uploads through the API produce correct embedding-based mappings.

The 489 tests do not cover probabilistic model behavior, detection accuracy and waste estimation calibration are evaluated through separate benchmarking frameworks, not pass/fail unit assertions.

7.2 Operational Data Collection

Following the initial system deployment, operational video footage was collected from returned meal carts at the airline partner's post-flight handling facility. This footage was processed through an automated extraction pipeline that performs interval sampling, blur quality filtering (Laplacian variance thresholding to reject motion-blurred frames), and perceptual deduplication (structural similarity indexing to eliminate near-duplicate frames). The pipeline produced over 15,000 unique candidate frames from the raw footage, a substantial reduction that ensures the resulting dataset contains visually diverse, high-quality samples.

Review of the collected frames exposed data quality challenges absent from laboratory conditions. Many entree containers had lids replaced by passengers, creating opaque occlusion that no camera-based sensing can penetrate, a data completeness constraint rather than a model accuracy issue. Side dish lids produced ambiguous translucency. Butter and yogurt foils had been resealed after consumption, introducing deceptive packaging states that compromise the telemetry pipeline's ability to distinguish consumed from unconsumed items. Tray compositions varied unpredictably: duplicate items, missing components, stacked containers, and cross-class service ware. These operational realities validated the system's manifest-default pattern (Section 5.4) as essential for maintaining data completeness when sensor coverage is partial.

7.3 Cost Analysis and Sync Pipeline Latency

The serverless architecture's cost profile was monitored against a $50/month budget threshold with alerts at 80% and 100%. During the development phase, cost analysis revealed that the Aurora PostgreSQL database stack consumed disproportionate idle costs relative to its utility. The feature-flagged stack design enabled its teardown without affecting other system operations, reducing monthly costs to well within the budget threshold. When manifest features required relational queries, the stack was redeployed, and the additional cost was justified by the feature's operational value.

The on-demand billing model for DynamoDB and the per-invocation billing for Lambda functions meant that the system's cost during low-traffic periods (no active flights) approached the minimum charges for always-on services (NAT Gateway, RDS Proxy) rather than scaling with provisioned capacity. This cost profile is well-suited to startup-stage deployments where traffic is intermittent and unpredictable.

The synchronization pipeline latency, from tray event finalization on the edge device to record availability in the dashboard, is dominated by network availability rather than processing time. Once connectivity is established, the end-to-end upload latency (device to API Gateway to DynamoDB) is measured in hundreds of milliseconds. The write-time aggregation pattern ensures that newly uploaded events are reflected in dashboard statistics within the next auto-refresh cycle (at most 30 seconds after upload). For fleet-scale deployments, storage projections indicate that DynamoDB handles sustained write throughput of thousands of items per second on on-demand billing, and the API Gateway and Lambda functions scale horizontally to handle concurrent uploads from all devices.

8. DISCUSSION

8.1 Edge-Native Versus Cloud-First: A Structural Argument

The decision to process inference entirely on-device, rather than streaming frames to a cloud service, is not a performance optimization, it is a structural necessity dictated by the operating environment. By processing on-device and transmitting only structured telemetry (~300 bytes per tray), the system reduces upstream bandwidth consumption by over 99.9% compared to frame streaming. This bandwidth reduction makes synchronization feasible even over congested airport Wi-Fi post-flight, and the store-and-forward protocol ensures that data is never lost even if synchronization is delayed by hours.

8.2 Six Stacks Versus Monolith: Cost Control Through Modularity

The six-stack architecture introduces deployment complexity, cross-stack references must be managed, deployment ordering must be respected, and stack teardown requires verification that no other stack depends on the resources being removed. A monolithic single-stack deployment would be simpler operationally. However, the monolithic approach creates an all-or-nothing cost commitment: the entire infrastructure is either deployed (with full cost) or not deployed (with no system).

The six-stack architecture enables selective cost control. During the months when the relational database was unnecessary, its stack was torn down, reducing costs substantially. During local development, the CloudFront CDN stack was not deployed. The budgets stack deploys to a different region entirely and has a lifecycle independent of the application stacks. This granularity proved essential for a startup-stage deployment operating within a strict monthly budget.

8.3 Embedding Similarity Versus String Matching

The embedding approach incurs an API call cost for each manifest upload (computing embeddings for menu items and class anchors). A manually curated string matching table would have zero marginal cost per lookup. However, the total cost of ownership must account for maintenance: every seasonal menu change, every new route with regional variations, and every language variation requires manual table updates under the string matching approach. The embedding approach handles these variations automatically, "Pollo con Arroz," "Chicken Rice," and "Hainanese Chicken with Fragrant Rice" all map to the main course class without any table maintenance. For an airline operating hundreds of routes with quarterly menu rotations, the maintenance cost of string matching tables would rapidly exceed the API costs of embedding similarity.

8.4 Write-Time Versus Query-Time Aggregation

The write-time aggregation pattern shifts computational cost from reads to writes. Each tray insertion requires an additional atomic update operation on the aggregate record, approximately doubling the per-write cost. However, the number of writes is bounded by operational reality (a device produces at most a few hundred tray events per day), while the number of reads scales with the number of dashboard users and the refresh interval. With auto-refresh every 30 seconds, even a single dashboard user generates 2,880 read operations per day. Under query-time aggregation, each of these reads scans the full dataset; under write-time aggregation, each is a single-record retrieval. The break-even point occurs almost immediately, once the dataset exceeds a few hundred records, write-time aggregation is more cost-effective than query-time aggregation for any non-trivial dashboard usage pattern.

8.5 Limitations

Several limitations constrain the system's current capabilities.

While serverless architectures eliminate provisioned capacity waste, they do not eliminate all idle costs. Always-on services such as NAT Gateways (required for Lambda functions in VPC to access the internet), RDS Proxy (required for connection pooling), and Cognito User Pools incur charges regardless of traffic volume. For startup-stage deployments with limited budgets, these baseline costs must be managed through selective stack deployment and teardown, which introduces operational complexity.

Operational data collection revealed that a substantial proportion of main dish containers are returned with lids placed back on by passengers. This is not a detection pipeline failure, the container is detectable regardless of lid state, but a data completeness challenge. The system cannot estimate waste levels when the food is not visible. The manifest-default pattern provides a fallback (reporting undetected waste states as "fully consumed"), but this introduces the bias discussed in Section 5.5. Addressing this limitation may require operational changes (instructing crew to remove lids before scanning) rather than technical solutions.

As described in Section 5.5, the manifest-default pattern systematically understates waste when the detection pipeline's recall is imperfect. The 10% manual sampling protocol planned for the ground pilot will quantify this bias, but until that data is collected, the magnitude of the understatement remains unknown.

8.6 Security and Compliance

The system implements a defense-in-depth security posture spanning multiple layers. Authentication uses Cognito with configurable TOTP MFA, the choice of TOTP over SMS-based MFA is deliberate, as TOTP works offline on aircraft (it requires only a synchronized clock, not cellular connectivity). Authorization enforces JWT validation on every API request, with the WAF providing an additional layer of rate limiting and managed rule protection. Encryption protects data at rest (KMS encryption on S3, DynamoDB, and RDS) and in transit (TLS on all API communications). Input validation enforces server-side constraints on all accepted fields, independent of client-side validation. Upload controls use presigned URLs with UUID filenames (preventing path traversal), content-length limits (preventing oversized uploads), and content-type allowlists (preventing non-image uploads). Privacy protection performs face blurring, EXIF stripping, and device integrity verification at the edge, before any data leaves the device. Audit logging writes append-only records with IAM-enforced immutability, providing a tamper-evident trail for regulatory compliance.

This security posture aligns with GDPR Article 25 (Data Protection by Design and by Default), which requires that data protection measures are integrated into the processing system's design rather than applied as afterthoughts.

8.7 Broader Implications for Regulated Industries

The SkyWaste architecture demonstrates several principles applicable beyond aviation waste management. First, edge-native processing is the only viable approach for IoT sensing in environments where cloud connectivity is unreliable, expensive, or unavailable, a description that applies to maritime vessels, underground mines, remote agricultural sites, and disaster response scenarios. Second, infrastructure-as-code modularity provides cost management capabilities that are essential for startup-stage IoT deployments, where the cost of idle infrastructure can exceed the cost of active computation. Third, store-and-forward synchronization with crash resilience is a reusable pattern for any IoT system where data loss is unacceptable and network availability is intermittent. Fourth, semantic embedding similarity provides a scalable approach to vocabulary mismatch problems in supply chain and provisioning contexts, where item descriptions vary across languages, regions, and time periods.

Several lessons from the presented system generalize to other startup-stage IoT deployments. Feature flags on infrastructure stacks are as valuable as feature flags in application code, the ability to deploy or tear down expensive infrastructure components on demand provides financial flexibility that fixed deployments cannot match. Write-time aggregation should be the default pattern for IoT dashboards with auto-refresh, the engineering cost of maintaining aggregate records at write time is repaid within days of dashboard operation. Store-and-forward synchronization should be designed from the start, not retrofitted, retrofitting crash resilience onto a system designed for reliable connectivity is architecturally expensive and error-prone. Privacy-preserving mechanisms must operate at the edge, before data enters any persistent storage, post-hoc anonymization creates a window of vulnerability and complicates compliance arguments.

9. CONCLUSION AND FUTURE WORK

9.1 Summary of Systems Engineering Contributions

This chapter has presented the systems engineering of a resilient edge-cloud IoT architecture for aviation food waste telemetry. The architecture addresses a fundamental constraint, the impossibility of cloud-dependent data collection in connectivity-constrained aircraft environments, through a three-plane topology that separates edge sensing, cloud analytics, and model improvement into independently operable domains.

The Edge Plane demonstrates that a complete inference pipeline can operate autonomously on commodity tablet hardware, producing structured telemetry within 67 milliseconds per frame with no cloud dependency. The store-and-forward synchronization protocol, drawing on DTN principles, provides crash-resilient data persistence with opportunistic, non-blocking upload and exponential backoff retry. Privacy-preserving mechanisms (face blurring, metadata stripping, device integrity verification) ensure compliance with data protection regulations at the point of capture.

The Cloud Plane demonstrates that six modular infrastructure-as-code stacks, deployed and managed independently, provide the flexibility needed for startup-stage cost control while supporting production-grade data ingestion, storage, and analytics. The write-time aggregation pattern eliminates the cost scaling problem inherent in query-time aggregation for auto-refreshing dashboards. Append-only audit records with IAM-enforced immutability provide the tamper-evident logging required for regulatory compliance.

The manifest integration subsystem demonstrates that text embedding similarity can bridge the vocabulary mismatch between free-text catering descriptions and fixed detection taxonomies, eliminating the maintenance burden of manually curated lookup tables while supporting seasonal menu changes through date-range validity.

9.2 Commercial Validation

The system's commercial viability is supported by a formal Letter of Intent from a major airline partner, providing three forms of validation. It formalizes the data-sharing agreement required for operational data collection, the bottleneck for all subsequent system improvement. It establishes the legal framework for processing waste telemetry under the partner's data governance policies. And it provides the commercial credibility needed for follow-on funding discussions. This agreement transitions the work described in this chapter from academic research to commercial deployment, with a defined pathway through ground pilot and operational pilot phases.

9.3 Future Directions

Several future directions extend the architecture beyond the pilot phase, framed as IoT fleet management challenges.

Alerting and threshold rules. The current system is reactive, operations staff must monitor dashboards to identify problematic patterns. Configurable threshold rules would trigger proactive notifications (for example, when waste exceeds a target percentage across three consecutive flights on the same route), shifting the operational model from monitoring to exception management. From an IoT perspective, this is an edge-to-cloud alerting pipeline, where the cloud backend evaluates incoming telemetry against configurable rules and pushes notifications through existing messaging channels.

Over-the-air model deployment. The ability to push updated models to deployed devices without requiring application updates is essential for continuous improvement. This requires version-controlled model storage in the cloud, automatic update checks by edge devices when connectivity is available, a hot-swap loading mechanism that replaces the active model without interrupting operation, and a rollback mechanism that allows the backend to revert devices to a previous model version if performance degradation is detected. From an IoT fleet management perspective, this is a firmware-over-the-air (FOTA) pattern applied to ML model artifacts rather than application binaries.

Model versioning and rollback. A comprehensive versioning framework would track which model version each device is running, monitor per-version performance metrics through the device telemetry already collected, and enable centralized rollback decisions when a new model version underperforms. This builds on the existing device telemetry infrastructure (inference latency, confidence distributions, thermal state) to provide fleet-wide model lifecycle management.

Lightweight adaptation for seasonal updates. Low-rank adaptation techniques enable model fine-tuning by training only small parameter matrices (~5 MB) alongside frozen base weights, preventing catastrophic forgetting while remaining small enough for over-the-air distribution. From an IoT fleet management perspective, these adapters are lightweight firmware patches that can be distributed to hundreds of devices with minimal bandwidth consumption, enabling the inference pipeline to adapt to seasonal menu changes without full model replacement.

REFERENCES

Bailey, J. (2022, December 7). Airlines burn $4 billion of untouched food and drink every year, here's why. Simple Flying. https://simpleflying.com/airlines-4-billion-untouched-food-drink/

Card, S. K., Moran, T. P., & Newell, A. (1983). *The psychology of human-computer interaction*. Lawrence Erlbaum Associates.

Chen, J., & Ran, X. (2019). Deep learning with edge computing: A review. *Proceedings of the IEEE, 107*(8), 1655–1674. DOI: 10.1109/JPROC.2019.2921977

Fall, K. (2003). A delay-tolerant network architecture for challenged internets. In *Proceedings of the 2003 Conference on Applications, Technologies, Architectures, and Protocols for Computer Communications (SIGCOMM)* (pp. 27-34). ACM. DOI: 10.1145/863955.863960

Few, S. (2006). *Information dashboard design: The effective visual communication of data*. O'Reilly Media.

Feyerke, A. (2013). Designing offline-first web apps. A List Apart. https://alistapart.com/article/offline-first/

Grigorik, I. (2013). *High performance browser networking*. O'Reilly Media.

Grous, A. (n.d.). Sky high economics: Quantifying the commercial opportunities of passenger connectivity. London School of Economics. https://www.lse.ac.uk/business/consulting/assets/documents/sky-high-economics-chapter-one.pdf

Hashimoto, M. (2016). Infrastructure as code. In *Proceedings of the 2016 ACM SIGPLAN International Conference on Object-Oriented Programming, Systems, Languages, and Applications (OOPSLA)*. ACM.

IATA. (2019, August). IATA cabin waste handbook. https://www.iata.org/contentassets/821b593dd8cd4f4aa33b63ab9e35368b/iata-cabin-waste-handbook.pdf

Jonas, E., Schleier-Smith, J., Sreekanti, V., Tsai, C.-C., Khandelwal, A., Pu, Q., Shankar, V., Carreira, J., Kraska, T., Gonzalez, J. E., Popa, R. A., Stoica, I., & Patterson, D. A. (2019). Cloud programming simplified: A Berkeley view on serverless computing. arXiv preprint, arXiv:1902.03383. https://arxiv.org/abs/1902.03383

Kleppmann, M. (2017). *Designing data-intensive applications*. O'Reilly Media

Metcalfe, R. M., & Boggs, D. R. (1976). Ethernet: Distributed packet switching for local computer networks. *Communications of the ACM, 19*(7), 395–404. DOI: 10.1145/360248.360253

Monarch, R. M. (2021). *Human-in-the-loop machine learning: Active learning and annotation for human-centered AI*. Manning Publications.

Nielsen, J. (1993). *Usability engineering*. Academic Press. DOI: 10.1016/B978-0-08-052029-2.50007-3

Nygard, M. T. (2018). *Release it! Design and deploy production-ready software* (2nd ed.). Pragmatic Bookshelf.

Reimers, N., & Gurevych, I. (2019). Sentence-BERT: Sentence embeddings using Siamese BERT-networks. In *Proceedings of the 2019 Conference on Empirical Methods in Natural Language Processing (EMNLP)* (pp. 3982-3992). https://arxiv.org/abs/1908.10084

Satyanarayanan, M. (2017). The emergence of edge computing. *Computer, 50*(1), 30–39. DOI: 10.1109/MC.2017.9

Shahrad, M., Fonseca, R., Goiri, I., Chaudhry, G., Bancy, P., Cober, T., Fontoura, M., Fonseca, R., & Bianchini, R. (2020). Serverless in the wild: Characterizing and optimizing the serverless workload at a large cloud provider. In Proceedings of the 2020 USENIX Annual Technical Conference (ATC). USENIX.

Shapiro, M., Preguica, N., Baquero, C., & Zawirski, M. (2011). Conflict-free replicated data types. In *Proceedings of the 13th International Conference on Stabilization, Safety, and Security of Distributed Systems (SSS)* (pp. 386-400). Springer. DOI: 10.1007/978-3-642-24550-3_29

Shi, W., Cao, J., Zhang, Q., Li, Y., & Xu, L. (2016). Edge computing: Vision and challenges. *IEEE Internet of Things Journal, 3*(5), 637–646. DOI: 10.1109/JIOT.2016.2579198

Wang, S., Qu, Y., Shen, J., & Araki, T. (2023). Entity alignment with semantic embeddings: A survey. *ACM Computing Surveys, 56*(2), 1–38. DOI: 10.1145/3604560

Wickham, H. (2014). Tidy data. *Journal of Statistical Software, 59*(10), 1–23. DOI: 10.18637/jss.v059.i10

Chapter 5
Real-Time IoT and Edge AI Framework for Hospital Visitor Access Control and Smart Door Automation

Cheuk Hin Yeung
http://orcid.org/0009-0001-5516-9909
The Hong Kong Polytechnic University, Hong Kong

Cong Wu
http://orcid.org/0000-0002-8652-3501
Liaoning Technical University, China

ABSTRACT

The Hong Kong public medical system is facing a critical manpower crisis while frontline nurses are spending up to one-third of their time on manual visitor management. This project proposes a Smart Door Automation System leveraging an AIoT architecture to streamline real-time access control. The system integrates a React.js self-service portal with a decentralized edge-computing module powered by a Raspberry Pi 4. To ensure patient privacy, identity verification—incorporating Tesseract OCR for HKID extraction and ArcFace-based facial matching, which is executed locally at the ward entrance. Experimental results have demonstrated an end-to-end latency of 13.47 seconds and an OCR accuracy of about 90%. Automating verification and maintaining secure audit trails can alleviate the administrative burden on staff while providing safeguarding services to patients and a privacy-aware blueprint for future Smart Hospitals.

DOI: 10.4018/979-8-2600-0701-3.ch005

1. INTRODUCTION

1.1 Background

Healthcare systems around the world are under increasing pressure from the dual challenges of a rapidly ageing population and a diminishing clinical workforce. According to the United Nations, the global population aged 60 and over will reach 2 billion by 2050, indicating that a shift in how medical facilities manage high-traffic environments is necessary to accommodate this demographic, as traditional management methods are increasingly inadequate.

Hong Kong's medical system has been facing similar pressures of overload and understaffing, with reports stating that there would be shortages of 1,383 nurses and 1,669 nurses by 2025 and 2030, respectively (Lam, 2022). Public hospitals have been working under heavy service demand, while the government and the Hospital Authority had undergone a healthcare reform in early 2025 (Secretary for Health, 2025). A meticulous examination presents an excellent opportunity to upgrade the hospital wards' infrastructure, thereby improving the hospitalisation flow for both staff and patients.

This report proposes the project, known as HOSPITAL VISITOR DOOR AUTOMATION SYSTEM USING AIoT FOR REAL-TIME ACCESS CONTROL, which is directly related to the identified problem, since its focus will be on the visitor registration process, verification of identification, and matching process at the ward's gate area. The combination of a self-service visitor portal, the backend approval mechanism, and the edge OCR & face recognition module will allow lowering the workload associated with managing visitors while ensuring proper traceability and enabling access control at the hospital ward. Shifting identity checking to the place where visitors enter also falls into the scope of using AIoT technology for decreasing latency, protecting data privacy, and real-time decision-making.

1.2 Problem Statement

Although many aspects of our lives have moved towards digitalisation, many public hospital wards still operate on manual, paper-based processes, with few moving to instant-messaging-based registration. Manual confirmation of visitors' appointments creates several inefficiencies. First, frontline nurses are frequently burdened by non-clinical administrative tasks, such as verifying visitors' identities and manually logging entry times. A considerable amount of nursing time, up to one-third, is currently allocated to non-clinical work, which includes managing and logging visitor traffic (Grosso et al., 2021). Second, manual access control makes auditability of visits uncertain. Previous outbreaks, such as SARS and COVID-19, demonstrated that minimising cross-contamination is also necessary in a healthcare setting (Wong et al., 2020). The inability to accurately track and limit visitors could introduce risk to patients in hospital wards.

There is a lack of existing systems for an automated, secure, and auditable access control system that verifies visitor identities while reducing the workload on hospital staff, showing a clear and urgent need for such systems.

1.3 Project Aim / Objectives

The primary aim is to design and implement a "Smart Door Automation System". The system should leverage AIoT technologies to provide real-time, autonomous, and contactless access control for hospital wards. Three objectives are then defined as below to reach the project aim.

1. Automatic Visitor Registration: By designing and developing a user-friendly, self-service visitor registration system, which will digitise the check-in process and eliminate the need for manual paperwork, hence reducing the time staff need to dedicate to visitor management.
2. Strengthen Identity Verification: A robust and automated identity verification module will be implemented to utilise Optical Character Recognition (OCR), and card-to-holder face matching on Hong Kong Identity (HKID) cards, ensuring the accuracy of identity information against the physical card bearer.
3. Enforce Controlled Access and Auditing: Develop and integrate an automated authorisation workflow. Following this workflow, the system will validate visitor credentials and timeslots against patient visiting permissions before granting them access to designated patient wards. The system should also maintain a tamper-proof digital audit trail of events for back tracing cross-contamination.

1.4 Contributions

This report proposes several significant contributions to hospital wards in three ways: A privacy-preserving edge AI framework, a Modular AIoT Architecture, and a new integrated workflow with high auditability.

1. Privacy-Preserving Edge AI Framework: Unlike traditional systems that upload sensitive identity data to the cloud, this project implements a "Privacy-by-Design" approach. The system architecture is designed to reduce and minimise the transmission count of sensitive and personal data, such as HKID images. By processing the image locally on the Raspberry Pi, and sending only encrypted results in text to the database, the system can reduce the risk of man-in-the-

middle attacks and comply with strict patient data privacy standards. This contribution aims to address the critical data sovereignty requirements in medical environments.

2. Modular AIoT Architecture: The project introduces a modular architecture that separates the system into 4 modules: the frontend interface, the RESTful API backend, the AIoT edge module, and the persistence layer. This provides scalability and allows more development to this project in the future. For example, this system can be integrated into the current "HA go" application, or different kiosk types with different authorisation methods could be developed without rewriting the backend logic.
3. Integrated Workflow and Auditability: The proposed system will bridge the gap between visitor online registration and physical security. By connecting the dots between a web-based reservation system and a hardware-controlled door lock, the system provides a comprehensive end-to-end solution for real-time visitor auditing and infection control, helping reduce the need for clinical staff intervention.

1.5 Report Organization

The remainder of this report is structured as follows: Chapter 2 provides a comprehensive literature review of current visitor management systems, OCR technologies, and face recognition algorithms. Chapter 3 details the research methodology, focusing on the system architecture, database design, and the implementation of the AIoT modules. Chapter 4 presents the experimental results, including performance benchmarks for the OCR and face matching components. Chapter 5 offers a critical discussion of the system's practical implications and limitations. Finally, Chapter 6 concludes the report and outlines directions for future work.

2. LITERATURE REVIEW

2.1 Smart Healthcare, Smart Hospital and AIoT

The evolution of hospitals to smart hospitals is driven by the integration of Artificial Intelligence of Things (AIoT). The high connectivity and automation features of AIoT support clinical decision-making, as interconnected devices collect real-time data to enhance doctors' and nurses' decision-making and operational efficiency. IoT integration in Hospitals is also considered one of the key factors in increasing operational efficiency (Anas et al., 2025). Furthermore, the Hong Kong Hospital Authority has also identified adopting technology to enable smart care as its strategic plan in 2022 (Hospital Authority, 2022). One of the major drivers of this technological shift is the global shortage of healthcare workers. As nurse shortages become a systemic issue, hospitals look toward automated technological solutions to ease staff burdens (Almarri et al., 2025).

Nurses and clinicians devote approximately 50% of their working time to doing administration tasks including documentations and checking of medical devices (Palan et al., 2025). Therefore, there is little time for taking care of patients, which increases work stress and adds to the problem of labor shortage. The integration of patients' individual information with IoT devices can reduce the need for manual documentation. (Javaid & Khan, 2021).

2.2 Visitor management systems for reducing staff workloads

Visitor management is the cornerstone of hospital security and infection control. Previously, manual visitor surveillance by nurses was both resource-intensive and prone to human error (van Mourik et al., 2017). The proposed system architecture is informed by successful precedents in visitor management systems. For instance, a previous deployment of a visitor registration system at Taipei Veterans General Hospital in 2020, which linked a registration form to a MySQL database and an entrance check-in interface (Hsu et al., 2020), successfully handled over 4,900 registrations within 20 days. A similar Electronic Visitor Management System, hosted in India called Pravesh (Karpaga Priya & Prabhune, 2024), has also shown success in reducing staff workload and the auditability of visitor traffic by using an online form-based registration system.

However, these existing systems show the fragmentation between data registration and ward access control, lacking integration between these systems. Furthermore, such systems lack the capability to automatically verify the real-time identity of a visitor, allowing the use of false credentials.

2.3 Low-cost computing for OCR capabilities

An automated and accurate identity verification is needed to ensure high-security access control. Optical Character Recognition (OCR) is a technology that is often used for digitising identity documents in real-time. Recent studies show that it is possible to scan information from ID cards easily, even when the card is being held by a person in non-ideal lighting conditions (Gupta et al., 2021). For edge-based usage of OCR, the computational cost of the OCR operation should also be considered. For the vast majority of uses, the Raspberry Pi system has enough computing power for real-time OCR implementation (Yanto et al., 2023). Another essential aspect in such an environment is choosing the proper OCR engine. Ramdhani et al.

(2021) tested the performance of three OCR engines, which were Foxit, PDF2GO, and Tesseract, in extracting information from files. Their results indicated that while Tesseract provided slightly worse results concerning recall rate compared to other OCR engines, it was the fastest of all engines in terms of average processing time at 0.044 seconds per document, compared to Foxit's 0.050 seconds. Furthermore, they highlighted that Tesseract allows for unlimited batch conversion and offline processing. For hospitals, this fact about Tesseract suits the privacy-by-design model well because files do not have to go beyond the local network.

2.4 Face recognition and verification process

Complementing OCR, a facial recognition of the visitor provides a secondary layer of authentication. Automating the identity verification process involves matching a live camera capture (a "selfie") with a scanned ID card photo, often formulated as a "cross-domain" face verification problem. This problem, which is more challenging than standard face recognition due to domain shift between live selfies and photographed ID documents, including differences in illumination, blur, noise, and acquisition conditions (Folego et al., 2016). Other studies have found that using a CNN with the Triplet Loss approach yields the best results compared to other existing solutions (Costa et al., 2021). J. Deng et al. (2019) introduced the ArcFace DCNN model, which introduces an "angular margin" that forces the network to learn more discriminative features, and applies the embedding space concept to face verification usages. The model introduces hypersphere-based embeddings and measures similarity using geodesic distance rather than the usual straight-line distance, which makes it outperform state-of-the-art models like VGG-Face.

2.5 Research Gap and Motivation

Despite the advancements in the field of AIoT and biometric verification, there are 3 critical gaps that are seen in the current literature.

First, there is the integration gap. Most academic publications show systems that address only visitor registration or biometric detection in isolation, showing a lack of research on unifying both by bridging the gap between most web-based reservation and door lock (or other physical hardware actuator) under a staff approval loop.

Second, there is the privacy and edge gap. Existing commercial solutions that match ID to faces rely heavily on cloud processing, which poses a significant privacy concern in medical settings, as patients' data, or the HKID data used, is critical information that could hurt the hospital's reputation if leaked. There is a lack of an edge-based framework that performs local identity verification to ensure data sovereignty.

Lastly, there is the operational gap. There is a lack of a solution that empowers clinical staff, such as nurses, to manage visit requests by incorporating role-based access control.

These three gaps motivate the establishment of the current project. By integrating edge-based OCR and face recognition with a cloud-managed staff portal, this project proposes a system that is secure, privacy-first, and provides a fully integrated access control workflow for clinical staff, addressing the operational realities of current hospital wards and future smart hospital wards.

3. ALGORITHMS AND MODULES IMPLEMENTATION

3.1 System Architecture Overview

The proposed Smart Door Automation System in this report is built on a decentralised AIoT architecture that balances the security and speed of local processing with a centralised management entry point for clinic staff. The following Context diagram (Figure 1) shows the system as a whole and the logical processes, detailing 4 main actors that could interact with the system: visitors, hospital staff, ward door hardware, and an external Instant Messaging (IM) notification service. To focus on the main data exchanges, the system's internal workings are left out. While visitors fill out their registration and identity information, they also get updates on their access status. Hospital staff log in to the system with their authentication credentials and decide who can get in. The system connects to the door hardware for access control and to the outside messaging service for alerts.

Figure 1. Context Diagram for the system

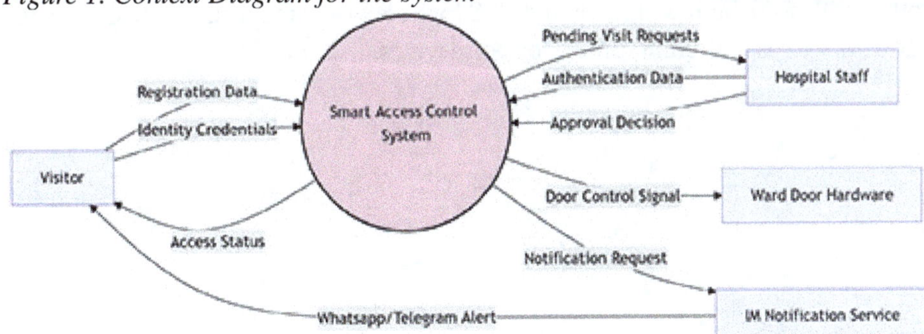

The Smart Access Control System is decoupled into four modular layers: the Edge Computing Layer, the Cloud Backend, the Web Frontends, and the Persistence Layer. This modular approach was chosen for 4 important reasons.

1. Latency optimisation: By doing tasks that are expensive in terms of processing power, like OCR and face matching, at the edge, the "round-trip" delay that is usually associated with cloud-based computing could be avoided, which would speed up the system as a whole.
2. Data privacy: Sensitive images, like still HKID card captures and face captures, never leave the edge AIoT module at the local ward entrance, upholding healthcare data sovereignty standards.
3. System Resilience: If the cloud backend experiences downtime, the edge devices can be updated to maintain a cached visitor list, eliminating the disruption caused by intermediate downtime of the backend, providing a fail-safe entry.
4. Future upgrades: This modular separation allows each component to be independently upgraded, and supports future hardware upgrades like performing horizontal or vertical scaling.

3.2 Detailed Module Descriptions

3.2.1 Web Frontend Portals (Visitor & Staff)

A web application built with React.js serves as the frontend portal for both visitors and clinic staff to access the system via their computer and smartphone devices. The web application is built as a mobile-optimised single-page application (SPA). The web application uses JSON Web Tokens (JWT) for session management, ensuring that only users who are authenticated are authorised to send requests to the server. Both portals were built with Tailwind CSS for a clean and modern responsive layout that works across different dimensions of devices.

Figure 2. Data Flow Diagram for the Visitor Frontend module

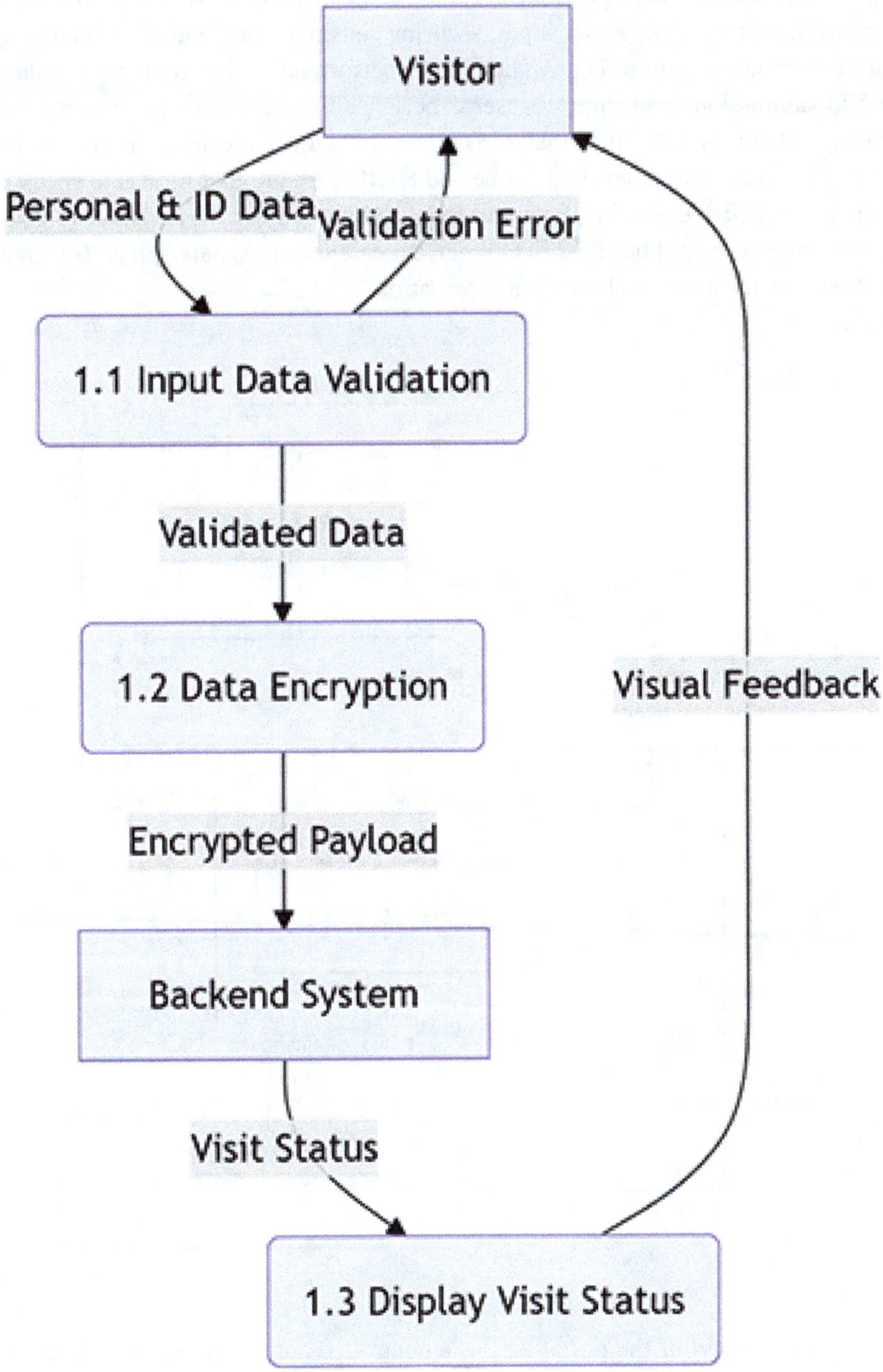

There are two parts of the portal. The first part allows potential visitors to register an account and request specific time slots for patient visits. The module is responsible for validating user input, securing sensitive data, and communicating with the backend system. Data validation is performed at the frontend to reduce invalid submissions and improve user experience, while encryption ensures that sensitive identity-related information is protected before transmission. This is also where the visitor will enter their Name and HKID card number for use to check in at the Edge AIoT Identity Verification Module. Optionally, in the registration steps, a phone number could be filled in to receive messages via WhatsApp or Telegram to receive status updates about their reservation.

Figure 3. Data Flow Diagram of the Staff Frontend module

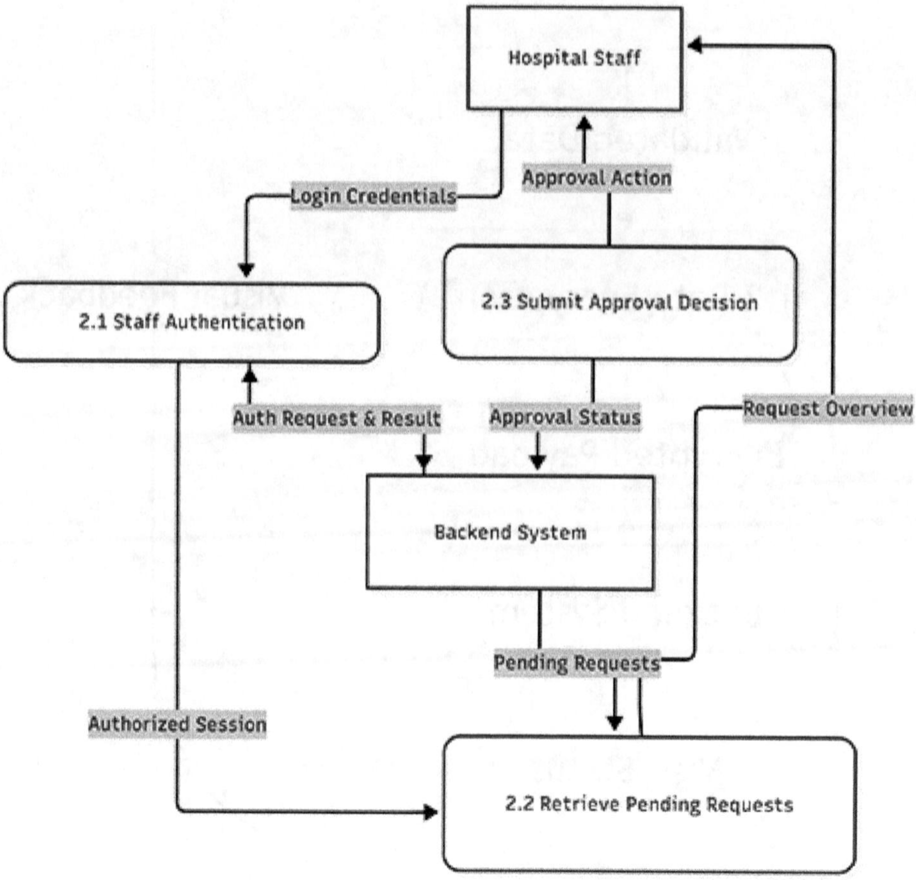

The second part of the portal acts as a controlled administrative interface, ensuring that only authorised personnel can access sensitive visitor information, and allows hospital nurses and administrators alike to have a real-time overview of any pending visit requests. Nurses can then approve or reject visits with a single click. Administrators can also perform user management and other maintenance tasks on this portal, like sending a manual notification message or amending information, such as patients who could be visited. By centralising approval workflows within this module, the system significantly reduces manual logging and coordination tasks traditionally handled by nursing staff. This automation directly contributes to workload reduction while maintaining oversight and accountability in visitor access management.

3.2.2 Edge AIoT Identity Verification Module

The Edge Module is a physical "kiosk" that would be deployed next to a hospital ward entrance. Powered by a Raspberry Pi 4 (4GB RAM), it is equipped with two high-resolution USB cameras and a screen. This module is the "gatekeeper" of the hospital ward.

Figure 4. Data Flow Diagram for the AIoT-based OCR and Face Recognition module

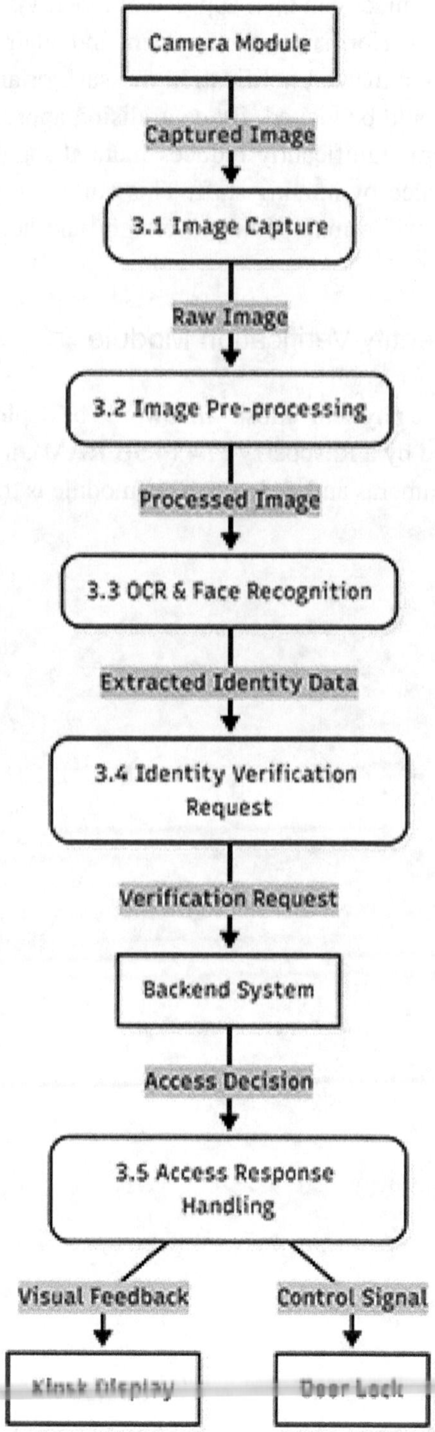

This edge module captures identity images, performs local image pre-processing, and extracts identity information using OCR and facial verification techniques. Leveraging OpenCV for extracting image data from the webcam's real-time video stream. The image is fed to the image preprocessing pipeline to ensure accurate data extraction from HKID cards. First, a perspective transform (4-point transform) is done to rectify the document's orientation to upright, then adaptive thresholding is used to eliminate shadows and reflections from lighting, resulting in a black and white image. The rectified image is then fed to the Tesseract OCR engine to extract text, like name, and HKID number from the document. Then, a second camera takes a "selfie" of the visitor to compare it to the ID photo that was taken using the DeepFace framework.

The system reduces latency and the transmission of sensitive raw image data to the backend by doing computationally intensive tasks at the edge. The backend checks the extracted identity data, and the access decision is immediately shown through visual feedback and physical door control. This design shows how AIoT principles can be used in the real world to make access control safe and real-time.

3.2.3 Backend API and Business Logic

The central nervous system that helps determine if the ward door should be unlocked lies within a RESTful API from Node.js/Express, where the backend system accepts requests from the Edge AIoT Identity Verification Module, then consults the databases to perform checks:

- Is the HKID registered in the database and has a visit request?
- Is the visit request approved by a staff member?
- Is the current system time within 60 minutes of the approved visiting window? (to allow leeway for visiting)

Figure 5. Data Flow Diagram of the Backend System

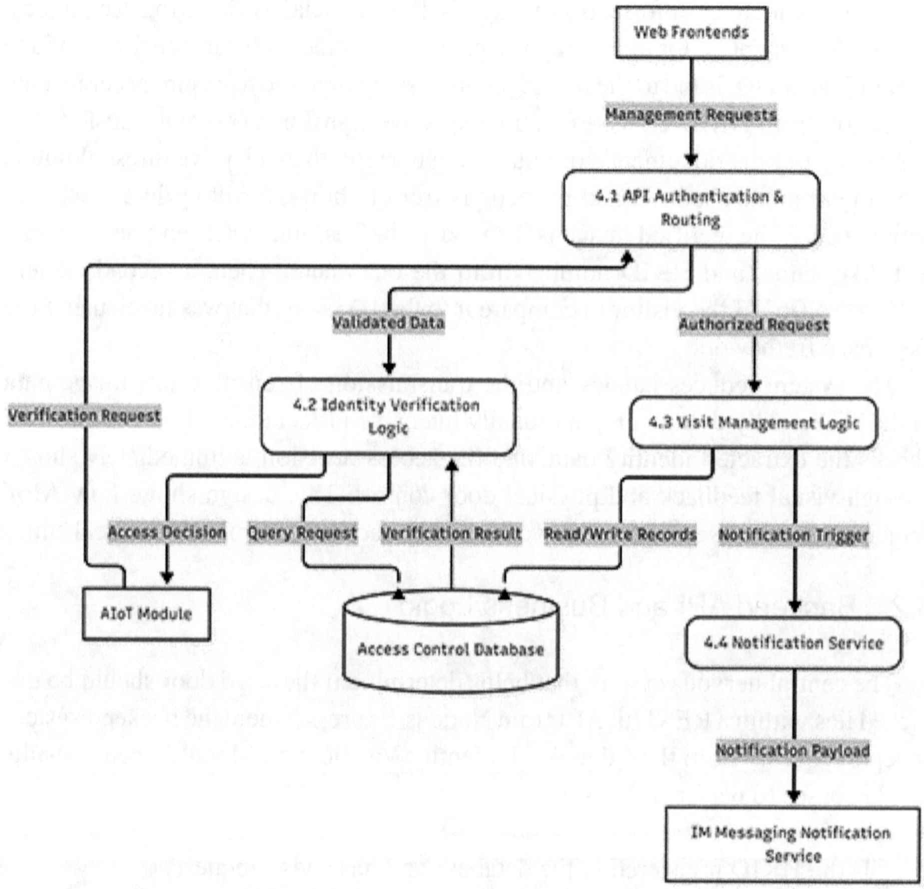

The backend system will not perform a door unlock signal until these conditions are satisfied. If all checks pass, then an MQTT signal will be sent to the door hardware actuator.

3.2.4 Database Design and Normalisation

Figure 6. The Entity relationship diagram for the current Database design

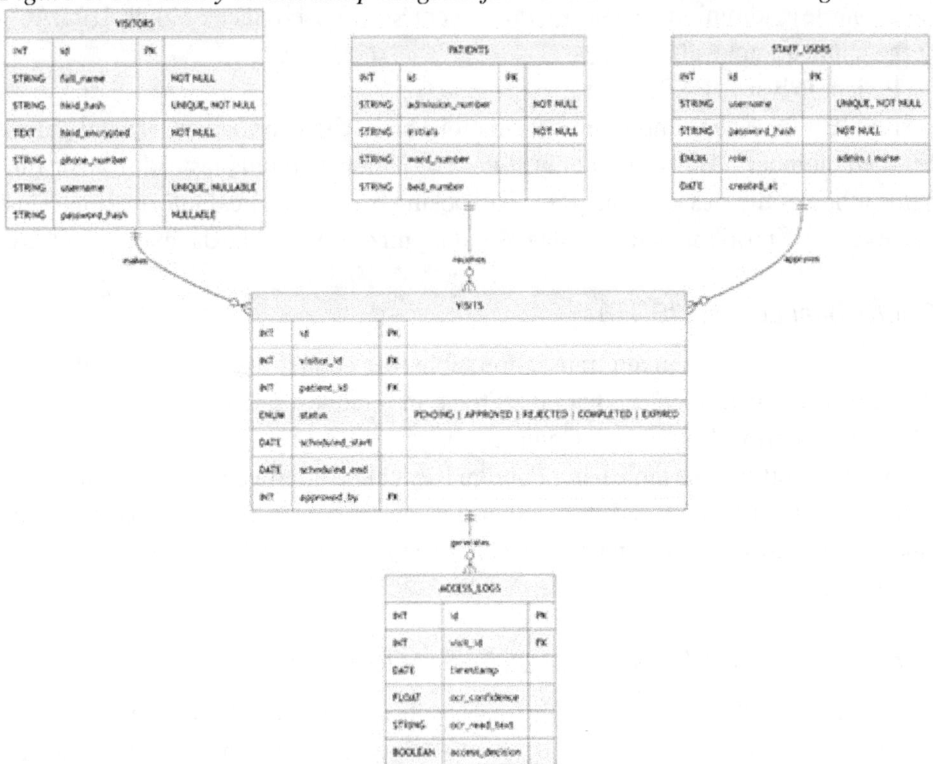

To support high-concurrency requests and data integrity, the system utilises a relational database following Third Normal Form (3NF). The database follows a normalised relational structure designed to ensure data integrity, privacy compliance, and accountability. The schema is divided into five core entities, with notable fields detailed below:

Identity Management (STAFF_USERS and VISITORS)

VISITORS: This is the primary database of all the visitors' identities. In compliance with privacy norms, HKID is stored in two ways: the first is hkid_hash, a one-way hash suitable for fast lookups, and the other is hkid_encrypted, an encrypted version of the text. Both temporary visitors and registered users can be stored in the table because the username and password_hash columns may contain null values.

STAFF_USERS: Stores credentials for hospital personnel. It utilises an ENUM field for role ("admin" or "nurse") to implement Role-Based Access Control (RBAC) at the application level.

Patient Privacy (PATIENTS)

PATIENTS: Keeps important information about where the patient is (ward_number, bed_number). The system only stores the patient's initials instead of their full name to lower the risk of a data breach exposing private medical information. This makes sure that patient identities stay pseudonymized even if the database is hacked.

Operational Logic (VISITS)

VISITS: This is the main transaction table that connects a Visitor to a Patient. It works as a junction entity that keeps track of a request's status using an ENUM (Pending, Approved, Rejected, Completed).

Accountability: The table has a Foreign Key called approved_by that connects it to STAFF_USERS. This makes sure that staff members could be held accountable, since each approved visit can be traced back to the nurse or administrator who gave the go-ahead.

Auditing and AI Performance (ACCESS_LOGS)

ACCESS_LOGS: A logging entity that is only for a certain VISIT. It keeps track of the exact time and the final access_decision (Allow/Deny) for each attempt to get in.

AI Diagnostics: This table is very important because it stores the ocr_confidence score and the ocr_read_text that the AIoT device made. This lets developers look at false positives and negatives and use real-world data collected during operation to retrain the OCR model, for use of further iterating the system for better performance.

3.3 Key Algorithms and Logic Flow

3.3.1 OCR and Document Rectification Algorithm

The system ensures that the OCR engine receives the highest quality input through a multi-stage algorithm:

Algorithm 1.

```
def preprocess_id_card(image):
# 1. Detection and Rectification
gray = convert_to_grayscale(image)
blurred = gaussian_blur(gray, (5, 5))
```

```
edges = canny_edge_detection(blurred)
rect_coords = find_largest_quadrilateral(edges)
rectified = four_point_transform(image, rect_coords)
# 2. Multi-stage OCR Enhancement (actual hkid_extractor logic)
# Upscale first to 2x resolution to help with small text
upscaled = rescale(rectified, scale_factor=2)
binary = adaptive_threshold(upscaled, method=GAUSSIAN)
# Erosion is critical to "thin out" bold HKID fonts for higher accuracy
processed = erode(binary, kernel_size=(2, 2))
return tesseract_extract_text(processed)
```

3.3.2 Face Verification Logic

The face matching process utilises the ArcFace model, known for its high accuracy in cross-domain verification. The algorithm computes the cosine similarity between the ID photo and the live feed:

Algorithm 2.

```
def verify_identity(id_face_img, live_visitor_img):
# ID photo enhancement: CLAHE + Upscale
enhanced_id_face = apply_clahe(id_face_img, clip_limit=2.0)
# Feature Extraction via ArcFace
embedding_id = deepface.represent(enhanced_id_face, model='ArcFace')
embedding_live = deepface.represent(live_visitor_img, model='ArcFace')
# Distance calculation (Lower = More Similar)
distance = calculate_cosine_distance(embedding_id, embedding_live)
# Allow some difference in face (can be changed)
if distance < 0.55:
return "MATCH_SUCCESS"
else:
return "MATCH_FAILURE"
```

3.4 Implementation Considerations

For a privacy-sensitive location such as a hospital ward, it is important to consider the system's data security. Several security measures have been integrated into the system design from day one to protect against external attacks and internal mismanagement.

Both the front end and the back end should clean up incoming and outgoing data to protect the system from malicious attacks and exploits. The Node.js backend has a strong Input Validation Layer that checks all incoming JSON payloads from both the web frontends and the Edge AIoT module. The validation logic uses strict schema checks to check the data type, set limits on string length, and filter values based on their range. Any request that doesn't follow the set schema because of errors in transmission or deliberate tampering is thrown away. This stops attacks like SQL Injection and Cross-Site Scripting.

3.5 Implementation Environment

The system development environment was standardised to ensure reproducibility:

- Hardware: Raspberry Pi 4 Model B, 2 USB Webcams (Auto-focus enabled).
- Server (Cloud): AWS EC2 Instance (t2.micro), Ubuntu 22.04 LTS.
- Server (Local Development): Ryzen 7 9700X, Windows 11 24H2
- Software Versions: Node.js v24.12, Python 3.12.9, React 19.2, Tesseract v5.5.0.

4. EXPERIMENTATION AND BENCHMARKING

4.1 Experimental Setup and Environment

To evaluate the performance and reliability of the Smart Access Control System, a controlled experimental environment was established to simulate a hospital ward entrance.

Hardware Configuration: The Edge AIoT Identity Verification Module with the Raspberry Pi 4 connected with two USB cameras: one with a changeable short focus high-definition USB camera for OCR, along with a secondary wide-angle camera with a longer focus range that captures live visitor feeds.

Software Environment: The backend was hosted on a local server environment running Node.js v24.12 on Windows 11 24H2 to minimise the network variables between the backend and the edge. The OCR on the Raspberry PI is powered by Tesseract v5.5, while the face recognition utilises the DeepFace Library.

4.2 Functional Validation

Table 1. Functional Validation of the system across 6 test cases

Test Case	Scenario Description	Expected Outcome	Observed Result	Status
TC1	Valid visitor with an approved time slot and a matching HKID.	Access Granted to the ward; Log recorded on the system.	Access Granted to the ward; Log recorded on the system.	Pass
TC2	Valid visitor but arriving 30 mins *after* the approved window.	Access Granted to the ward; Log recorded on the system.	Access Granted to the ward; Log recorded on the system.	Pass
TC3	Valid visitor but arriving 90 mins *after* the approved window.	Access Denied; Log recorded on the system.	Access Denied; Log recorded on the system.	Pass
TC4	Valid visitor but with a different person's HKID card.	Face Match Failure; Access Denied.	Face Match Failure; Access Denied.	Pass
TC5	Unauthorised staff trying to access the Admin Dashboard.	The login attempt fails, and no modifications to the system can be made.	The login attempt fails, and no modifications to the system can be made.	Pass
TC6	Visitor with an unapproved/pending registration.	Access Denied to the ward; Log recorded on the system.	Access Denied to the ward; Log recorded on the system.	Pass

4.3 Sub-module Performance Benchmarks

4.3.1 OCR Extraction Accuracy

The OCR extraction part of the Edge AIoT Identity Verification Module is individually tested using a sample of 3 HKID cards, each processed 5 times under different combinations of light brightness: Strong spotlight + Strong Environment light, Weak spotlight + Strong Environment Light, Strong spotlight + Weak Environment light, Weak spotlight + Weak Environment Light. This totals 60 OCR scans.

Table 2. OCR accuracy across different lighting conditions

Light Conditions	Successes	Failures	Success %
Strong spotlight + Strong Environment light	15	0	100%
Weak spotlight + Strong Environment Light	12	3	80%
Strong spotlight + Weak Environment light	15	0	100%
Weak spotlight + Weak Environment Light	14	1	93.33%

4.3.2 OCR tolerance to off-angle ID card placements

The OCR extraction part of the Edge AIoT Identity Verification Module is also tested for its tolerance against off-angle ID card placements. Upon testing, the system allows for nearly 30° of rotations clockwise and counter-clockwise. While the system currently does not allow putting the card rotated 180°, the UI provide by the Raspberry Pi 4 lets the visitor to see if they have put the card the correct way.

The four-point perspective transformation formula currently used by the module can only account for the geometry, but not correcting semantic orientation. No matter how the card is rotated, the card will be corrected to fit a flat, rectangular image. However, for a card with 180° turn, the physical top left of the card will become bottom right, resulting in a perfectly flat, upside down card, and a card turned 90° will become a portrait rectangle.

Nevertheless, the system could be improved in the future to accept 180° rotated card, or even all placements of the card given that it is faced up in the future, by doing 4 OCR scans on the possible rotations of the card, or by doing a feature detection on the card. For example, the face of the left most 1/3 of the card would be a great candidate for detecting the orientation of the card.

4.3.3 Face scanning distance allowance experiment

The face scanning part of the Edge AIoT Identity Verification Module is then tested for the distances that it can capture embeddings. With the 720P camera currently used for face captures, the distance of 40cm between the camera and the visitor nets the lowest cosine distance (~0.3) for ArcFace. The camera can reach acceptable cosine distance (<0.6) for ranges camera distance within 1 meters. The face matching fails (finding no faces from the camera feed) when the visitor stands further than 1.5 meters away. The cosine distance threshold is then set to 0.55 after the result of this experiment to strike the balance between the visitor's convenience and security of the system.

4.4 System Integration and Latency Analysis

The system is also tested for its "End-to-End" latency. For reproducible testing results, the "End-to-End" latency is defined as starting from the moment the ID card is placed down to the moment that the door receives the unlock signal via MQTT and the physical actuator clicks / LED light changes. The system latency is measured across 10 trials.

Table 3. Performance timing analysis for Edge AIoT Identity Verification Module

Operation	Time spent (s)
Image stabilisation wait	4.26
OCR pre-processing	0.25
Tesseract extraction	1.30
Camera switch	3.01
Face capture	4.47
ArcFace inference	0.13
Backend API round trip	0.05
Average End-to-End latency	13.47

Bottleneck Analysis: One of the bottlenecks that was observed was the camera switching, taking up more time than expected, amounting to nearly 3 seconds. The OCR scan also needs to confirm that the image is stable before performing the scan.

The result of the End-to-End test is an acceptable result, as the 13 seconds spent on a visitor allows more than 240 visitors per hour to the hospital ward. However, the latency can be further reduced.

4.5 End-to-End System Walkthrough

In order to showcase the interaction of these independent AIoT modules as well as demonstrate fluidity of the operation flow, below is provided a walkthrough of the typical scenario where the visitor will enter the premises by means of biometric authentication, which mirrors a standard hospital procedure, starting with clinical personnel authorisation and ending with a visitor access permission.

This step takes place on the React.js-based Staff Dashboard at the hospital nursing station. Once entered, clinical personnel must verify themselves using appropriate credentials. In order to protect patient and visitor data, upon successful authorisation, the Node.js backend returns a JSON Web Token (JWT). This token safely stores user roles (“nurse” or “admin”), implementing strict Role-Based Access Control (RBAC) mechanisms and allowing only appropriate users to gain access to further administrative panels.

Figure 7. Screenshot of the Clinical Personnel Authorisation page

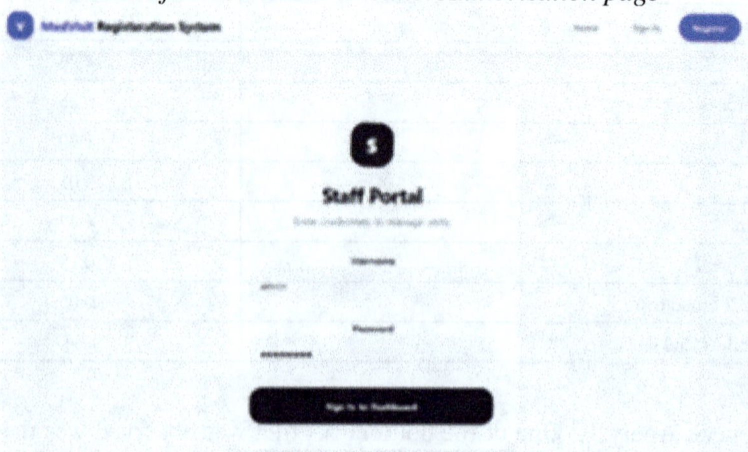

Figure 8. Screenshot of the Pending Visitor Registration Requests and Approval

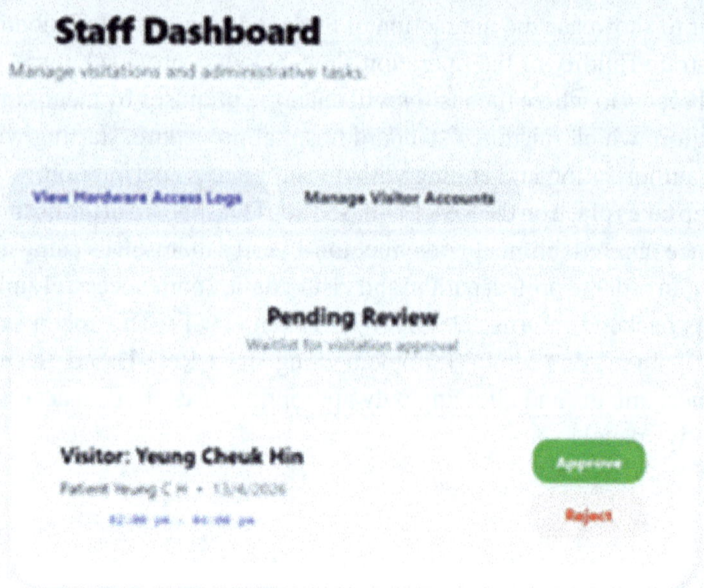

Figure 9. Screenshot of the Telegram Notification message that was sent to visitor

Next, the nurse opens the request overview panel, which automatically fetches all pending visitors' requests from the MySQL database. This panel is designed in such a way that all the necessary information needed to make an informed decision (the visitor's name, the requested time slot, and the patient they want to visit) is presented immediately to the user. When the clinic staff click "Approve," the front end makes a safe, asynchronous API call. Next, the backend changes the status of the visitor request to "Approved" and sends the message to the user through either Telegram or WhatsApp. This creates a smooth approval loop with no paperwork.

Figure 10. Screenshot of AIoT Kiosk UI at idle

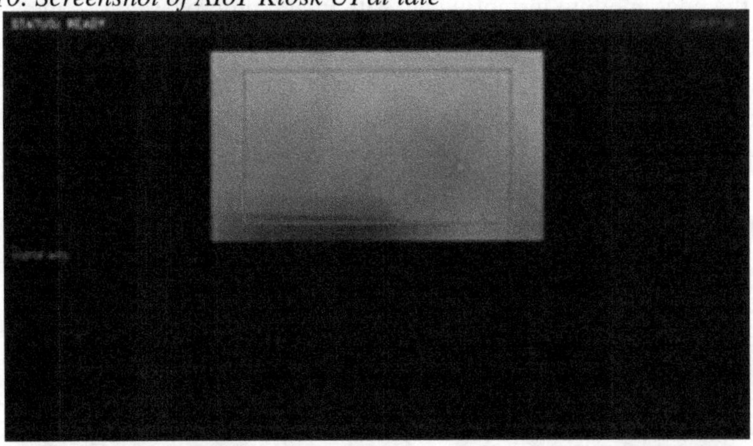

After receiving the approval message, the visitor arrives at the hospital ward entrance, where a Raspberry Pi 4-based physical Edge AIoT module is installed. The system is equipped with a screen that displays the system state, creating a convenient kiosk for visitors to use. The system works using idle loops, monitoring the camera for any activity in relation to HKID card detection.

Figure 11. Screenshot of AIoT Kiosk scanning a HKID card

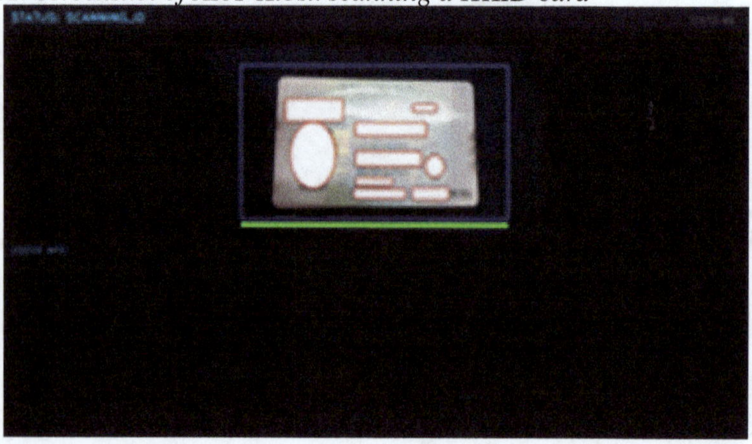

Once the HKID card of the visitor is placed in front of the first high-definition camera, the edge module instantly starts its image preprocessing sequence. After the stability meter fills up, the real-time edge detection and four-point perspective transform is performed in the background to create a flat image that accounts for off-centre positioning. Afterward, the Tesseract OCR is run to extract alphanumeric sequences from the processed image. Since sensitive HKID information gets converted into text and subsequently hashed at the edge end itself, it ensures that no biometric images pass over the hospital's private network.

Figure 12. Screenshot of AIoT Kiosk performing Biometric Facial Matching

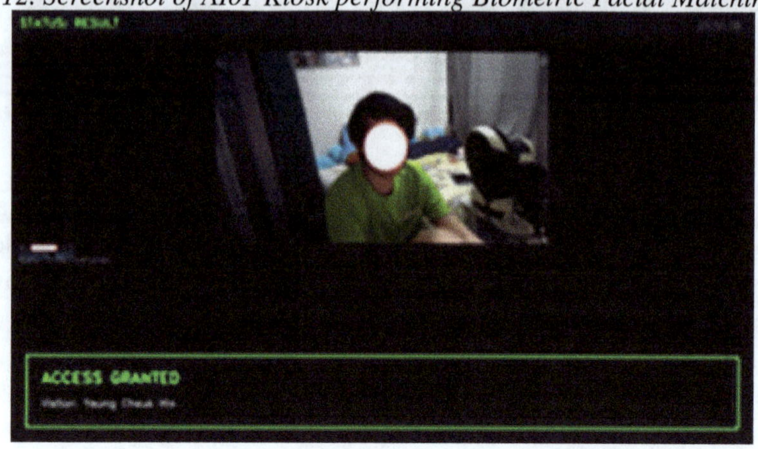

After a successful OCR process is completed, the second stage starts where the system shifts to using the wide-angle camera in place to capture a facial image. In this case, we have used the DeepFace library with ArcFace model to generate embeddings for the ID face and live face captured by the camera respectively. By performing cosine similarity comparison of the two vectors, it allows the edge machine to verify the physical identity of the person against the credentials presented, thus protecting against potential stolen/borrowed HKID.

Finally, after a successful matching, the Raspberry Pi sends the hashed HKID number and the True Boolean value to the Cloud Backend, which cross-verifies with the approved visits database in relation to time slots. If it is confirmed that the verified individual was granted permission, then it sends a Boolean True message to the Raspberry Pi. Afterwards, the Raspberry PI also sends a signal to via the MQTT protocol and releases the magnetic door lock, allowing the user to enter the room freely. Additionally, this triggers an immutable record creation in the ACCESS_LOGS table on the Cloud Database side.

4.6 Comparative Analysis: Manual vs Automated Workflow

A qualitative benchmark was performed to compare the traditional visitor logging process that is present in public hospitals.

Table 4. Comparison between manual and automated visitor registration workflow

Feature	Manual Process	Automated AIoT System
Processing Time	2-5 minutes per visitor.	< 20 seconds per visitor.
Human Resource	Requires 1 nurse/clerk on duty.	Fully autonomous; staff for approval only.
Data Integrity	Prone to handwriting errors.	Digital extraction with high accuracy.
Infection Control	High (physical contact with pens/logs).	Contactless (Camera-based).
Auditability	Difficult to search past records.	Real-time searchable SQL database.

5. DISCUSSION

5.1 Interpretation of Results

Through experimental evaluation, the "Smart Door Automation System" demonstrates that applying edge-based AIoT for access control in a hospital ward is technically feasible. The successful implementation of three disparate technologies, which are web-based visitor registration, edge-based OCR identity extraction and biometric facial verification, is the core achievement of this study, leading to a unified, staff-mediated workflow.

The experiments done in Chapter 4 also validates the success of the system on reaching the 3 objectives defined in Chapter 1.3:

1. Automatic Visitor Registration: The test case TC1, TC2 and TC6 in Chapter 4.2 demonstrates that the system is capable of allowing visitor entry autonomously without staff/nurse involvement.
2. Strengthen Identity Verification: The OCR extraction accuracy of >90% in Chapter 4.3.1 along with test case TC4 in Chapter 4.2 demonstrates the system is capable of performing Identity Verification with high accuracy.
3. Enforce Controlled Access and Auditing: In Chapter 4.2, the test cases TC1, TC2, TC3, TC5, and TC6 show that the system makes log entries for all visit attempts, whether they are successful or not. Ensuring there is a way to trace the audit trail for people trying to enter the hospital ward.

Functional testing also demonstrates that the system can correctly identify approved visitors and triggers physical hardware blockers (door locks) within an average latency of 13.47 seconds, compared to taking around 2 to 5 minutes to register manually with the clinic staff, showing that the system satisfies the operational requirements in a high-traffic hospital ward environment.

5.2 Interpretation and Comparison with Literature

This report discusses the "Integration Gap," which occurs when visitor management and door access control are two separate systems, as noted in Chapter 2.5. The Taipei Veterans General Hospital put in place a digital registration system in 2020 to handle Covid-19 visitor traffic (Hsu et al., 2020). However, the system focused solely on entering visitor data and did not provide real-time access control for the hospital ward. Another similar system located in India called Pravesh used online form-based registration, providing auditability, but lacking in biometric verification and physical door integration. The "Smart Door Automation System" extends these two systems by integrating visitor registration data using RESTful APIs to a Raspberry Pi-controlled hardware interface.

Furthermore, to align with hospital patient privacy policies, the system's choice of Edge-based OCR avoids the risks of cloud-centric solutions that may expose data to interception. Processing HKID card images on the edge ensures that raw HKID images are processed and discarded locally. The successful implementation of ArcFace also demonstrates that biometric verification can be achieved on low-cost edge hardware, addressing the "privacy and edge gap".

5.3 Limitations of the Current Prototype

Despite the fundamental functionality being valid, several limitations must be stated as such when considering the real deployment of this system to a hospital.

Firstly, the environmental sensitivity of the Tesseract OCR engine remains a challenge due to the failure of recognition of laminated HKID cards in high-glare environments. Further improvements to the scanning environment could reduce scanning time, such as providing a covered spot for placing the HKID card, similar to the e-Channels at border control.

Secondly, the current face recognition module lacks "Liveness Detection", making it vulnerable to spoofing attacks using photographs of approved visitors. In a comprehensive survey done by Ming et al. (2020), standard 2D facial recognition systems powered by generic RGB cameras are highly susceptible to printed photo or video replay impersonation attacks. In a hospital setting, this vulnerability is especially significant, since unauthorised individuals such as family members of visitors have strong motivations to borrow identification cards and use easily obtainable photo of registered visitors to bypass access controls.

Possible countermeasures for this vulnerability could include using multiple cameras and multiple perspectives to verify the three-dimensional depth and liveness characteristics of the visitor, or by using remote PhotoPlethysmoGraphy (rPPG) technology, which analyses blood flow changes on the human skin's surface (Debnath & Kim, 2025). The face recognition module also exclusively compares the face on the printed ID with the visitor's face, without considering a possible forged ID card. For deployment to public hospitals, it would be wise to connect the system to the government ID database to check the ID validity.

Thirdly, the system currently doesn't account for multiple visitors entering at once. Under the current system, group visits (such as families) would be separate visit requests, which might introduce more work than the current system. The edge module also has no countermeasures for tailgating another valid visitor to enter the hospital ward.

Lastly, due to budget and procurement restraints, the system was only tested in a simulated laboratory environment. The absence of real-world stressors, such as entrance congestion (multiple people) at the entrance and fluctuating hospital network bandwidth, means the current performance metrics are merely preliminary.

5.4 Theoretical and Technical Contributions

This project contributes to the AIoT field by proposing a Decentralised Healthcare Access Framework. This provides a hybrid model where business logic is managed in the cloud, and biometric inference is executed at the edge, which is the optimal architecture for privacy-sensitive environments. This research also provides a modular blueprint for the adaptation of other medical areas, such as medication storage rooms or intensive care units (ICUs), where auditability is maximised and response latency is minimised.

5.5 Recommendations for Future Deployment

To make sure that the system works well in public hospitals, it is best to add it to the existing mobile app "HA Go" so that visitors can easily schedule their visits. Also, the edge hardware should be updated to include infrared (IR) sensors. This will help OCR work better in low-light situations. Furthermore, a data retention and deletion policy should be detailed and implemented in code for the ACCESS_LOGS entries to comply with Data Protection Principle 2 (DPP2) and Section 26 of Hong Kong's Personal Data (Privacy) Ordinance (PDPO). ACCESS_LOGS and other related biometric data should be permanently deleted once they are no longer applicable for infection tracing purposes (for example, 31 days) to prevent legal liability.

Finally, a pilot study involving actual hospital staff should be conducted to evaluate the User Experience in different roles and identify any operational flaws within the approval workflow.

6. CONCLUSION AND FUTURE WORK

6.1 Summary of Contributions

This report presented a unified platform for hospital ward access control that integrates Edge AI, cloud, and Instant messaging notifications as a Smart Door Automation System. There are four key contributions that stand out:

Modular Architecture that preserves privacy: The first key contribution is a decentralised design comprising four layers that allows independent vertical scaling: Web Frontend, Cloud Backend, Edge AIoT module, and the Persistence Layer. This architectural approach is built on "Privacy-by-Design", which ensures that sensitive personal data is processed only at the edge, reducing the risk of data breaches while it is in transit and supporting independent maintenance of components.

Edge-Based Identity Verification Pipeline: The second key contribution is the implementation of Tesseract OCR and ArcFace on a Raspberry Pi 4 for localised identification and verification. This system can perform high-security tasks such as document rectification, text extraction, and facial matching on low-cost edge hardware, achieving an end-to-end latency of approximately 13.47 seconds and handling the visitor throughput of a typical hospital environment. The modular design also supports horizontal scaling, allowing hospital wards to operate multiple entrances to further increase the throughput.

Staff-Mediated Closed-Loop Workflow: The project goes beyond basic automation by building a full administrative stack on the front end of the system. A staff-mediated approval loop with real-time WhatsApp and Telegram API notifications makes sure that clinical staff always watch over automated access. This method successfully connects digital registration processes with existing physical security protocols, and it also saves clinical staff time by bringing all of the steps together on one platform.

Robust Data Integrity and Auditability: A principal contribution is the implementation of a normalised third normal form (3NF) relational database with integrated audit logging. This system maintains a tamper-proof security record for each entry attempt, providing hospital management with leading indicators that are critical for infection control and administrative optimisation. By using MySQL database as a base, existing SQL backup solutions could be applied to minimise the risk of data loss.

6.2 Technical and Operational Limitations

In summary, while the system is architecturally sound, its current use is limited by the camera's sensitivity to light and the ease with which face image can be faked for a simple RGB based camera. These limits show how hard it is to use complex AI models on low-power edge devices in open physical spaces right now.

6.3 Future Work and Deployment Roadmap

A phased roadmap will be put into place to help future development and turn the prototype into a medical infrastructure that is ready for production.

Phase 1: Q2 2027 - Hardware Optimisation and Anti-Spoofing

The first priority is upgrading the optical hardware to include Autofocus Macro Lenses and Infrared sensors to alleviate blooming on HKID cards caused by lighting interference. On the software side, Liveness Detection algorithms (blink detection, lightning reflection analysis or texture analysis) will be integrated into the

DeepFace pipeline to prevent the possibility of spoofing. Additionally, an auxiliary pre-processing neural network will be trained specifically on HKID holographic emblems to improve OCR extraction accuracy under high-glare conditions or to eliminate any spoofing with fake IDs.

Phase 2: Q4 2027 - Operational Pilot and Ecosystem Integration

The second milestone is a controlled pilot within a hospital ward in Hong Kong. This phase will be focusing on the following:

Integration with "HA Go": Building a safe connection to the Hospital Authority's current mobile app ecosystem so that visitors can use their already-verified profiles to sign up for profiles right away.

Multi-Entrance Scalability: Expanding the system to monitor multiple ward doors and windows seamlessly and simultaneously, utilising MQTT brokers for low-latency, multi-device synchronisation and real-time monitoring.

Alert Personalisation: Implementing personalised notifications that address visitors by name, utilising cognitive ergonomics principles to prevent 'alarm fatigue' for hospital staff in high traffic situations.

Phase 3: 2028 and Beyond - Continuous Learning at Scale

With enough operational data, Anomaly Detection models will be trained along with access logs to find unusual visitor patterns that could be signs of security threats or clusters of infections.

6.4 Concluding Remarks

The Hong Kong medical system has always faced a persistent manpower crisis that demands innovative and automated solutions. This research demonstrates that a well-designed AIoT framework can transform a manual hospital administration that lacks fail-safes and is manpower-hungry into a secure, auditable and contactless automated process. By replacing traditional logbooks with item-level biometric telemetry, the system provides hospitals with the tools needed to safeguard patient recovery and optimise staff workflow. The successful validation of the 13.47-second end-to-end latency and the implementation of a privacy-aware edge architecture validate that the "Smart Hospital" workflow is feasible with current AIoT technology. As the healthcare industry goes digital, systems like the one shown above will be very important for protecting both patients and medical staff. They will also help turn incident response from reactive to proactive, data-driven safety management.

REFERENCES

Almarri, A., Hunaiti, Z., & Manivannan, N. (2025). A review of smart healthcare: Concept, drivers, characteristics, and challenges. *Hospitals*, *2*(4), 26. DOI: 10.3390/hospitals2040026

Anas, A. S., Alchamdani, , Syam, S., Arwan, , & Firmansyah, . (2025). The role of IoT and emerging technologies in shaping smart hospitals. *Medicor: Journal of Health Informatics and Health Policy*, *3*(2), 96–109. DOI: 10.61978/medicor.v3i2.1083

Costa, F., Melo, M. V., Gadelha, I., Fôlego, G., Gambaro, L., & Rodrigues, A. (2021). Self-portrait to ID document face matching: CNN-based face verification in cross-domain scenario. *Anais do XVII Workshop de Visão Computacional (WVC 2021)*, 31-36. DOI: 10.5753/wvc.2021.18885

Debnath, U., & Kim, S. (2025). A comprehensive review of heart rate measurement using remote photoplethysmography and deep learning. *Biomedical Engineering Online*, *24*(1), 73. Advance online publication. DOI: 10.1186/s12938-025-01405-5 PMID: 40542336

Deng, J., Guo, J., Xue, N., & Zafeiriou, S. (2019). ArcFace: Additive angular margin loss for deep face recognition. *2019 IEEE/CVF Conference on Computer Vision and Pattern Recognition (CVPR)*, 4685-4694. DOI: 10.1109/CVPR.2019.00482

Folego, G., Angeloni, M. A., Stuchi, J. A., Godoy, A., & Rocha, A. (2016). Cross-Domain Face Verification: Matching ID Document and Self-Portrait Photographs. /arXiv.1611.05755DOI: 10.48550

Grosso, S., Longhini, J., Tonet, S., Bernard, I., Corso, J., Marchi, D., Dorigo, L., Funes, G., Lussu, M., Oppio, N., Grassetti, L., Pais Dei Mori, L., & Palese, A. (2021). Prevalence and reasons for non-nursing tasks as perceived by nurses: Findings from a large cross-sectional study. *Journal of Nursing Management*, *29*(8), 2658–2673. DOI: 10.1111/jonm.13451 PMID: 34369615

Gupta, M. K., Shah, R., Rathod, J., & Kumar, A. (2021). SmartIdOCR: Automatic detection and recognition of identity card number using deep networks. *2021 Sixth International Conference on Image Information Processing (ICIIP)*, 267-272. DOI: 10.1109/ICIIP53038.2021.9702703

Hsu, C., Yuan, E. J., Fen, J., Lee, C., Ming, J., Chen, T., Lee, W., & Chen, S. (2020). Deployment of a computerized ward visitor registration system in coronavirus disease 2019 epidemic: Experiences of a large academic medical center in Taiwan. *Journal of the Chinese Medical Association*, *83*(6), 561–565. DOI: 10.1097/JCMA.0000000000000327 PMID: 32502118

Javaid, M., & Khan, I. H. (2021). Internet of things (IoT) enabled healthcare helps to take the challenges of COVID-19 pandemic. *Journal of Oral Biology and Craniofacial Research, 11*(2), 209–214. DOI: 10.1016/j.jobcr.2021.01.015 PMID: 33665069

Karpaga Priya, P. S., & Prabhune, G., A. (. (2024). Transforming hospital crowd management: A case study on Pravesh electronic visitor management system. *BOHR International Journal of Internet of things. Artificial Intelligence and Machine Learning, 3*(1), 40–44. DOI: 10.54646/bijiam.2024.27

Lam, G. (2022). Shortage of nurses in Hong Kong: The challenges ahead. *Asian Education and Development Studies, 12*(1), 89–102. DOI: 10.1108/AEDS-08-2021-0179

Ming, Z., Visani, M., Luqman, M. M., & Burie, J. (2020). A survey on anti-spoofing methods for facial recognition with RGB cameras of generic consumer devices. *Journal of Imaging, 6*(12), 139. DOI: 10.3390/jimaging6120139 PMID: 34460536

Palan, A., Nadoda, M., & Sajwan, P. (2025). A comprehensive framework for smart hospitals using IoT and artificial intelligence. *International Journal of Advanced Research in Computer and Communication Engineering, 14*(12). Advance online publication. DOI: 10.17148/IJARCCE.2025.141204

Ramdhani, T. W., Budi, I., & Purwandari, B. (2021). Optical character recognition engines performance comparison in information extraction. *International Journal of Advanced Computer Science and Applications, 12*(8). Advance online publication. DOI: 10.14569/IJACSA.2021.0120814

United Nations. (n.d.). *World population projected to reach 9.8 billion in 2050, and 11.2 billion in 2100.* https://www.un.org/en/desa/world-population-projected-reach-98-billion-2050-and-112-billion-2100

Van Mourik, M. S., Perencevich, E. N., Gastmeier, P., & Bonten, M. J. (2017). Designing surveillance of healthcare-associated infections in the era of automation and reporting mandates. *Clinical Infectious Diseases, 66*(6), 970–976. DOI: 10.1093/cid/cix835 PMID: 29514241

Wong, S., Leung, M., Tong, D. W., Lee, L. L., Leung, W. L., Chan, F. W., Chen, J. H., Hung, I. F., Yuen, K., Yeung, D. T., Chung, K., & Cheng, V. C. (2020). Infection control challenges in setting up community isolation and treatment facilities for patients with coronavirus disease 2019 (COVID-19): Implementation of directly observed environmental disinfection. *Infection Control and Hospital Epidemiology, 42*(9), 1037–1045. DOI: 10.1017/ice.2020.1355 PMID: 33280617

Yanto, W., Nayan, N. M., & Sulaiman, R. S. (2023). Implementation of optical character recognition (Ocr) based on text image using raspberry Pi with Arduino to write text. DOI: 10.2139/ssrn.4500046

Chapter 6
Real-Time Edge AIoT Fridge System for Inventory Tracking and Expiry-Aware Monitoring

Yan Wut Lam
http://orcid.org/0009-0007-9675-0840
The Hong Kong Polytechnic University, Hong Kong

Cong Wu
http://orcid.org/0000-0002-8652-3501
Liaoning Technical University, China

ABSTRACT

A simple misremembering can trigger household food waste: people can forget about food that is already stored, they can have something that was lost somewhere behind other goods, and they cannot see the printed expiry dates before it is too late. In this project, an in-depth AIoT smart-fridge prototype is introduced that considers the problem as a workflow and not as a single recognition task. An ESP32-S3 is used to sense door openings, a Jetson Orin to sense items, their position, and expiry zone, a Raspberry Pi to compare inventory records with predicted date parts, and a mobile application to display the retained value over a local MQTT-based network. The contribution does not claim to resolve all the cases of fridges. Instead, it is a practical integration of sensing, edge vision, local messaging, persistence, and readable inventory display for a small set of expirable items.

DOI: 10.4018/979-8-2600-0701-3.ch006

1. INTRODUCTION

1.1 Background and Motivation

Home food waste usually originates in the usual daily habits as opposed to the apparent failure. Individuals fill the fridge, forget about what they had left there, and lose sight of what is to be used the most. Date labels are beneficial, yet they fail to eliminate ambiguity since labels may be ignored, misinterpreted, or viewed as strict rules where household judgement may be more complex (Newsome et al., 2014; Zielińska et al., 2020; Wilson & Miao, 2025).

A good place to tackle this issue is the refrigerator since there are numerous little food-management choices that occur there. Products are piled, shoved, pushed aside, and covered with some of the goods. Evidence from date-label knowledge and food-waste behaviour studies still indicates that a lack of understanding regarding storage information is a factor that leads to unneeded wastage (Liegeard et al., 2020; Szenderak et al., 2025).

This uncertainty has already been reduced with suggested digital food-management tools, such as smart refrigerators, connected packaging, mobile inventory systems, and expanded IoT kitchen environments (Golshany et al., 2025; Rajan & Wani, 2025). Lots of these methods remain high-level, however, extensive, or are interrupted by a small model work. This project thus has a more limited path through developing a more specific AIoT process on selected expirable goods stored in a single fridge-like cabinet.

Figure 1. Prototype hardware and test setup used for the integrated smart-fridge demonstration

1.2 Key Challenges

This prototype starts by considering the fridge as a challenging setting to integrate computer vision and AIoT. The initial difficulty is the item identification. Four reaction items that can be stored in a household fridge include fruit, beverages, packaged goods, leftovers, dairy commodities, and prepared meals. The appearance within an individual category can be diverse by brand, package, colour and size. A system that attempts to identify everything is prone to destabilising and be difficult to assess. In the case of a capstone prototype, where there is limited time, it is thus more reasonable to reduce the set of classes instead of promising infinite recognition.

The second difficulty is the interpretation of expiry-date. Food packaging is more difficult to read than more general OCR since the system must first locate the appropriate region then determine whether the visible day, month and year are a sensible expiry result or not. Printed dates can be small, illegible, rounding about packaging, partially obscured, or combined with non-relevant production codes. There are resourceful studies dedicated to this issue, and a fridge presents additional challenges since packages are usually considered at awkward angles and with an uneven lighting (Seker & Ahn, 2022; Wang et al., 2023).

The third obstacle is the dynamics of a door session. A fridge is not an object-detection scene that is not moving. When a door is opened, objects can be inserted, deleted or moved rapidly. It implies that a working system cannot be based on individual frames. It has to make a choice on how to observe change on a single interaction and how to transform that observation into a revised state of inventory thereafter.

The fourth is obstacle of system integration. A useful smart-fridge prototype is not just a model on one device. It involves sensing, messaging, state persistence and user result. The correct processing window has to be triggered by door events, the image outputs have to be connected with the proper tracked items, the records should be updated regularly, and the final state should be readable. In the absence of this expert annotation, the constructive household system will not be created even with a strong recognition model.

1.3 Aim and Scope

To address these limitations, the project will assume the implementation of a consistent AIoT prototype incorporating door sensing, camera-based item observations, expiry-related logic, the local messaging, handling persistent inventory, and the provision of a basic mobile inventory display. It is structured around door based sessions as the most significant changes to the fridge occur when the door is open and objects are being transferred. A single session with that time period results in a clean boundary on what the system observes and subsequent inventory reconciliation are easier to justify.

The areas are purposely narrow. The item detector is trained on three demonstration classes; banana, orange, and noodle. The expiry pipeline is not guaranteed but favourable. When the evidence is weak the backend does not render an expiry field resolved as required by a full date. The mobile app is read-only and is concerned more with clarity of presentation on the retained inventory state other than user correction. These limits are chosen, as they would make the project focused on what is/can be implemented, tested, and discussed in a capstone environment.

The architecture is also local in nature. The demo Wi-Fi hotspot and the amqtt broker are on a separate Raspberry Pi, with the session-time vision work on the Jetson Orin, the SQLite inventory and date prediction managed by the backend Raspberry Pi, and the whatsoever runs on a SQLite inventory and date prediction in the mobile app connected to the retained state on the same local network. This makes the workflow acceptable to the discussion of edge-computing instead of transforming the project into a cloud-dependent smart-home platform.

1.4 Contributions

Practical integration is the key contribution of the project. Various studies already offer some helpful fragments of the larger issue, such as the ideas of smart-kitchen, food-recognition, and expiry-date data, along with edge-based computing support of connected devices (Liegeard & Manning, 2020; Seker and Ahn, 2022; Kong et al., 2023). It is more unusual to find a small working prototype which integrates these notions into one fridge oriented workflow of tracking the various expirable items which one is interested in.

To be more precise, the project has four contributions. To begin with, it proposes a session based design where door events are used to determine the unit of observation. Second, it uses item-detection and expiry-region-detection on the Jetson such that compact session summaries and the optimal ROI of the expiry of each track seen should be published. Third, it connects these outputs to a Raspberry Pi backend that balances the state of inventory, buffers against out-of-order messages, foretells prospective nullable date parts, and caches probable resolved expiry dates. Fourth, it displays the outcome in easy mobile format that enables the inventory state that is retained to be seen on a single page.

1.5 Report Structure

The remainder of this report is organised as follows. Section 2 reviews work on digital food management, fridge and product recognition, expiry-date recognition, and edge-based AIoT systems. Section 3 explains the implemented modules and how they cooperate in the prototype. The later sections will expand the benchmark results, discussion, and final conclusions using the uploaded benchmark summaries and images.

2. RELATED WORKS

This project is based on multiple research domains, but there is no one area that completely can address the issue of household fridge on its own. Earlier studies have considered digital food-management frameworks, fridge item identification, expiry item identification, buyer product datasets and edge-based IoT frameworks. There is something valuable in each of the strands, but the literature remains largely task separate. That is why, this section considers the most topical strands of work and then reveals the gap filled in by the contemporary prototype.

2.1 Smart Kitchen and Household Food Management

It is evident from research on household food waste that technical support tools are crucial, as waste is influenced by everyday practice rather than solely by spoilage. Liegeard and Manning (2020) examine intelligent applications like food management tools, smart refrigerators, and connected packaging and contend that digital systems can assist households in making better disposal and storage choices. This review is helpful because it shows the issue as a practical one involving money, trust, and user acceptance, rather than just a sensing issue. The framing is significant in this context, as a smart-fridge prototype is only valuable if it facilitates everyday storage decisions, rather than generating technical output that has minimal household relevance.

The utilization of mobile applications is one of the most direct methods of promoting household behavior. Johansen (2022) looked at smartphone apps that try to cut down on food waste and discovered that app-based support can change people's behavior, but only if the users are interested and find the app easy to use. The mobile application described by Mastorakis et al. (2024) enables users to record food items, associate expected expiry information, and receive reminders. These studies demonstrate the value of user-facing tools even in the absence of complete automation. However, they also show a weakness that led to this project: many solutions still rely on entering data by hand a lot. This results in friction during daily use and undermines the functionality of a "smart" system.

The discourse on connected domestic environments has been expanded beyond applications in recent smart-kitchen reviews. According to Golshany et al. (2025), IoT-enabled kitchen technologies are an expanding application area that connects storage systems, refrigerators, and other appliances to more general food-management objectives. Similarly, Rajan and Wani (2025) examine smart food and packaging technologies that could enhance household monitoring and traceability. This writing helps us understand why a prototype of a connected fridge is important at this time. However, much of it remains broad in scope. It frequently addresses kitchen ecosystems, packaging innovations, or high-level opportunities, as opposed to a compact fridge-centered prototype that illustrates a single, unambiguous workflow.

2.2 Food Recognition in Fridge and Product Contexts

Although the most evident technical component of a smart-fridge system is food recognition, the literature also demonstrates why it is challenging to translate vision models into actual refrigerator conditions. Dai (2024) addresses refrigerator food recognition directly and highlights the practical difficulty of clutter, shadow, reflection, and changing viewpoint. Scenes in a fridge are different from settings where objects are easier to find because things may be partially hidden, uneven lighting can cast shadows, and packaging may be shiny or uneven. This is crucial for the ongoing project as it reinforces the notion that the refrigerator is not merely a reduced version of an online food image dataset.

Research on the recognition of retail products has also extensively documented the concept of visual similarity. Wei et al. (2022) developed RPC as a fine-grained, large-scale retail product checkout dataset that is intended for the accurate identification of packaged products. es in demonstrating how the task becomes challenging very quickly when there are many and visually similar categories. Wang et al. (2023) conducted additional research on smart retail SKU checkout via an enhanced residual network, demonstrating that densely arranged and highly similar products continue to pose a challenge in commercial environments. These studies are useful because they help us think about packaged foods like cup noodles, where category meaning is linked to how they look and how they are packaged by brand. Despite this, retail checkout is still different from refrigerator checkout. Repeated door sessions, household storage habits, and expiration interpretation are not the focus of retail datasets.

This distinction matters for system design. It's not necessary for a household prototype to solve unlimited recognition in order to be useful. If the class set allows for consistent observation of specific expirable items, it can be a good engineering choice. Therefore, a more modest approach is supported by the fridge and retail literature: acknowledge that visual difficulty increases rapidly with class scale and concentrate first on a class scope that can be trained and demonstrated clearly.

2.3 Expiry Date Recognition

Expiry date recognition is another significant area of related research, as a refrigerator system that is functional should not merely register the existence of an item. We already know that this task is more than just OCR from earlier work. Flowers and Rebedea (2020) suggested a deep-learning-based method for reading expiration dates on packaging. In this case, it was shown that small, distorted, or noisy date text needs to be handled in a way that is unique to that domain. Gong et al. (2021) also suggested a two-step process: first, the use-by-date region is found, and then it is recognized. That staged design is significant because it keeps interpretation and localization apart rather than addressing the entire image.

One of the best direct resources for the current project is offered by Seker and Ahn (2022). With their ExpDate framework, they created a dataset and a method that explicitly model day, month, and year instead of just extracting raw characters. In the case of a smart fridge prototype, this is especially important since the useful output is not just a string of text. It is an expiry date that can be attached to an inventory item. ExpDate is still more like understanding dates at the package level than at the full fridge scene. The pictures don't fully show how the food is stored because it could be tilted, shiny, partially hidden, or only briefly visible during a door-op.

More recent work continues to push toward lightweight and mobile-friendly expiry support. In a mobile-oriented prototype, Dospinescu et al. (2025) integrate OCR and image classification, while Peng et al. (2025) outline an effective expiration-date recognition framework for mobile applications. It is clearer now that low-cost deployment is the best option. However, they also tend to assume that the package or date label can be captured deliberately. The issue starts earlier in a fridge prototype: the system must first separate an item from a cluttered scene before determining whether any visible area is likely to have any useful expiration information.

Scene-text literature in general helps to show why a staged design makes sense. Wang et al. (2023) survey deep-learning approaches for text detection and recognition and show how irregular text, complex backgrounds, and varied layouts continue to challenge end-to-end recognition. In this case, that supports the choice to skip full-scene OCR and instead find an area on an item crop where the expiration date is likely to be before trying to predict the date.

2.4 Edge-Based AIoT and Research Gap

The last piece of related work is about system architecture. A smart fridge prototype is more than just a problem with vision. It is also an IoT workflow that connects sensing, messaging, local inference, persistence, and user presentation. Both Yu et al. (2018) and Kong et al. (2023) conduct a survey of edge computing for the Internet of Things and lay out the benefits of local processing for latency-sensitive and device-connected applications. Their discussion is relevant here because a fridge session is short and event-driven. Processing item observations locally and sharing small results is more useful than relying on a cloud service for each step of the process.

When combined, the research shows a gap. Although studies on food management and smart kitchens explain the importance of digital support, they frequently concentrate on manual tools or general systems. The work on fridge recognition shows how hard the environment is to see, but the focus is usually only on recognition. The work with an expiration date gives you good datasets and staged methods, but not in a messy fridge workflow. Studies on edge computing support local processing, but they don't give a fridge-specific way to do it. What is lacking is a compact, well-thought-out prototype that unifies these concepts around a single, obvious domestic chore.

That is the space this project addresses. It does not suggest a new model for general-purpose recognition, a new dataset, or a full smart-fridge platform for sale. Instead, it integrates multiple pre-existing concepts into a targeted prototype: a vision stage for specific items, a staged expiry pipeline, a session trigger, a local backend for inventory reconciliation, and a basic mobile interface for showing the retained state. Because of this, the contribution is more useful than just theoretical.

3. ALGORITHMS/MODULES IMPLEMENTATION

This part describes the prototype as a group of modules that work together, not as a single smart-fridge program. Upon the launch of a session by the door-open event, the vision stage observes visible items, the backend reconciles the result with stored inventory records, and the mobile app displays the retained state to the user. Because the prototype's value is derived from the integration of sensing, edge vision, messaging, and state handling, a modular perspective is more aligned with the project's objectives than a monolithic description.

Figure 2. Prototype system architecture for the deployed local smart-fridge pipeline

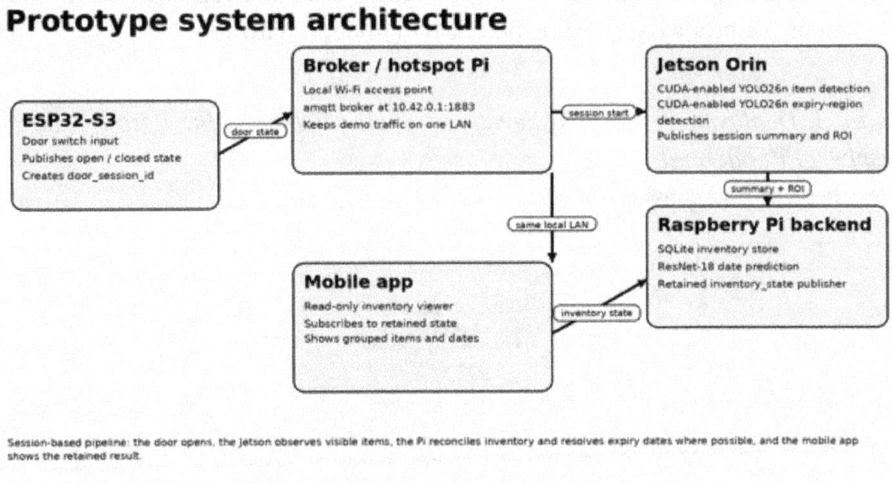

3.1 Overall Architecture and Device Roles

The implemented prototype uses a local deployment in which all hardware connects to the same Wi-Fi network hosted by a separate Raspberry Pi. That same device runs the amqtt broker at 10.42.0.1:1883, which keeps demo traffic on one LAN and avoids unnecessary dependence on external services. Within this network, the ESP32-S3 acts as the session trigger, the Jetson Orin performs session-time vision processing, the backend Raspberry Pi manages inventory and date prediction, and the mobile app reads the latest retained inventory state. This layout reflects an edge-computing design because the useful work stays close to the devices that create and consume the data.

The architecture also separates fast visual processing from persistent state management. The Jetson was chosen for the live vision stage because it can use CUDA-enabled GPU acceleration, and the deployed Jetson programs and Jetson-side benchmarks were run with CUDA enabled. That makes repeated frame-based inference, tracking, and crop selection much more suitable on the Orin than on a Raspberry Pi. By contrast, the Raspberry Pi backend is better placed to maintain SQLite records, apply date-resolution rules, and publish retained state updates. This division keeps each module narrower and easier to reason about while also reducing the chance that one slow stage will bottleneck the whole pipeline.

Figure 13 makes the hardware split more explicit. The local broker Pi keeps messaging repeatable for the demonstration network at 10.42.0.1, the Jetson handles the CUDA-enabled visual workload, and the Raspberry Pi backend focuses on persistence and lightweight structured prediction. This division of labour is one of the reasons the prototype remains practical on modest hardware.

Figure 3. Deployment split and device rationale for broker Pi, Jetson Orin, and Raspberry Pi backend

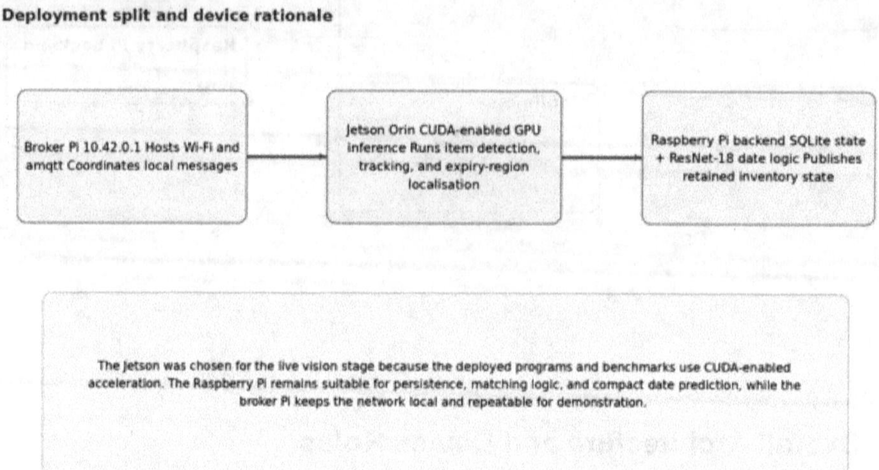

Table 1. Key MQTT topics used by the prototype

Topic	Publisher	Purpose
fridge/door/state	ESP32-S3	Starts and ends one door-based session.
fridge/vision/session/summary	Jetson Orin	Publishes the final tracked-item summary for a session.
fridge/vision/roi/expiry/{door_session_id}/{session_track_id}	Jetson Orin	Publishes one selected 224 x 224 expiry ROI per visible track.
fridge/inventory/state	Pi backend	Publishes the retained inventory view consumed by the mobile app.

3.2 Session Trigger and Communication Logic

The session trigger defines when the system should pay attention to the fridge. Instead of analysing the scene continuously, the prototype uses door-based sessions. When the door opens, the ESP32-S3 publishes a message containing state=open and a door_session_id. When the door closes, it publishes the corresponding closed event. This gives the rest of the system a clear processing boundary and reflects the way fridge contents usually change in practice.

The choice of MQTT fits this event-based design. Modules do not need to block one another while processing, and the retained inventory topic allows the mobile app to obtain the latest state immediately after subscribing. The topic structure is deliberately small, which makes debugging easier and keeps the report focused on message meaning rather than message variety. The prototype also preserves two identifiers across modules: door_session_id carries session-level meaning, while session_track_id carries item-level meaning within a session. This avoids pretending that stable identity across all future openings has already been solved.

Figure 10 summarises the operational sequence across one door session. The diagram is useful because the prototype is best understood as a bounded event pipeline rather than as a permanently active smart-home service. It also clarifies why the retained inventory state is published only after the session closes and reconciliation has been completed.

Figure 4. End-to-end fridge session flow from door trigger to updated retained inventory state

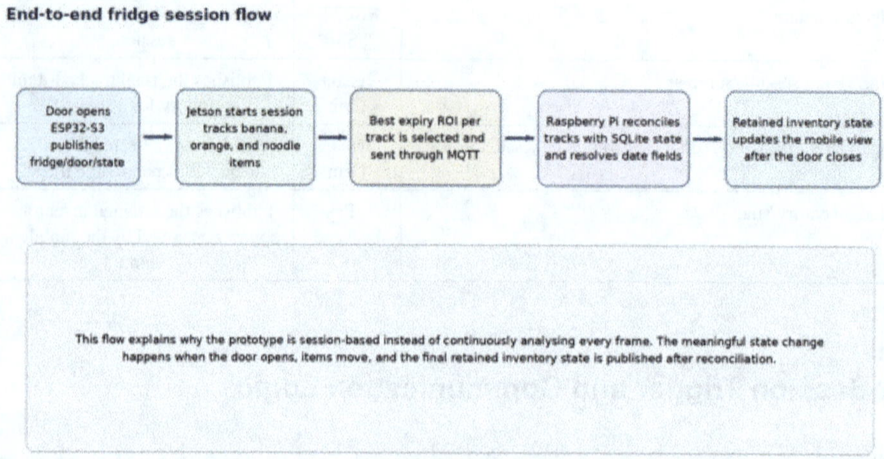

Table 2. Device and module responsibilities

Module	Runs on	Main responsibility
Door trigger	ESP32-S3	Detect door state changes and publish session start/end events.
Vision processor	Jetson Orin	Run item detection, tracking, expiry-region detection, and final session publication.
Inventory backend	Raspberry Pi	Persist items in SQLite, reconcile sessions, run date prediction, and publish retained state.
Mobile viewer	Android app	Display grouped items, expiry status, and class-level summaries to the user.

3.3 Jetson Item-Detection and Tracking Module

The Jetson Orin handles session-time visual processing because it is the most suitable device in the prototype for repeated inference on live camera frames. In the implemented build, the item-observation stage uses an Ultralytics YOLO26n model running on 1280 x 720 webcam input with inference performed at image size 640, and the Jetson-side programs use CUDA-enabled execution. This is one of the main reasons the vision stage sits on the Orin rather than on a Raspberry Pi. The model is trained independently from the expiry-region detector and the date model, which keeps the role of each model clear.

The item dataset is deliberately shaped around the final demonstration classes: banana, orange, and noodle. Banana and orange images are drawn primarily from LVIS, while the noodle class is built from custom webcam photographs of the exact demo product. This training design is important because the noodle product is the most deployment-specific class. Final balancing is anchored on noodle image and annotation counts, all noodle examples are kept, and other classes are down-sampled accordingly. In addition, negative images with empty label files are included so that the detector learns when none of the target classes are present.

The Jetson module does more than detect classes. During an active session it tracks items, keeps the initial centroid and last visible centroid for each track, crops item regions, scans those crops for likely expiry regions, and stores only the best expiry candidate per track. This structure avoids publishing every transient crop over MQTT. Instead, the Jetson publishes one compact session summary plus, where available, one selected expiry ROI for each visible track at session finalisation.

The uploaded Nano.py code shows several safeguards that make this stage more reliable. The tracker must provide the session_track_id, so fallback identifiers are not invented. Initial-track windows and final-visibility handling are measured in elapsed time rather than frame counts, which is more stable when frame rate varies. The session is finalised only when the matching door_session_id closes, and the published summary includes only tracks with a meaningful initial or final centroid. These small rules help maintain item-level meaning across the rest of the pipeline.

3.4 Expiry-Region Selection on Item Crops

The staged expiry pipeline is shown separately in Figure 12 because it explains an important design choice in the implementation. The Jetson keeps the visual task narrow by locating one useful date region, while the Raspberry Pi resolves structured components and can still leave uncertain fields null. This reduces message size and avoids treating every visible string on the package as equally meaningful.

Expiry handling on the Jetson is implemented as a second vision stage applied only to item crops. Rather than reading dates directly from the full fridge image, the system first narrows the problem to the detected item and then runs a second YOLO26n model at image size 416 to localise the most likely expiry region. This two-stage design reduces clutter, unrelated printed text, and background noise before any date prediction begins.

The expiry-region detector is built from a remade version of ExpDate Dataset A. The dataset creation rule is important: when multiple date candidates appear in one image, the latest valid parsed date is chosen as the expiry annotation and exported as the single target box. This keeps the model aligned with the project's needs because the aim is not to detect every printed date-like string, but to locate the region that is most likely to correspond to the expiry date a household user actually cares about.

Figure 5. Expiry-date resolution pipeline from fridge frame to structured day, month, and year output

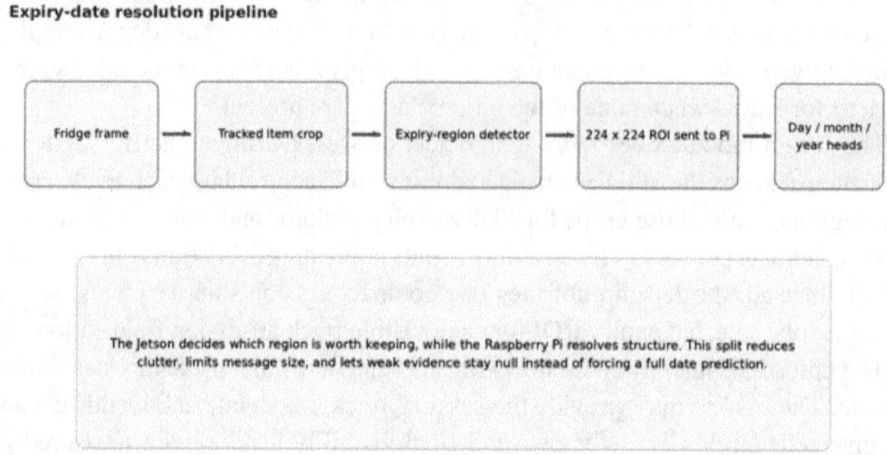

A separate reconciliation diagram is included in Figure 11 because the backend logic is harder to infer from screenshots alone. The figure makes clear that the prototype does not claim perfect permanent identity. Instead, it compares class and approximate position, updates stored items when a reasonable match exists, creates a new item when no match is suitable, and marks older records as removed when they are no longer supported by the session result.

Figure 6. Inventory reconciliation logic on Raspberry Pi for matching, creation, and removal decisions

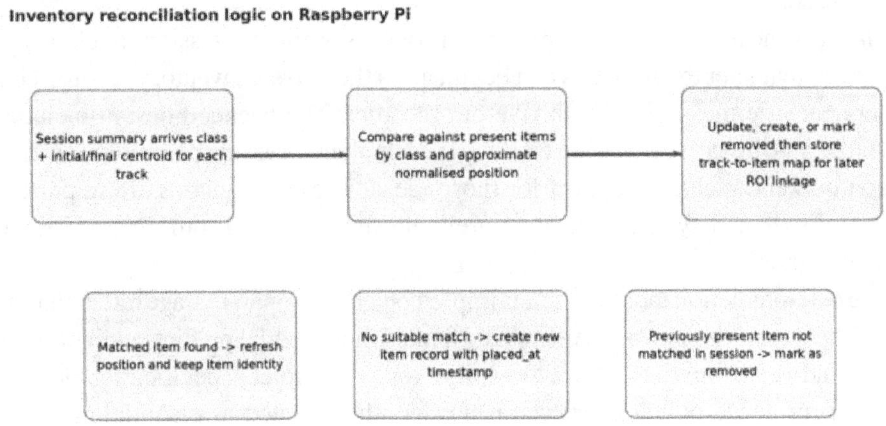

At runtime, the Jetson scores candidate expiry ROIs within each item crop and keeps only the best one. The selected region is converted into a square 224 x 224 JPEG before publication. Restricting the output to one kept ROI per visible track reduces message volume and prevents the backend from repeating visual work that has already been done during the session. It also provides a cleaner hand-off to the date-prediction stage on the Raspberry Pi.

3.5 Inventory Reconciliation and Date Prediction on Raspberry Pi

The Raspberry Pi backend acts as the state-management centre of the prototype. Its main tasks are to persist inventory records, reconcile completed session results against the current stored state, connect expiry ROIs to the correct item records, run date prediction, and publish the retained inventory view. The uploaded Pi.py uses SQLite, which is appropriate here because the prototype needs a lightweight local database that survives across sessions without introducing unnecessary infrastructure.

Session reconciliation is based on class and approximate normalised position rather than on a promise of permanent long-term visual identity. When a session summary arrives, the backend compares the completed tracks with the current set of present items. A suitable match updates an existing record, an unmatched track creates a new item, and an item that is no longer matched can be marked removed.

This approach is modest, but it is reasonable for a capstone prototype whose goal is to track practical inventory change rather than solve identity persistence in the general case.

The backend also stores a mapping from {door_session_id, session_track_id} to item_id so that later expiry ROIs can be linked to the correct inventory record. This matters because messages may arrive out of order. The backend design includes ROI buffering, processed-session tracking, and duplicate-session protection, and the Python backend includes support for those ideas. These safeguards are important because distributed systems become unreliable very quickly if ordering and repetition are ignored.

The date-prediction module itself is implemented as a separate stage rather than as part of the Jetson vision work. A ResNet-18 multi-head model predicts nullable day, month, and year from the 224 x 224 expiry crop. The project specification defines distinct confidence policies for each head: month is treated as essential and kept stricter, while day and year can remain null after filtering. If month is unresolved, no expiry date is formed. If month is present but day or year is missing, the backend may infer the missing part under explicit rules, such as using the last valid day of the month or inferring a plausible year from the placed_at date so that the resolved expiry does not fall before placement.

The separation between ROI selection and date resolution makes the backend easier to explain. The Jetson decides which crop is worth keeping, while the Pi decides whether the predicted components are strong enough to become an inventory date. That is a good fit for the prototype because it keeps visual selection, prediction, and business rules distinct instead of hiding them in one opaque stage.

Table 3. Model training and deployment summary

Model	Training source	Runtime role	Key deployment setting
YOLO26n item detector	LVIS banana/orange data + custom noodle photos + negatives	Detect and track banana, orange, and noodle during active sessions.	Jetson Orin, imgsz 640, 1280 x 720 source frames
YOLO26n expiry-region detector	Remade ExpDate Dataset A with one chosen expiry box per image	Locate the best expiry ROI inside each item crop.	Jetson Orin, imgsz 416, publishes 224 x 224 ROI
ResNet-18 multi-head date model	ExpDate Dataset B with nullable day/month/year labels	Predict day, month, and year components for the selected ROI.	Pi backend, month required before storing expiry_date_iso

3.6 Mobile Viewing Module

The mobile app is implemented as a read-only inventory viewer because the project aims first to make the retained result understandable. It subscribes to the retained inventory-state topic and groups items by class so that a user can quickly see what is present, what date has been attached, and which items are due soon. This is a sensible design choice for a prototype because it keeps the mobile side clear and does not introduce write-back complexity before the observation pipeline itself has been validated.

The uploaded screenshot shows a layout based on class cards, item rows, and simple status labels such as due soon. That presentation style fits the household use case well. A fridge tool does not need the user to inspect raw tracking data or date-model confidences. It needs to turn the pipeline output into a readable inventory view.

Figure 7. Mobile app inventory view showing grouped items, expiry status, and position summaries

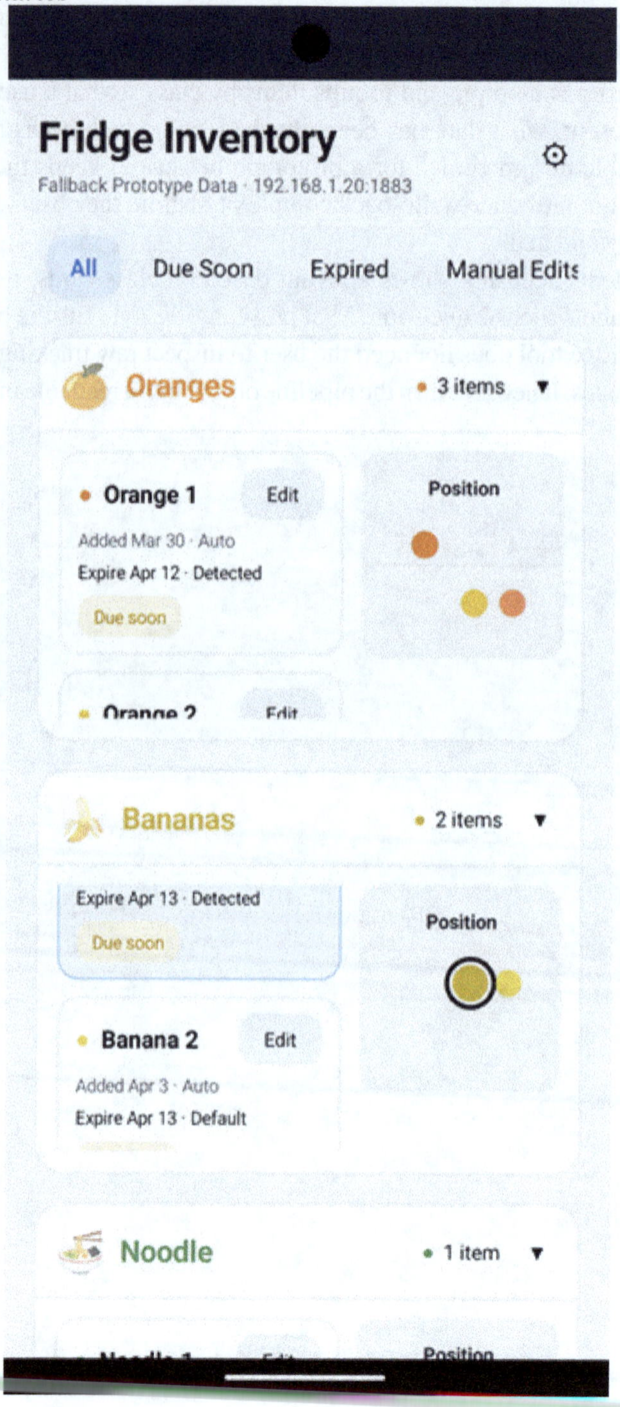

3.7 Implementation Outcome

Taken together, the implemented modules form a focused algorithmic pipeline for one practical household task: observe selected items during a door-open interaction, keep the most relevant expiry evidence, reconcile that evidence into stored inventory records, and publish a readable retained state. The strength of the design is not that any one model is novel on its own. It is that the modules are organised so their outputs remain useful to the next stage.

In summary, the most important implementation outcome is modular coherence. The ESP32 defines the session boundary, the Jetson produces structured observations, the Raspberry Pi turns those observations into inventory state, and the mobile app makes the result understandable. This is the foundation on which the next benchmark and discussion sections can be built.

4. EXPERIMENTATION AND BENCHMARK

4.1 Benchmark Design and Evidence Base

This section evaluates the prototype module by module instead of forcing one headline score for the whole smart-fridge pipeline. The benchmarks ask four practical questions: can the item detector separate the chosen demo classes, can the expiry stage isolate useful date regions, can the date model decode those regions accurately enough to update inventory records, and is the edge stack responsive enough for live use. Because the uploaded benchmark packages were produced on more than one machine, each result is interpreted as evidence for a specific module rather than as one fully standardised end-to-end laboratory run.

Table 4 summarises the evidence used in this build. All figures in this section were regenerated from the uploaded summary.json or CSV files, except Figure 5, which uses a live screenshot from the current prototype. This keeps the section grounded in benchmark outputs.

Table 4. Benchmark sources and workloads

Benchmark area	Main evidence	Hardware / context	Workload	Purpose
Item detector	per_class_metrics.csv, summary.json, live screenshots	Supporting offline CPU run plus current demo feed	63 frames, 3 classes	Check class separation and live-scene usability
Expiry-region detector	summary.json, per_item_metrics.csv, sample grid	Offline dataset evaluation and Jetson Orin runtime	658 test images; 375 measured webcam items	Check ROI localisation quality and live-session latency
Date predictor	summary.json, per_item_predictions.csv, speed runs	Pi 5 accuracy run plus Pi 5 and Orin speed runs	509 test crops; 504 speed images	Check date accuracy and hardware fit
Transport layer	round_trip_results.csv, summary.json	Orin client and Pi-host broker/server	500 runs plus 50 warmup per payload	Compare message cost for door, summary, and ROI payloads

4.2 Item Detector Results

The item detector is judged here as a deliberately narrow three-class model rather than as a general food recogniser. That is the correct standard for this prototype because the demo only needs stable observation of banana, orange, and noodle items during one door-open session. On the uploaded dataset-mode benchmark, the model processed 63 measured frames with an approximate throughput of 24.24 frames per second, a median latency of 39.69 ms, and a p95 latency of 55.10 ms. Across the three classes, the mean precision was 95.68%, the mean recall was 82.36%, the mean mAP50 was 91.62%, and the mean mAP50-95 was 83.21%. Those values are strong enough for a bounded demo, especially because the detector does not need to classify a large open-world food vocabulary.

Figure 4 shows the class-wise breakdown. The noodle class is the strongest, with 95.41% recall, 99.09% mAP50, and 94.71% mAP50-95. This is expected because the noodle class was trained with custom images of the exact demonstration product. Orange also remains reliable, while banana is the weakest class because recall drops to 71.43% even though precision stays high at 97.39%. In practical terms, the detector is usually correct when it decides a banana is present, but difficult banana views are still easier to miss than the other two classes.

Figure 8. Item detector class-wise evaluation regenerated from the uploaded per-class benchmark metrics

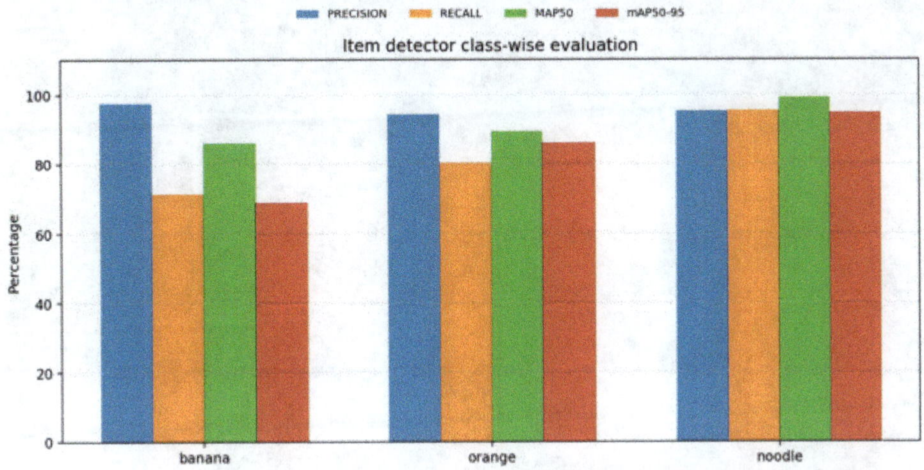

The quantitative benchmark above was not run on the final Jetson deployment device, so it should be interpreted as supporting accuracy evidence rather than as a direct deployment guarantee. Figure 5 therefore complements the offline metrics with a live fridge-scene screenshot from the current prototype. It shows that the detector can separate multiple target classes in one cabinet-style scene and still keep useful boxes when items are partly stacked or placed on different shelves. This is the important behaviour for the report, because the backend only needs a reliable session summary, not perfect frame-by-frame object permanence.

Figure 9. Live item detector output from the current fridge-scene prototype

4.3 Expiry-Region Localisation and Date Prediction

Expiry handling is more central to the smart-fridge workflow than raw OCR alone because the system first needs to decide whether a useful date region exists at all. The expiry-region detector performed well on the Dataset A test split, which contained 658 images. Its precision reached 90.69%, recall 85.25%, mAP50 91.64%, and mAP50-95 78.45%. These numbers suggest that the model usually localised the correct date region and retained enough box quality to support the next stage. Figure 6 is helpful here because it shows the detector operating across different packaging materials, label styles, and viewing angles rather than only on one product family.

Runtime on the Jetson Orin was also comfortably within demo range. The dataset-speed run reported a median latency of 33.03 ms over 658 images, while the webcam-mode run reported a median latency of 39.30 ms over 375 measured items. These Jetson-side runs used CUDA-enabled execution, which is consistent with the design choice to place the live vision stage on the Orin instead of on a Raspberry Pi. The webcam result is more important for the project because it better reflects active use with camera capture, cropping, and scene variation. Even in that setting, the model stayed in the tens-of-milliseconds range, which is fast enough to select one useful expiry crop per visible track during a session without dominating the whole pipeline.

Figure 10. Expiry-region detector sample predictions on Dataset A test examples

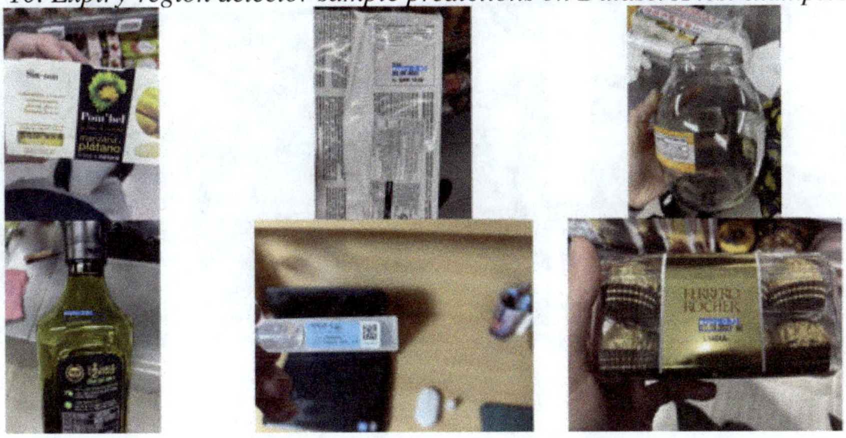

The date predictor produced the strongest numerical result in the current benchmark set. On the Dataset B test split, evaluated on the Raspberry Pi 5 over 509 crops, day accuracy reached 99.41%, month accuracy 99.61%, year accuracy 98.82%, and exact full-date accuracy 98.43%. In count terms, that means only 8 crops out of 509 were not exact matches. A look through the per-item prediction file shows that the remaining errors were usually partial mistakes such as year ambiguity or a nullable day component rather than a complete collapse of all three heads. This is important for the prototype because a household inventory tool benefits most from reading the month and year correctly, while a single uncertain component can still be left unresolved rather than forced into a false full date.

Figure 11. Date-component prediction accuracy on Dataset B test split

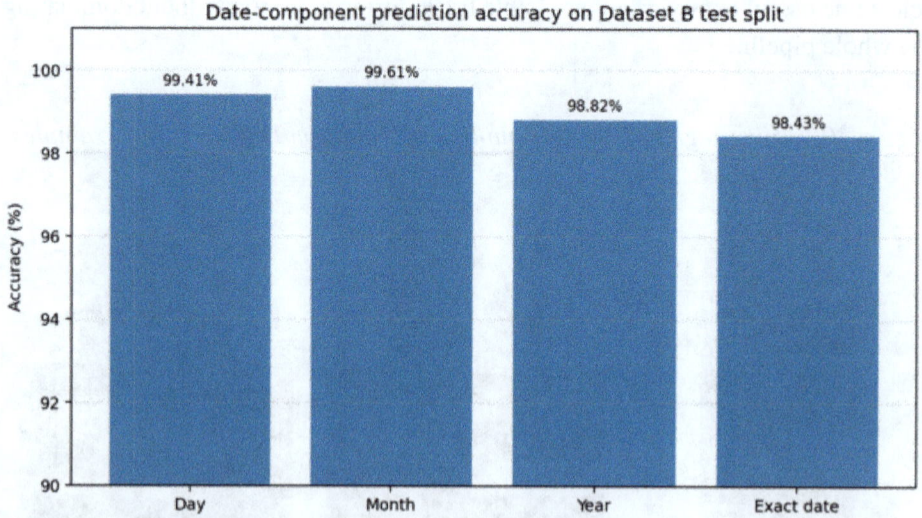

Hardware speed creates a clearer division of labour inside the prototype. The same date model had a median latency of 17.84 ms on the Orin but 265.30 ms on the Pi 5, which makes the Pi about 14.9 times slower for this task. Even so, the Pi 5 result is still acceptable for the chosen system design because date prediction happens after ROI selection and is linked to visible tracks rather than run continuously on every frame. In other words, the fridge prototype uses the faster Orin for the visually busy session-time work, while the Pi 5 performs slower but manageable post-session reasoning and state updates.

4.4 Runtime and Transport Behaviour

Figure 8 brings the available latency results together. Two points stand out. First, all Orin-side vision tasks in the uploaded benchmark set remain in the tens-of-milliseconds range, which suits a live camera pipeline. Second, the Pi 5 date stage is clearly the slowest module, yet its placement after ROI selection keeps it from becoming a frame-rate bottleneck. The offline item-detector timing sits in a similar latency band to the Orin expiry detector, but because that run came from a different computer it should only be used as a supporting reference.

Figure 12. Observed inference latency across available benchmarks

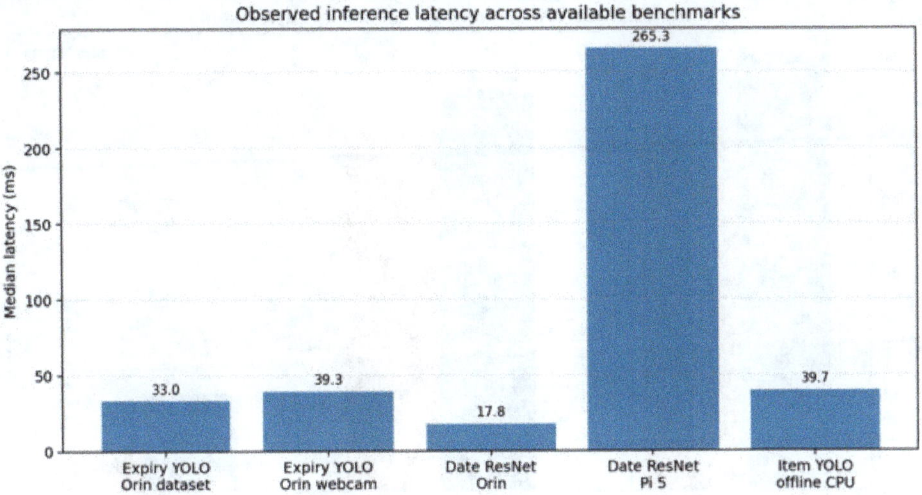

The transport benchmark compared MQTT and HTTP for door-state messages, session summaries, and expiry ROI transfers. Each case used 500 measured runs after 50 warmup runs, and every run succeeded. HTTP was leaner in this controlled setup. Median round-trip latency was 50.43 ms versus 60.56 ms for door, 50.41 ms versus 55.69 ms for summary, and 6.35 ms versus 13.64 ms for ROI. Approximate network cost per successful round trip followed the same pattern: 1.22 KiB versus 2.85 KiB for door, 1.69 KiB versus 3.76 KiB for summary, and 7.42 KiB versus 15.05 KiB for ROI.

Figure 13. Transport benchmark median latency by payload type

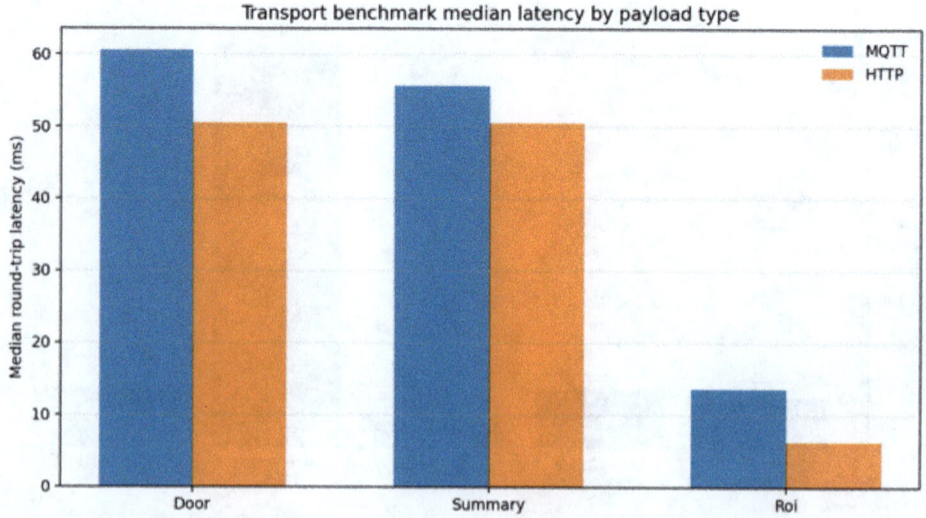

Table 5. Transport benchmark round-trip comparison

Payload	MQTT median (ms)	HTTP median (ms)	MQTT bytes/msg (KiB)	HTTP bytes/msg (KiB)
Door	60.56	50.43	2.85	1.22
Summary	55.69	50.41	3.76	1.69
ROI	13.64	6.35	15.05	7.42

These numbers do not mean HTTP should replace MQTT in the prototype. The benchmark is a tightly controlled request-response test, so it naturally rewards the leaner one-request exchange. The smart-fridge system still benefits from publish-subscribe messaging, retained inventory state, and loose coupling between the ESP32, Jetson, Pi backend, and mobile app. In that sense, HTTP won the microbenchmark for raw efficiency, but MQTT still fits the integrated demo better.

4.5 Benchmark Summary

Taken together, the benchmark results support the report's central claim that the prototype is feasible as a practical edge workflow rather than only as a collection of disconnected models. The strongest evidence is the combination of high date accuracy, solid expiry-region localisation, and live-session latency that stays within a workable range on the Orin. At the same time, the benchmarks also show the current limits of the system. The item detector still needs a final quantitative run on the deployment hardware, and the transport comparison should be read as a module-level microbenchmark rather than as a verdict on the full architecture. These limits define the next iteration and prepare the ground for the discussion section.

Table 6. Consolidated benchmark summary

Module	Key numerical result	Interpretation
Item detector	Mean precision 95.68%, mean mAP50 91.62%, mean mAP50-95 83.21%	Strong enough for the narrowed three-class demo; noodle is the most reliable class
Expiry-region detector	mAP50 91.64%, mAP50-95 78.45%, Orin webcam median 39.30 ms	ROI selection is accurate enough and fast enough for session-time use
Date predictor	Exact accuracy 98.43%, Pi 5 median 265.30 ms, Orin median 17.84 ms	Very accurate; Pi 5 is slower but acceptable when used after ROI filtering
Transport layer	0 failures in all measured runs; HTTP lower raw cost than MQTT	MQTT remains justified by retained state and publish-subscribe integration needs

5. DISCUSSION

5.1 What the Results Mean for the Prototype

The benchmark results show that the project works best when it is judged as an integrated household prototype rather than as a claim to solve general food recognition. The strongest finding is not a single headline score. It is the way several moderate-sized components fit together into one event-driven workflow. The item detector is accurate enough for the chosen classes, the expiry-region stage can isolate likely date areas, the date model can resolve most labelled crops reliably, and the transport layer keeps messages moving quickly enough for a live demonstration. In that sense, the evidence supports the project aim stated earlier: a useful smart-fridge system depends on coordination between sensing, inference, state management, and presentation, not only on raw model performance. This interpretation is also consistent with wider smart-kitchen and food-management literature, which shows that digital tools are most valuable when they reduce practical uncertainty in everyday storage decisions rather than when they merely add another interface layer (Liegeard & Manning, 2020; Mathisen & Johansen, 2022; Mastorakis et al., 2024).

A second important point is that the prototype benefits from a deliberately narrow design. The three-class item detector is not a weakness by itself. It is an engineering decision that keeps the system within a realistic student-project scope while still supporting a clear demonstration of the overall idea. The literature on fridge recognition and retail product recognition helps explain why this matters. Visual clutter, partial occlusion, and package similarity make wide-class recognition difficult even in stronger retail settings, so expecting stable open-ended food recognition in a domestic fridge would be unrealistic at this stage (Dai, 2024; Wei et al., 2022; Wang et al., 2023). By restricting the class set and focusing on session-based updates, the project trades breadth for reliability. The benchmark results suggest that this trade-off was sensible.

5.2 Strengths of the Current Design

The main strength of the system is modular coherence. Each device has a clear job, and the uploaded code reflects that separation. The ESP32-S3 defines when a session starts and ends. The Jetson handles the heavier live vision tasks. The Raspberry Pi backend persists state, reconciles tracks, and resolves expiry information. The mobile app then exposes the retained result in a readable way. This division reduces unnecessary duplication and makes the system easier to explain, test, and revise. It also matches the general logic of edge computing, where latency-sensitive processing is kept close to the data source instead of being pushed into a remote cloud pipeline (Yu et al., 2018; Kong et al., 2023; Stefanou et al., 2025).

A further strength is that expiry interpretation is treated as a staged problem. Earlier research has already shown that date recognition is not the same as ordinary OCR because localisation and date reasoning must both be handled carefully (Florea & Rebedea, 2020; Gong et al., 2021; Seker & Ahn, 2022). The present prototype applies that lesson well. The Jetson first narrows the problem to a candidate expiry region on an item crop, and only then does the Pi model attempt day, month, and year prediction. This reduces noise from irrelevant packaging text and fits the structure of the ExpDate-inspired datasets used in training. The high date-component accuracy reported in the benchmark section suggests that this staged design is more than theoretically neat. It is practically effective for the selected test data.

5.3 Limitations and Threats to Validity

The current build also has clear limitations, and these should be stated plainly. First, the prototype is evaluated on a controlled scope. The item detector only covers three classes, the fridge scene is relatively stable, and the benchmark datasets are not the same as a long-term field deployment with many brands, changing lighting, reflective packaging, and arbitrary user behaviour. That means the report can support a claim of prototype feasibility, but not a claim of broad household readiness. The same applies to expiry understanding. Although the Pi date model performs strongly on Dataset B, real domestic packaging can still contain curved prints, damaged labels, and competing date-like strings that may be harder than the benchmark crops. The literature on date recognition and scene text supports this caution, especially when labels are small or visually irregular (Seker & Ahn, 2022; Wang et al., 2023; Peng et al., 2025).

Second, some benchmark evidence comes from more than one machine context. The item detector accuracy benchmark was not run on the final Jetson deployment device, and the date model was benchmarked on both the Orin and the Pi 5 for speed. The Jetson-side vision runs discussed in this report were executed with CUDA enabled, which matters when interpreting their latency. This does not invalidate the results, but it changes how they should be read. Accuracy numbers remain useful as model evidence, while device-specific latency numbers remain useful as deployment evidence. The report should therefore avoid collapsing those into one misleading claim about end-to-end performance. A prototype like this is better defended through a layered argument: model quality, device timing, and architectural fit each matter, but they are not identical forms of evidence.

Third, the user-facing evaluation is still thin. The mobile app screenshot suggests that the retained inventory state can be presented clearly, but the project has not yet included a formal usability study with repeated household use. That matters because food-management tools succeed only when users can understand and trust the information being shown (Liegeard & Manning, 2020; Mathisen & Johansen, 2022). In future work, it would be useful to test whether people actually find the grouped inventory cards, due-soon labels, and expiry displays more helpful than simpler reminder-based approaches.

5.4 Deployment Implications and Future Design Choices

Even with those limitations, the project offers several useful lessons for later development. The first is that local edge processing is justified here. The session-based design keeps the amount of data manageable, reduces dependence on internet availability, and makes the demo more resilient in a classroom or lab setting. Broader Edge AI literature similarly argues that local deployment is attractive when latency, bandwidth, privacy, and device autonomy matter together rather than separately (Abdelmoniem et al., 2024; Stefanou et al., 2025). For a household fridge tool, those conditions are highly relevant.

The second lesson is that communication protocol choices should be judged by system role, not by isolated timing alone. The benchmark showed lower round-trip latency for HTTP in a tightly controlled request-response test, but the prototype still benefits from MQTT because retained state, asynchronous publication, and loose coupling fit the multi-device workflow better. That is especially important once session summaries, expiry crops, backend updates, and mobile subscriptions all need to coexist without direct polling. In other words, the transport benchmark is useful precisely because it clarifies that the fastest isolated request is not always the best overall architecture.

A final lesson concerns trust and privacy. A household prototype may appear low risk compared with medical or industrial IoT, but it still collects visual information from a private domestic space. That makes local processing, limited scope, and conservative data retention sensible design choices. Future extensions should therefore include clearer policies for what images are stored, how long derived inventory records remain available, and how the system handles authentication or broker access. Recent IoT survey work continues to treat privacy, security, and device trust as central design concerns rather than optional extras (Kumar & Prashar, 2021; Dritsas & Trigka, 2025).

6. CONCLUSION AND FUTURE WORK

6.1 Conclusion

The goal of this report was to show that a smart-fridge project can be defended as a single edge-computing workflow instead of a bunch of separate vision demos. The final build supports that claim. The project illustrates a comprehensive local pipeline that incorporates door events to define observation sessions, the Jetson Orin to perform live item and expiry-region inference, the Raspberry Pi backend to reconcile inventory state and predict date components, and the mobile app to present the retained result in a readable manner. This is illustrated by the utilization of benchmark outputs, screenshots, and expanded code. Thus, the prototype accomplishes its main objective: it shows that a single AIoT architecture can be utilized to track, process, and display specific food items and evidence related to their expiration dates without relying on a cloud-heavy design.

The report's main conclusion is that integration matters more than any single component metric. The prototype works because each part holds the others up. The ESP32 gives the system a natural session boundary. The item detector narrows the fridge scene to a manageable class set. The expiry-region stage avoids forcing OCR onto the whole frame. The Pi backend turns detections into a stateful inventory view. Finally, the app shows the result in a more useful way than just showing bounding boxes or single predictions. Benchmarks reinforce this conclusion. The expiry pipeline is reliable, the item model is sufficiently accurate for its intended use, and the timing and transport outcomes are still consistent with a live demo workflow.

At the same time, the project needs to be understood at the appropriate scale. This is a prototype for specific expirable items, not a fully functional domestic appliance. That narrower claim is still valuable. Prior research frequently treats expiry recognition, food management apps, smart kitchen systems, and edge computing architecture as distinct subjects (Liegeard & Manning, 2020; Seker & Ahn, 2022; Kong et al., 2023). This project contributes by bringing those strands together into one working fridge-centred example. That is, the report doesn't say that the problem with smart fridges is fixed. This proves that a useful direction focused on the home is possible and can be explained using the resources that are already available for student projects.

6.2 Future Work

The first priority for future work is broader and more realistic item coverage. The current three-class detector is good for a demo, but the next generation of the system should be able to handle more types of packaged and fresh food while still working reliably in fridges with different levels of clutter, obstructions, and lighting. The expansion should be driven by data rather than solely architectural considerations. More instances of challenging packaging views, more negative scenes, and the final camera position would probably increase robustness more than just selecting a larger model. Similar care should be taken when handling expired items. To test the system under the conditions it is intended to withstand, future datasets should include more real-world household examples, particularly labels with blur, glare, curvature, and partial visibility.

A second priority is stronger end-to-end evaluation. This report gives good evidence at the module level, but a future version should also measure whole-session behavior in a more organized way. For example, it should look at the time it takes to go from the start of a session to the end of it, the percentage of sessions that are resolved correctly, the number of unmatched tracks, and how often this happens over several days of use. Small user studies would also strengthen the project considerably. For example, participants could be asked whether the app helped them notice food that should be consumed soon, whether they trusted the inferred expiry output, and whether the grouped card view was easy to understand. It has been shown that usability and perceived usefulness are very important in deciding whether or not food management apps are actually used (Mathisen & Johansen, 2022; Mastorakis et al., 2024) in research.

A third priority is trustworthiness beyond raw accuracy. Clearer data-retention decisions, a more explicit privacy narrative, and a safer broker configuration will be necessary as the prototype approaches actual domestic use. Security and access control remain critical in IoT systems, despite the fact that t processing already alleviates the need for images to leave the local environment (Dritsas & Trigka, 2025). Depending on the trade-off between latency, privacy, and maintenance load, future iterations may take into account more cooperative edge designs, lightweight model compression, or selective cloud support only for challenging situations (Abdelmoniem et al., 2024).

Overall, expanding the system right away is not the smartest thing to do next. In the same fridge context, it is to increase its credibility by providing more accurate data, robust end-to-end testing, more transparent user evaluation, and more stringent operational safeguards. From a successful capstone project, the prototype could become a more reliable smart-home inventory assistant.

REFERENCES

Abdelmoniem, A. M., Jaber, M., Anwar, A., Zhang, Y., & Gao, M. (2024). Towards a Decentralized Collaborative Framework for Scalable Edge AI. *Future Internet*, *16*(11), 421. DOI: 10.3390/fi16110421

Dai, X. (2024). Robust deep-learning-based refrigerator food recognition. *Frontiers in Artificial Intelligence*, *7*, 1442948. DOI: 10.3389/frai.2024.1442948 PMID: 39697798

Dospinescu, O., Grigorcea, G.-L., & Lefter, B.-I. (2025). A mobile approach to food expiration date determination using OCR and on-cloud image classification. *Applied System Innovation*, *8*(6), 176. DOI: 10.3390/asi8060176

Dritsas, E., & Trigka, M. (2025). A Survey on Cybersecurity in IoT. *Future Internet*, *17*(1), 30. DOI: 10.3390/fi17010030

Florea, V., & Rebedea, T. (2020). Expiry date recognition using deep neural networks. *International Journal of User-System Interaction*, *13*(1), 1–17. DOI: 10.37789/ijusi.2020.13.1.1

Golshany, H., Ni, Y., Yu, Q., & Fan, L. (2025). IoT-enabled smart kitchen technologies and their impact on food storage, preparation, and culinary experiences: A systematic review. *Food Research International*, *213*, 116557. DOI: 10.1016/j.foodres.2025.116557 PMID: 40436590

Gong, D., Jiang, Y., Sun, L., Xu, J., & Wu, Y. (2021). A novel unified deep neural networks methodology for use-by-date recognition in retail food package image. *Signal, Image and Video Processing*, *15*(3), 449–457. DOI: 10.1007/s11760-020-01764-7

Kong, L., Tan, J., Huang, J., Chen, G., Wang, S., Jin, X., Zeng, P., Khan, M., & Das, S. K. (2023). Edge-computing-driven Internet of Things: A survey. *ACM Computing Surveys*, *55*(8), 174. Advance online publication. DOI: 10.1145/3555308

Kumar, T. B., & Prashar, D. (2021). Review on Efficient Food Waste Management System Using Internet of Things. *International Journal of Current Research and Review*, *13*(6), 143–150. DOI: 10.31782/IJCRR.2021.13603

Liegeard, J., & Manning, L. (2020). Use of intelligent applications to reduce household food waste. *Critical Reviews in Food Science and Nutrition*, *60*(6), 1048–1061. DOI: 10.1080/10408398.2018.1556580 PMID: 30633547

Mastorakis, G., Kopanakis, I., Makridis, J., Chroni, C., Synani, K., Lasaridi, K., Abeliotis, K., Louloudakis, I., Daliakopoulos, I. N., & Manios, T. (2024). Managing household food waste with the FoodSaveShare mobile application. *Sustainability (Basel), 16*(7), 2800. DOI: 10.3390/su16072800

Mathisen, T. F., & Johansen, O. J. (2022). The impact of smartphone apps designed to reduce food waste on improving healthy eating, financial expenses and personal food waste: Crossover pilot intervention trial studying students' user experiences. *JMIR Formative Research, 6*(9), e38520. DOI: 10.2196/38520 PMID: 36053667

Newsome, R., Balestrini, C. G., Baum, M. D., Corby, J., Fisher, W., Goodburn, K., Labuza, T. P., Prince, G., Thesmar, H. S., & Yiannas, F. (2014). Applications and perceptions of date labeling of food. *Comprehensive Reviews in Food Science and Food Safety, 13*(4), 745–769. DOI: 10.1111/1541-4337.12086 PMID: 33412704

Peng, H., Bayon, J., Recas, J., & Guijarro, M. (2025). Efficient expiration date recognition in food packages for mobile applications. *Algorithms, 18*(5), 286. DOI: 10.3390/a18050286

Rajan, S. S., & Wani, K. M. (2025). A review of smart food and packaging technologies: Revolutionizing nutrition and sustainability. *Food and Humanity, 4*, 100593. DOI: 10.1016/j.foohum.2025.100593

Seker, A. C., & Ahn, Y. H. (2022). A generalized framework for recognition of expiration dates on product packages using fully convolutional networks. *Expert Systems with Applications, 204*, 117310. DOI: 10.1016/j.eswa.2022.117310

Stefanou, E. E., Bitilis, P., Bouchouras, G., & Kotis, K. (2025). Collecting, Integrating and Processing IoT Sensor Data on Edge Devices for PD Monitoring: A Scoping Review. *Applied Sciences (Basel, Switzerland), 15*(19), 10541. DOI: 10.3390/app151910541

Szenderák, J., Fróna, D., & Rákos, M. H. (2025). Expiration date label knowledge and food waste reduction: A synthesis and critical evaluation of policy recommendations. *Environmental Challenges, 20*, 101269. DOI: 10.1016/j.envc.2025.101269

Wang, C., Huang, C., Zhu, X., Li, Z., & Zhao, L. (2023). Smart retail SKUs checkout using improved residual network. *Scientific Reports, 13*(1), 22512. DOI: 10.1038/s41598-023-49543-x PMID: 38110520

Wang, X.-F., He, Z.-H., Wang, K., Wang, Y.-F., Zou, L., & Wu, Z.-Z. (2023). A survey of text detection and recognition algorithms based on deep learning technology. *Neurocomputing, 556*, 126702. DOI: 10.1016/j.neucom.2023.126702

Wei, X.-S., Cui, Q., Yang, L., Wang, P., Liu, L., & Yang, J. (2022). RPC: A large-scale and fine-grained retail product checkout dataset. *Science China. Information Sciences*, *65*(9), 197101. Advance online publication. DOI: 10.1007/s11432-022-3513-y

Wilson, N. L. W., & Miao, R. (2025). Food waste, date labels, and risk preferences: An experimental exploration. *Applied Economic Perspectives and Policy*, *47*(3), 1029–1057. Advance online publication. DOI: 10.1002/aepp.13507

Yu, W., Liang, F., He, X., Hatcher, W. G., Lu, C., Lin, J., & Yang, X. (2018). A Survey on the Edge Computing for the Internet of Things. *IEEE Access : Practical Innovations, Open Solutions*, *6*, 6900–6919. DOI: 10.1109/ACCESS.2017.2778504

Chapter 7
Session-Based Inventory Tracking on ESP32 With MQTT and Edge AI

Chi Hang Lau
http://orcid.org/0009-0009-5306-3770
The Hong Kong Polytechnic University, Hong Kong

Po Chun To
http://orcid.org/0009-0005-9711-7137
The Hong Kong Polytechnic University, Hong Kong

ABSTRACT

This report presents a local smart-fridge prototype that connects a door-trigger node, a Jetson Orin vision service, a Raspberry Pi backend, and a Vue-based web interface through MQTT. Each fridge interaction is treated as one session and is linked across the full pipeline by a shared door_session_id. In the current prototype, the Orin records while the door is open and processes the session after the door closes. The Pi stores results in SQLite, updates the inventory state, and serves the user interface. The web application shows inventory, alerts, review tasks, and an optional recipe feature. The evaluation uses real session artifacts, backend logs, database rows, and an MQTT door-node validation archive. The results show add-item and remove-item handling, safe unknown-item fallback, empty-session handling, and a support-image refinement example for egg classification. The main contribution is a traceable local workflow that avoids writing uncertain detections into inventory until the user confirms them.

DOI: 10.4018/979-8-2600-0701-3.ch007

1. INTRODUCTION

1.1 Background

Tracking food inside a home refrigerator sounds simple, but in practice it is difficult to do well. A manual list on paper or on a phone is easy to understand, but it only works if the user remembers to update it every time an item is added or removed. Many people do not keep that habit for long. A camera-based smart-fridge system looks like a useful solution because it can reduce manual work. However, the vision problem inside a fridge is not easy. Items overlap each other, hands block the view, shelves create shadows, and packaging materials can reflect light or hide the food. Dai (2024) shows that even dedicated refrigerator food-recognition research still struggles under realistic conditions.

Because of that, this chapter does not treat the smart-fridge problem as only an image-classification problem. Instead, it treats it as a systems-integration problem. The main goal is to build a local prototype that can detect one fridge interaction, process it as one session, update local inventory safely, and show the result through a usable interface. In simple terms, the chapter asks whether sensing, messaging, vision, storage, and user review can work together in a reliable way.

This system view is important because many real errors happen between components rather than inside one component. Even if a model gives a reasonable prediction, the overall system can still fail if the door event is missed, if the wrong session is matched, if the backend saves a weak prediction as if it were certain, or if the user has no way to correct an uncertain result. MQTT is useful here because it lets the trigger node, the vision service, and the backend exchange clear messages without depending too tightly on each other (OASIS Open, 2019). A local-first design is also sensible because it reduces network dependency and keeps household images on local devices instead of sending them to remote cloud services.

The prototype therefore has four main parts. First, a door-trigger node starts and ends each session. In the current prototype, this is often tested on ESP8266 hardware, while ESP32-S3 remains the final hardware direction. Second, a Jetson Orin listens for the door events and treats the time between door open and door close as one vision session. Third, a Raspberry Pi acts as the main application hub by running the MQTT broker, backend API, SQLite database, and inventory logic. Fourth, a Vue-based web application provides the user interface on a touch display and, if needed, from another browser on the same LAN. The shared link across the whole pipeline is door_session_id. The same ID appears in MQTT payloads, session summaries, backend logs, and database rows, so the full workflow can be traced clearly.

For that reason, this chapter is written honestly as a controlled local prototype study. It does not claim production deployment, universal food recognition, or large benchmark-level accuracy. Instead, it shows a practical engineering workflow that combines physical triggering, edge AI processing, local persistence, and manual confirmation when the model evidence is weak.

1.2 Aim

The main aim of this chapter is to present a local smart-fridge prototype that can detect one fridge interaction as one session, process that session with edge AI, update a local inventory state, and display the result through a web application with manual review support.

This aim is divided into the following objectives:

1) Design a reliable door-trigger mechanism using firmware that supports the current ESP8266 prototype and the final ESP32-S3 direction.
2) Use MQTT to connect the trigger node, the vision service, and the backend with low coupling.
3) Persist authoritative state locally in SQLite on the Raspberry Pi.
4) Provide a web interface that shows inventory, activity, alerts, and review tasks.
5) Demonstrate add, remove, unknown, and empty-session behaviour through collected prototype cases.

1.3 Scope and Contributions

The scope of this chapter is intentionally limited. It assumes one fridge, one fixed camera view, one local network, and a small set of supported item classes. It does not attempt full expiry-date OCR, large-scale dataset collection, cloud-native deployment, or strong cross-household generalization. These limits are appropriate for this chapter because they keep the discussion realistic and make the final claims easier to justify.

Within that scope, the main contributions are:

1) A session-based smart-fridge architecture controlled by door open and door close events.
2) A practical MQTT contract that links embedded sensing, edge vision, backend processing, and user interface state.
3) A local-first persistence design in which SQLite on the Raspberry Pi acts as the authoritative application state.
4) A safe unknown-item review path that prevents weak detections from silently corrupting inventory.

5) A small support-image refinement workflow that shows how later sessions can improve for difficult classes such as egg.

1.4 Paper Organization

The remainder of this chapter follows the required chapter structure. Section 2 reviews related work on MQTT-based IoT systems, food recognition, session-level vision reasoning, and uncertainty-aware design. Section 3 explains the research methodology, system architecture, module design, and data model. Section 4 presents the experimental results using figures, screenshots, UML diagrams, message examples, logs, and database evidence. Section 5 discusses what the results mean, the main design trade-offs, and the system limitations. Section 6 concludes the chapter and suggests future work.

2. RELATED WORKS

2.1 MQTT and Local IoT Integration

MQTT is widely used in IoT systems because it is lightweight and works well for event-driven communication between devices. The MQTT 5.0 specification uses a publish-subscribe model, so a device that sends a message does not need direct knowledge of the device that receives it (OASIS Open, 2019). This is useful in the presented system because the door node only needs to publish a state change, while the Jetson Orin and Raspberry Pi can subscribe and respond in different ways. For a student prototype, this loose coupling makes the system easier to build and easier to maintain.

MQTT also suits smart-home applications because small JSON messages are enough to describe events such as open, closed, or session_summary. The retained-message feature is also useful because the backend can publish the latest inventory state for later subscribers. In practice, open-source tools such as Eclipse Mosquitto and Eclipse Paho make MQTT practical on small devices like a Raspberry Pi and easy to integrate with Python services (Eclipse Foundation, n.d.-a, n.d.-b).

At the same time, MQTT is only a messaging layer, not a complete system solution. It moves messages, but it does not guarantee security, validation, or correct application logic on its own. NIST guidance on IoT cybersecurity makes it clear that device identity, access control, and data protection still need explicit design choices (National Institute of Standards and Technology, 2019). This is relevant to the present work because the prototype functions well on a local network, but a more complete deployment would still need stronger broker security and topic control

2.2 Food Recognition and Refrigerator Vision

Food recognition has been studied for many years, and Food-101 is one of the best-known benchmark datasets in this area (Bossard et al., 2014). Datasets like Food-101 are useful because they show that food classes can still be difficult even in benchmark settings. A refrigerator environment is harder again. Items may be partly hidden, stored in plastic, placed at unusual angles, or affected by changing light when the door opens. Dai (2024) discusses these same challenges and shows that robust refrigerator food recognition is still difficult in realistic scenes.

This literature helps explain why the presented system does not make broad accuracy claims. A smart-fridge system can look impressive in a small demonstration but still perform poorly in real use. For that reason, this chapter focuses more on presenting a reliable workflow around the model than on claiming that one model can solve the whole recognition problem. That is why the system works with session-based reasoning, saves evidence images, and includes a review path when the predicted class is uncertain.

2.3 Detection, Open-Vocabulary Reasoning, and Prompt-Based Recognition

The detection pipeline used in the presented system is based on the YOLO family. The original YOLO paper is important because it showed that object detection could be treated as a single end-to-end problem and still run fast enough for practical use (Redmon et al., 2016). Later work such as YOLOv4 improved the balance between speed and accuracy, which is useful for edge systems with limited resources (Bochkovskiy et al., 2020). These papers are not about smart fridges directly, but they explain why YOLO-style models are a sensible starting point for this type of system.

The presented system also uses prompt-oriented classification ideas instead of relying only on a fixed closed-set classifier. This connects to the wider development of vision-language models. CLIP showed that images and text can be aligned well enough to support strong zero-shot transfer across many classes (Radford et al., 2021). More recent work such as YOLO-World extends open-vocabulary reasoning into a fast detector setting, which is conceptually similar to using text prompts and support-image refinement in the presented system (Cheng et al., 2024). The chapter does not claim to reproduce any one published model exactly, but this literature supports the choice of prompt-based recognition.

2.4 Tracking and Session-Level Reasoning

A single image is not enough for a fridge workflow because the system must decide whether an item entered or left the fridge. That requires temporal information. For this reason, tracking-by-detection methods are relevant background. SORT is a common baseline that combines detections with lightweight motion and assignment logic for online tracking (Bewley et al., 2016). ByteTrack later showed that keeping lower-confidence detections can improve track continuity, especially when objects are partly blocked or briefly weak in confidence (Zhang et al., 2022).

This chapter is not presented as a tracking research contribution. However, its session summaries clearly depend on track-level reasoning. Fields such as count_change, entered_or_left, initialPosition, and finalPosition only make sense if the system can follow an object through time. This is why the system reasons over a full session instead of trusting one frame by itself.

2.5 Uncertainty, Reject Option, and Human-in-the-Loop Design

One of the most important design choices in the presented system is that the system does not always force an automatic label into the database. This choice is supported by two related areas of research. First, calibration research shows that modern neural networks can be overconfident, so a high confidence score does not always mean the result is truly reliable (Guo et al., 2017). Second, reject-option and selective-classification research argues that it can be better for a model to abstain on uncertain cases than to make a wrong prediction that only looks precise (Geifman & El-Yaniv, 2019).

In real applications, this idea connects naturally to human-in-the-loop design. Mosqueira-Rey et al. (2023) explain that human input is often useful when models face ambiguity or incomplete context. A fridge inventory is not as safety-critical as healthcare or autonomous driving, but it is still a persistent state that the user depends on. If the system writes the wrong item into inventory and the user does not notice, trust in the application will slowly decrease. For that reason, the unknown-item review queue in the presented system is not just an extra screen. It is a core part of the system design.

2.6 Research Gap

The literature already provides strong ideas in several separate areas: lightweight IoT messaging, fast object detection, session-aware tracking, prompt-based recognition, and uncertainty-aware review. What is less common in small student prototypes is a clear end-to-end workflow that combines these ideas into one local and auditable system. The gap addressed in this chapter is therefore not the lack of recognition models. Instead, it is the lack of a small but complete workflow that starts from a physical event, moves through MQTT, produces evidence-rich session outputs, updates local state safely, and gives the user a review path when the evidence is weak. This chapter focuses on that workflow gap.

3. RESEARCH METHODOLOGY

3.1 System Architecture Overview

The prototype is split across several devices because each device is used for a different job. The current local trigger hardware is often a NodeMCU ESP8266, while the codebase and final hardware direction target ESP32/ESP32-S3. The Jetson Orin is used for GPU-backed vision, and the Raspberry Pi acts as the broker, backend, database, and UI hub (Espressif Systems, n.d.-a; NVIDIA, 2025; Raspberry Pi Ltd., n.d.) as shown in Table 1. This separation keeps the system modular and makes debugging easier.

Table 1. summarizes the device roles.

Component	Main responsibility	Main output
Door trigger node	Detect door open/close, create door_session_id, publish MQTT events	dumbfridge/door/state
Jetson Orin	Record the open-door interval and run GPU-targeted inference after close	dumbfridge/vision/session/summary
Raspberry Pi backend	Subscribe to MQTT, update SQLite, publish retained state, serve API and UI	DB rows, alerts, inventory state
Web application	Show inventory, logs, alerts, review queue, and optional recipe assistance	User-facing interface

The architecture is local-first. All traffic stays inside one LAN, and the database remains on the Raspberry Pi. This design also supports a more privacy-aware approach because fridge images do not need to be sent to external cloud services.

The physical prototype photos are also important because they show that the system was built as a real hardware setup rather than a purely simulated pipeline. The closed view makes the camera placement and enclosure structure clear, while the opened view shows the shelf layout and the kinds of items used during testing.

Figure 1. External view of the prototype shelf, showing the enclosure, camera mount, and nearby touch-display setup.

Figure 2. Internal view of the prototype shelf with the door open, showing representative items used during prototype sessions.

Figure 3. System context of the proposed smart-fridge prototype, showing device roles and user-facing access paths.

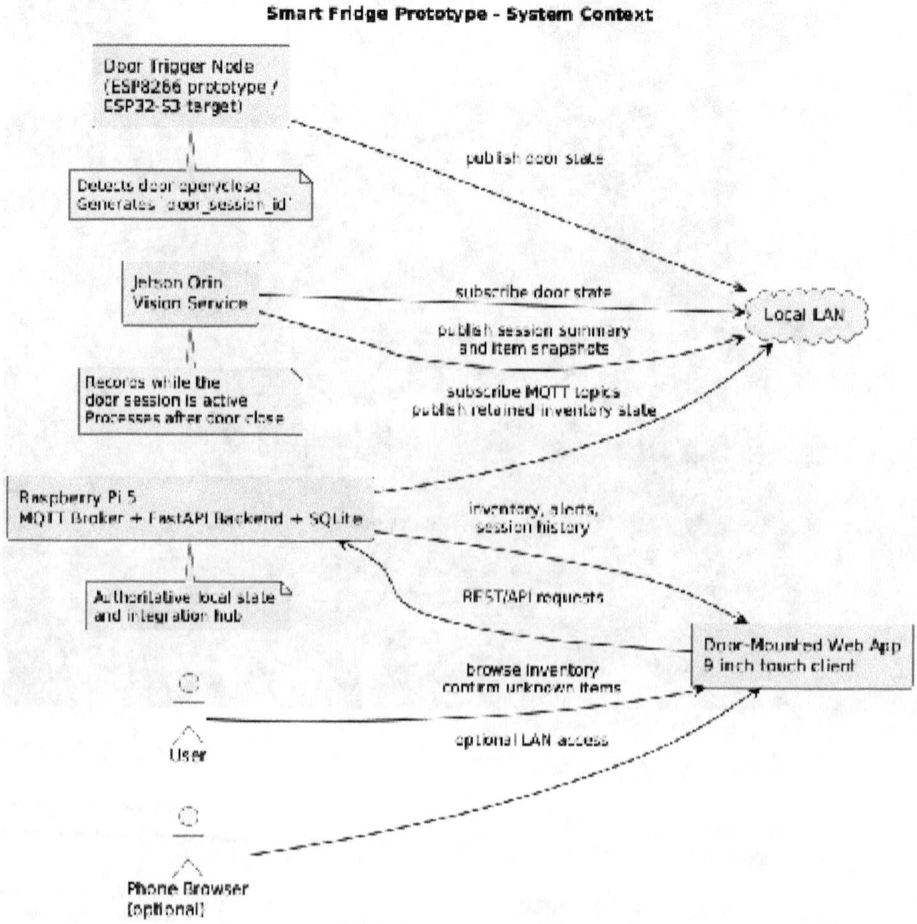

The diagram highlights one key design principle: device-level events use MQTT, while the user interface talks to the backend API instead of directly reading raw broker messages. This keeps the frontend more stable even if the device-level topic set changes later.

3.2 Technology Stack and Implementation Environment

The implementation stack was chosen for practical reasons rather than novelty. The trigger firmware is written to support both ESP8266 and ESP32-family boards, with ESP32-S3 as the final hardware direction and ESP8266 used often for local verification. The Jetson AGX Orin developer materials provide the basis for the vision environment, and the Raspberry Pi 5 is suitable as a compact local hub because it can run the broker, backend, database, and browser-facing interface on one device (Espressif Systems, n.d.-a, n.d.-b; NVIDIA, 2025; Raspberry Pi Ltd., n.d.).

For the application layer, the backend uses FastAPI and Uvicorn, while the frontend uses Vue 3 and PrimeVue (FastAPI, n.d.; Uvicorn, n.d.; Vue.js, n.d.; PrimeTek, n.d.). SQLite is used as the local persistence layer because it is lightweight, file-based, and easy to package inside an edge prototype (SQLite Consortium, n.d.). On the messaging side, the prototype follows the MQTT 5.0 specification and uses common tools such as Mosquitto and Paho (OASIS Open, 2019; Eclipse Foundation, n.d.-a, n.d.-b) as shown in Table 2.

Table 2. summarizes the software stack.

Layer	Technology	Role in the system
Embedded firmware	ESP8266 / ESP32 / ESP32-S3 firmware	Door sensing and event publish
Messaging	MQTT + Mosquitto + Paho	Device-to-device communication
Vision service	Jetson Orin + YOLOE pipeline on cuda:0	Session recording and after-close inference
Backend API	FastAPI + Uvicorn	Persistence and application logic
Database	SQLite	Authoritative local state
Frontend	Vue 3 + PrimeVue	Touch-friendly web interface with optional recipe extension

3.3 Module Descriptions

3.3.1 Door Trigger Node

The door node has a deliberately narrow role. It detects door state changes, generates a door_session_id, adds a timestamp, and publishes the payload to the MQTT broker. It does not perform recognition or inventory reasoning. The source is written to support both ESP8266 and ESP32-family boards, and the current local prototype is often verified on NodeMCU ESP8266 hardware even though the final hardware direction is ESP32-S3.

The payload is intentionally small and stable. A representative message is:


```
{
  "state": "open",
  "door_session_id": "20260408T103203Z-472229",
```

```
  "timestamp": "2026-04-08T10:32:03Z"
}
```


This contract is enough to define the start of a session, and a matching closed event with the same session ID defines the end. The MQTT test archive in Section 4 later shows that the live device produced this structure consistently during a three-cycle validation run.

The physical switch installation is shown below. This image links the software description to the sensing hardware used to create the session trigger.

Figure 4. Door-switch placement used by the prototype door-trigger node for session triggering.

3.3.2 Jetson Orin Vision Service

The Jetson Orin is the main processing node for the vision pipeline. It subscribes to the door-state topic, keeps the camera ready, records the session while the door is open, and performs inference after the door closes. The current configuration targets the GPU on cuda:0, but the normal prototype path is still after-close rather than true real-time inference. This makes the system easier to audit because each session leaves a clean artifact set instead of only temporary streaming output. A future version can rework the Orin side into a real-time GPU pipeline when throughput and stability are improved.

This artifact set is one of the strongest elements discussed in this chapter. Each processed session can contain a session_summary.json, an overview image, a best-candidate crop, extra thumbnails, and sometimes a video capture. This makes the system much easier to evaluate because the evidence can be matched directly to one session ID.

The session summary is also richer than a simple final label. It records the final class and confidence together with movement direction, count change, text-stage guess, support-image match, and whether manual confirmation is needed. This is especially useful for uncertain cases because it separates intermediate evidence from the final decision.

3.3.3 Raspberry Pi Backend and Web Application

The Raspberry Pi hosts the main application layer. It subscribes to MQTT topics, validates and stores session summaries, updates item lots in SQLite, creates alerts when needed, and republishes the latest inventory state. The backend therefore acts as the authoritative state manager rather than a passive relay. This matters because a smart-fridge interface is only trustworthy if it reflects durable state instead of temporary broker traffic.

The web application is the main user-facing surface. It provides inventory cards, recent activity, alert views, and a queue for uncertain or unknown items that require confirmation. It also includes an optional inventory-grounded recipe recommendation and recipe chat extension, although that feature is secondary to the core contribution of this chapter. The screenshots used in this chapter show that the system is not only a device demonstration; it is also an application prototype with a touch-oriented interface.

Figure 5. Main web interface showing grouped inventory cards and summary information.

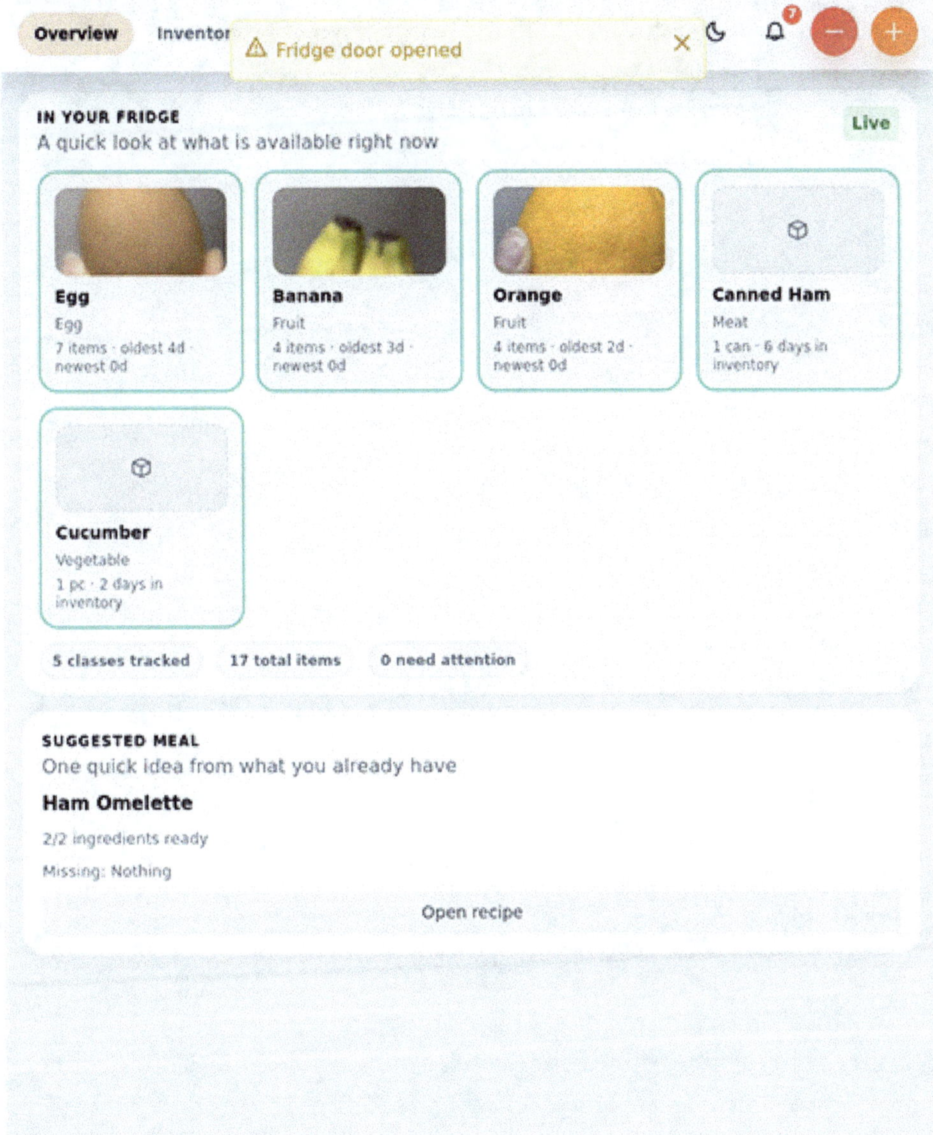

Figure 6. Review queue used to surface uncertain detections for manual confirmation.

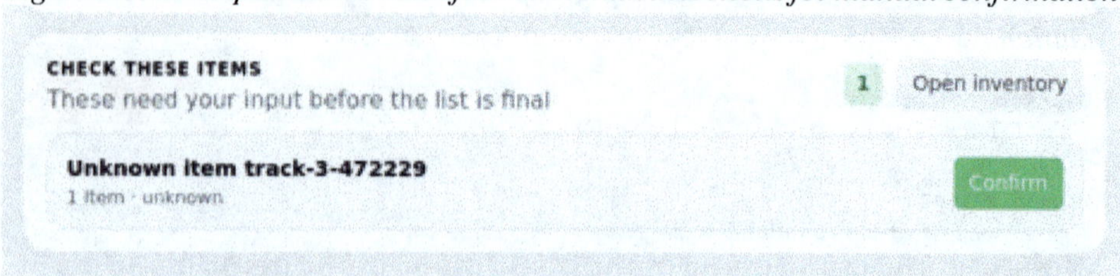

Figure 7. System and alert page used for monitoring and operational review.

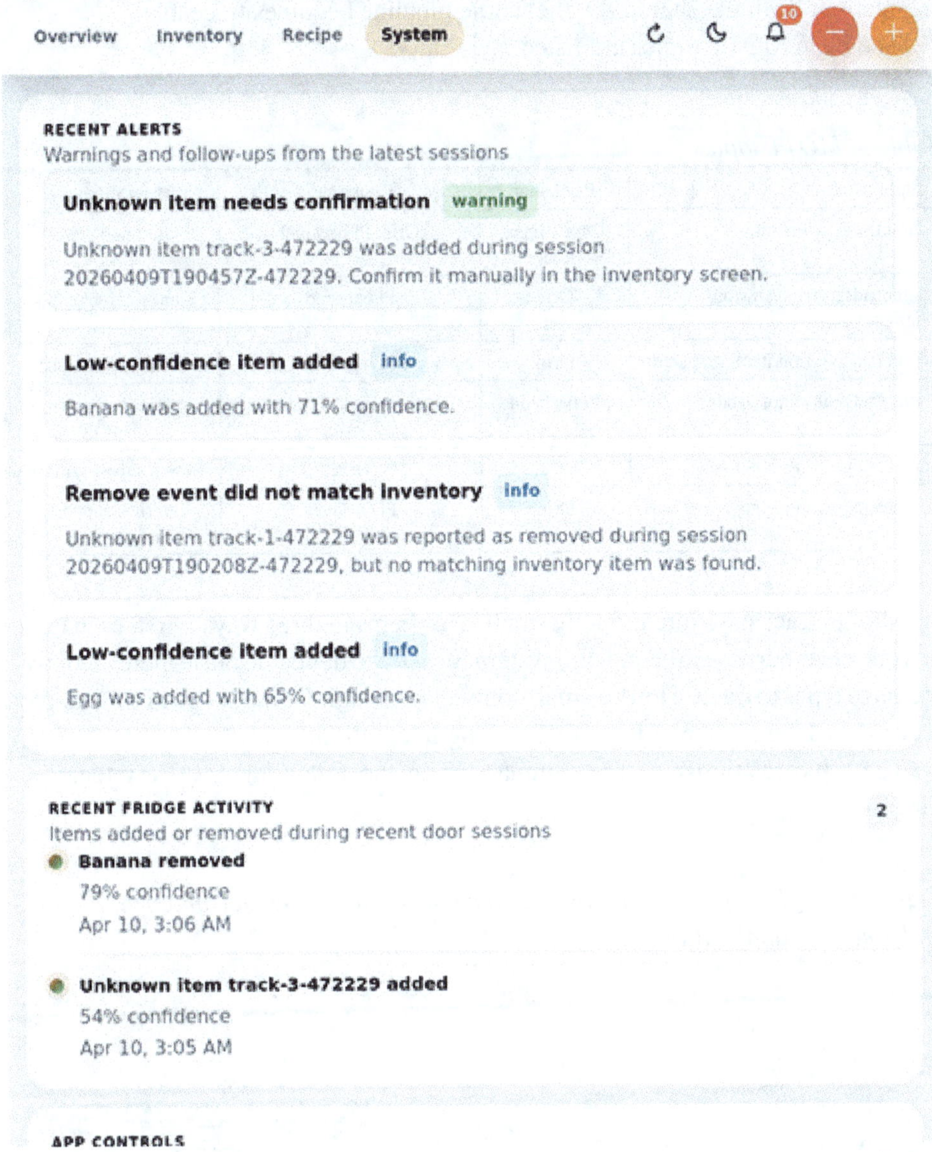

3.4 Message Contract and Session Identity

The single most important design choice in the presented system is the use of door_session_id as the end-to-end join key. Without a stable session ID, the backend would have to guess which open and close events belong together, and the chapter would have to infer which images, logs, and database rows refer to the same fridge interaction. With the shared ID, the whole pipeline becomes traceable.

The main MQTT topics are listed in Table 3.

Table 3. MQTT topics

Topic	Publisher	Consumer	Main purpose
dumbfridge/door/state	Door trigger node	Orin, Pi backend	Start and end a session
dumbfridge/vision/session/summary	Orin	Pi backend	Final session output
dumbfridge/vision/item_snapshot	Orin	Pi backend	Candidate thumbnails
dumbfridge/inventory/state	Pi backend	Debug tools and optional subscribers	Retained inventory snapshot
dumbfridge/orin/support/save	Pi backend	Orin	Save a support image
dumbfridge/orin/support/save_ack	Orin	Pi backend	Confirm support-image save

The contract stays intentionally small. This helps in three ways. First, each topic has one clear purpose. Second, the system is easier to debug because there are fewer message types to trace. Third, a small contract is enough for the scope of this chapter and reduces the chance of topic drift during integration.

The topic relationships are summarized in the following MQTT integration diagram.

Figure 8. MQTT topic interactions between the door trigger, Jetson Orin, Pi back-end, and web application.

3.5 Database Design

SQLite on the Raspberry Pi acts as the authoritative state store. This choice matches the local-first design of the system described in this chapter. A file-based database is easy to deploy, inspect, and back up, and it is more than sufficient for the scale of this prototype (SQLite Consortium, n.d.). The active system database contains the tables that matter most for the chapter: items, session_logs, processed_sessions, alerts, and item_images.

The following schema excerpt captures the most relevant structures:

```
CREATE TABLE items (
    id INTEGER PRIMARY KEY AUTOINCREMENT,
    name TEXT NOT NULL,
    food_type TEXT NOT NULL,
    quantity REAL NOT NULL DEFAULT 1,
    status TEXT NOT NULL DEFAULT 'review',
    confidence REAL NOT NULL DEFAULT 0,
    door_session_id TEXT NOT NULL DEFAULT '',
    temp_track_id TEXT NOT NULL DEFAULT '',
    support_save_status TEXT NOT NULL DEFAULT 'not_request-
ed'
);
CREATE TABLE session_logs (
    id INTEGER PRIMARY KEY AUTOINCREMENT,
    door_session_id TEXT NOT NULL DEFAULT '',
    event_type TEXT NOT NULL,
    item_name TEXT NOT NULL DEFAULT '',
    direction TEXT NOT NULL DEFAULT '',
    confidence REAL NOT NULL DEFAULT 0,
    event_at TEXT NOT NULL
);
CREATE TABLE processed_sessions (
    door_session_id TEXT PRIMARY KEY,
    payload_hash TEXT NOT NULL DEFAULT '',
    processed_at TEXT NOT NULL
);
```

The database is still small, but it already captures repeated usage. The current live counts are 15 rows in items, 368 rows in session_logs, 56 rows in processed_sessions,

12 rows in alerts, and 14 rows in item_images. These counts show that the system has been used beyond one isolated demonstration.

The logical data model is shown below as a rendered ER-style diagram.

Figure 9. Logical application data model showing the main entities used for inventory state, session history, alerts, and stored images.

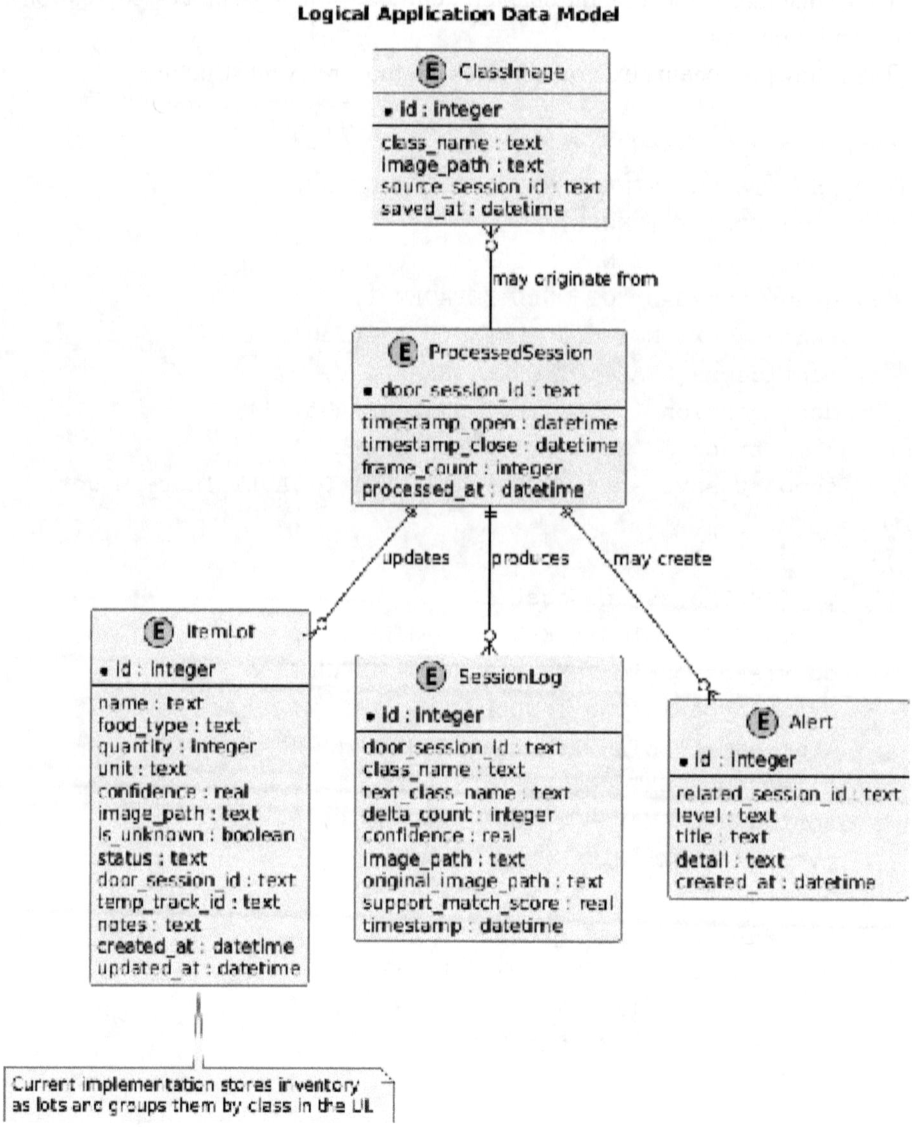

3.6 Workflow Algorithm and Session Lifecycle

From a methodological point of view, the system is best understood as a session pipeline:

1) The ESP node publishes a door-open event and creates a new `door_session_id`.
2) The Orin begins recording the scene while the door is open.
3) The user adds or removes an item.
4) The ESP node publishes a door-close event using the same session ID.
5) The Orin stops recording, performs tracking and classification, and builds a session summary.
6) The Pi backend stores the session, updates inventory if appropriate, and creates alerts for uncertain cases.
7) The web interface shows the updated inventory or a review task.

The end-to-end sequence is captured below.

Figure 10. End-to-end interaction sequence for a normal fridge session from door open to inventory update.

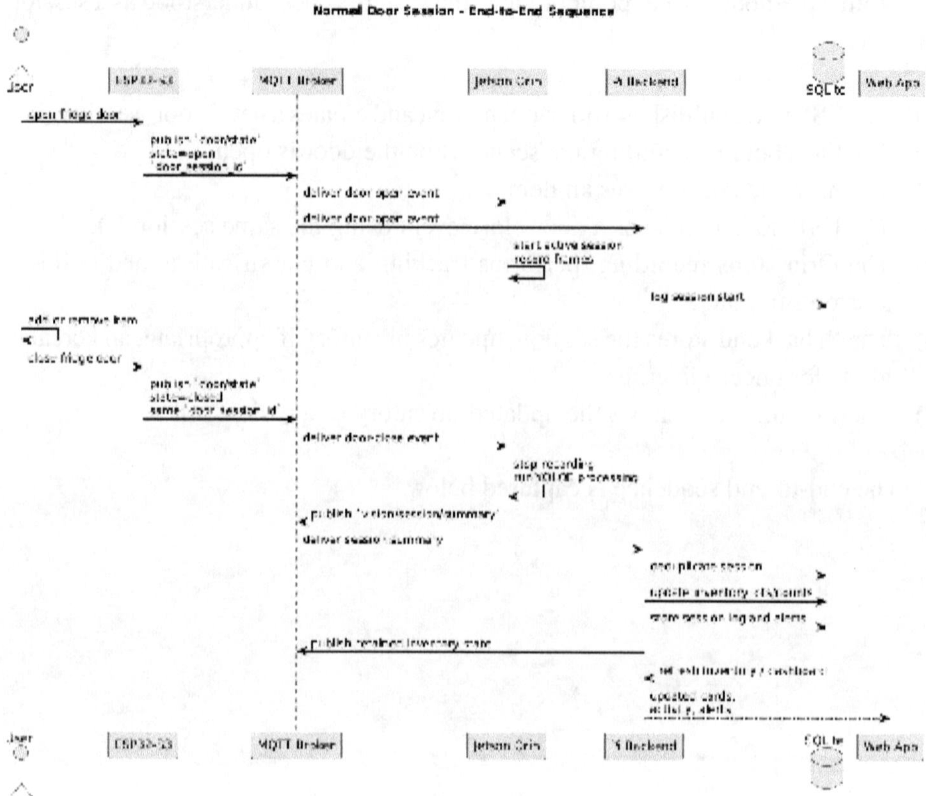

This workflow explains why the chapter emphasizes traceability so strongly. The system only succeeds if every important step leaves evidence that can be checked later in artifacts, logs, or the database. For the same reason, the experimental section uses selected real sessions instead of making broad claims with weak support.

The behavioural logic can be summarized with the following session-lifecycle diagram.

Figure 11. Behavioural state model of a fridge session, including normal completion, empty-session handling, and review flow.

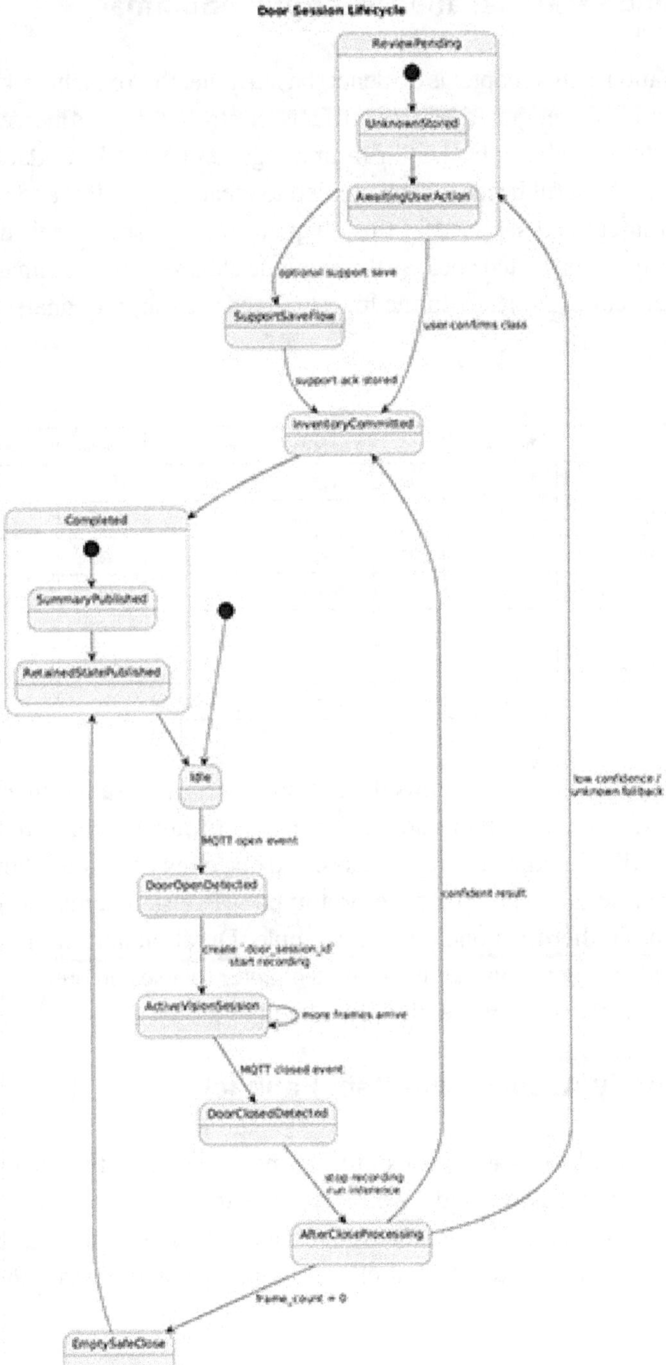

4. EXPERIMENTAL RESULTS

4.1 Evaluation Material and Descriptive Summary

The evaluation in this chapter is evidence-based rather than benchmark-based. The strongest sources are integrated session artifacts, controlled egg-refinement sessions, the active system database, backend runtime logs, and MQTT validation records. These sources are useful because they are tied to clear session IDs and timestamps.

The main integrated set used in this chapter contains seven well-documented sessions. These were selected because they include clear session_summary.json files and enough matching evidence in the log or database to support analysis.

Table 4.

Session ID	Duration (s)		Final class	Count change	Direction	Needs confirmation
20260409T181812Z-472229	9	10	egg	+1	entered	false
20260409T182658Z-472229	6	15	unknown	+1	entered	true
20260409T182911Z-472229	8	23	egg	+1	entered	false
20260409T184321Z-472229	14	37	unknown	-1	left	true
20260409T184820Z-472229	0	0	-	-	-	-
20260409T190457Z-472229	13	34	unknown	+1	entered	true
20260409T190606Z-472229	14	34	Banana	-1	left	false

The summary numbers from this dataset are descriptive rather than statistical. Across the seven sessions, there are 153 processed frames, four add events, two remove events, three unknown-review cases, and one empty session. These numbers are useful because they show that the prototype was tested across several meaningful outcomes rather than one perfect example. Duration and frame-count plots were generated during analysis, but the main chapter focuses on outcomes and case evidence because those are more directly related to the system story.

4.2 Case Study A: Unknown-Item Fallback

The unknown fallback case is one of the strongest results in this chapter because it demonstrates safe behaviour under weak evidence. In session 20260409T190457Z-472229, the system detected an entered item, but it did not commit to a confident food label. Instead, it stored the final class as unknown and marked the result for confirmation.

Figure 12. Distribution of add, remove, unknown-review, and empty-session outcomes in the integrated evaluation set.

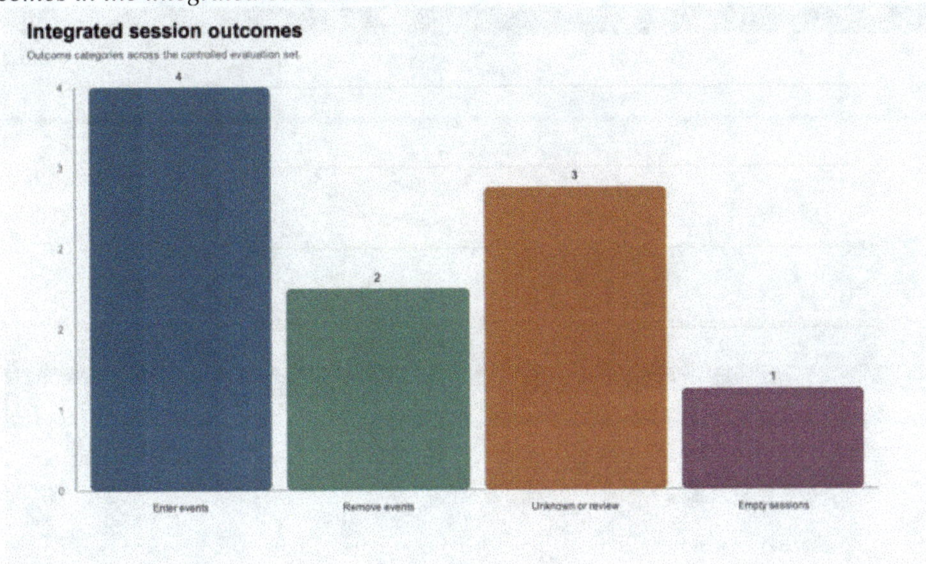

Figure 13. Overview image for the unknown-item fallback session `20260409T190457Z-472229`.

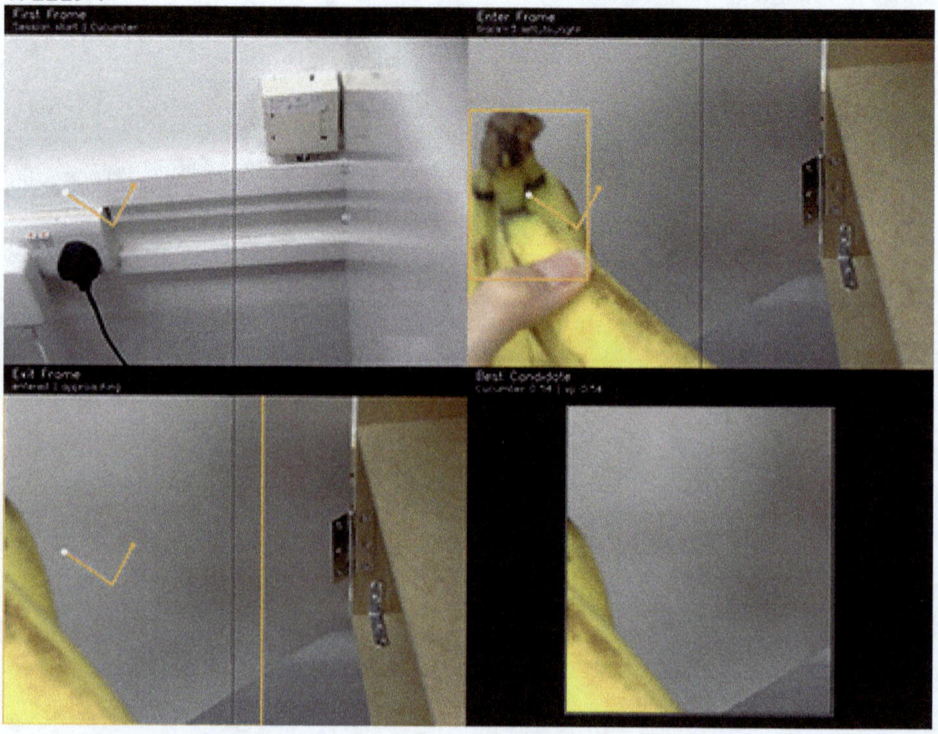

The session summary shows why this happened. The text stage suggested banana with low confidence, while the support-image match leaned toward Cucumber but was still weak. The final result therefore became:


```
{
  "class_name": "unknown",
  "confidence": 0.541,
  "count_change": 1,
  "entered_or_left": "entered",
  "text_class_name": "banana",
  "text_confidence": 0.2428,
  "support_match_class_name": "Cucumber",
  "support_match_score": 0.541,
  "needs_confirmation": true,
  "decision_source": "manual_confirmation"
}
```

This is good behaviour for a stateful application. A weaker design might have chosen the strongest available label and silently added it to inventory. The backend log confirms that the safer path was taken:

```
[backend] door state handled session=20260409T190457Z-472229 state=open
[backend] door state handled session=20260409T190457Z-472229 state=closed
[backend] session summary handled session=20260409T190457Z-472229 processed=1 entered=1 left=0 unknown=1
```

The database also contains a matching processed_sessions row, a session_item log, an items row named Unknown item track-3-472229, and an alert telling the user that confirmation is required. Taken together, these artifacts show a complete evidence chain from vision output to persisted application state.

4.3 Case Study B: Successful Remove-Item Event

The remove-item case shows the opposite pattern. In session 20260409T190606Z-472229, the system confidently identified a Banana leaving the fridge and passed that result cleanly through the pipeline.

Figure 14. Overview image for the remove-item session `20260409T190606Z-472229`.

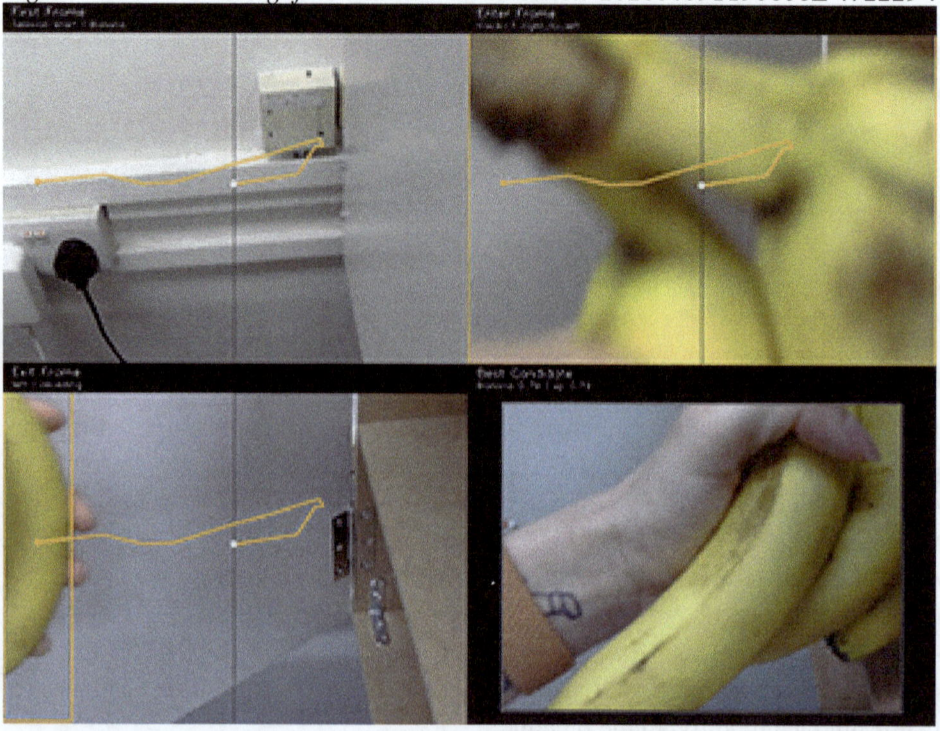

The relevant fields in the session summary are:


```
{
  "class_name": "Banana",
  "confidence": 0.793,
  "count_change": -1,
  "entered_or_left": "left",
  "text_class_name": "banana",
  "text_confidence": 0.6083,
  "support_match_class_name": "Banana",
  "support_match_score": 0.793,
  "needs_confirmation": false,
  "decision_source": "visual_prompt"
}
```


Here the automatic path is justified because the evidence is consistent. The label, support match, and direction all agree, and the backend log records left=1 and unknown=0. A matching item_images row and processed_sessions row in SQLite show that the session was stored correctly. This case matters because it proves that the system can apply a negative inventory change, not only additions.

4.4 Case Study C: Empty-Session Safe Handling

The empty session 20260409T184820Z-472229 is not visually dramatic, but it is technically important. The session summary reports frame_count = 0 and an empty tracked_items list. The backend still handled the session safely and recorded processed=0 entered=0 left=0 unknown=0. This is useful evidence because it shows that the pipeline does not force a result when there is no meaningful visual data. For an appliance-like workflow, safe no-data handling is part of correctness.

4.5 Case Study D: Egg Support-Image Refinement

The egg case is the clearest machine-learning refinement example in this chapter. It should be interpreted carefully. The result does not prove one-shot learning in a broad sense. Instead, it shows a practical support-image workflow that improved a later decision for a difficult class.

Based on the recorded egg-case evidence, the sequence was:

1) A brown egg was first misclassified as `orange`.
2) The best crop from that session was saved as egg support.
3) A later white egg was initially misclassified as `apple` by the text stage.
4) The visual-prompt stage matched the saved support image and corrected the final class to `egg`.

The baseline session ended with class_name = orange and confidence = 0.594, while the refined session ended with class_name = egg, text_class_name = apple, and support_match_class_name = egg with a score of 0.7897. This comparison is meaningful because the text stage was still wrong in the later session. The improvement came from the support-image stage, not from pretending that the original classifier had suddenly become reliable by itself.

4.6 MQTT Door-Node Validation

The door-node validation archive gives a focused check of the MQTT trigger behaviour. The summary reports three expected cycles and three observed complete cycles on the topic dumbfridge/door/state. The archive also reports passed = false, but the raw capture explains why: two unrelated retained inventory-state messages were included in the broader topic capture and failed the door-message validator.

```
{
  "expected_cycles": 3,
  "observed_complete_cycles": 3,
  "total_messages": 8,
  "bad_messages": 2,
  "passed": false
}
```

Figure 15. Baseline egg case in which a brown egg was initially classified as `orange`.

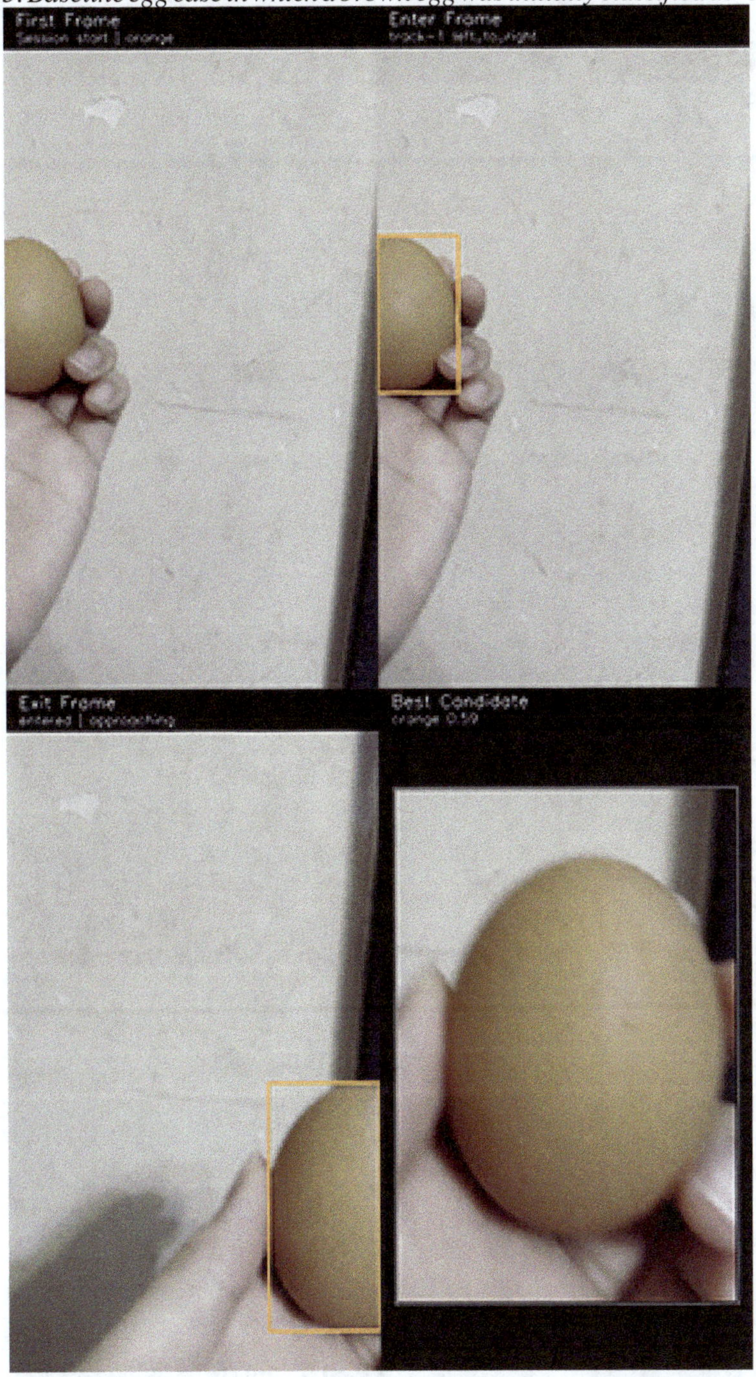

Figure 16. Later white-egg case in which support-image refinement corrected the final class to `egg`.

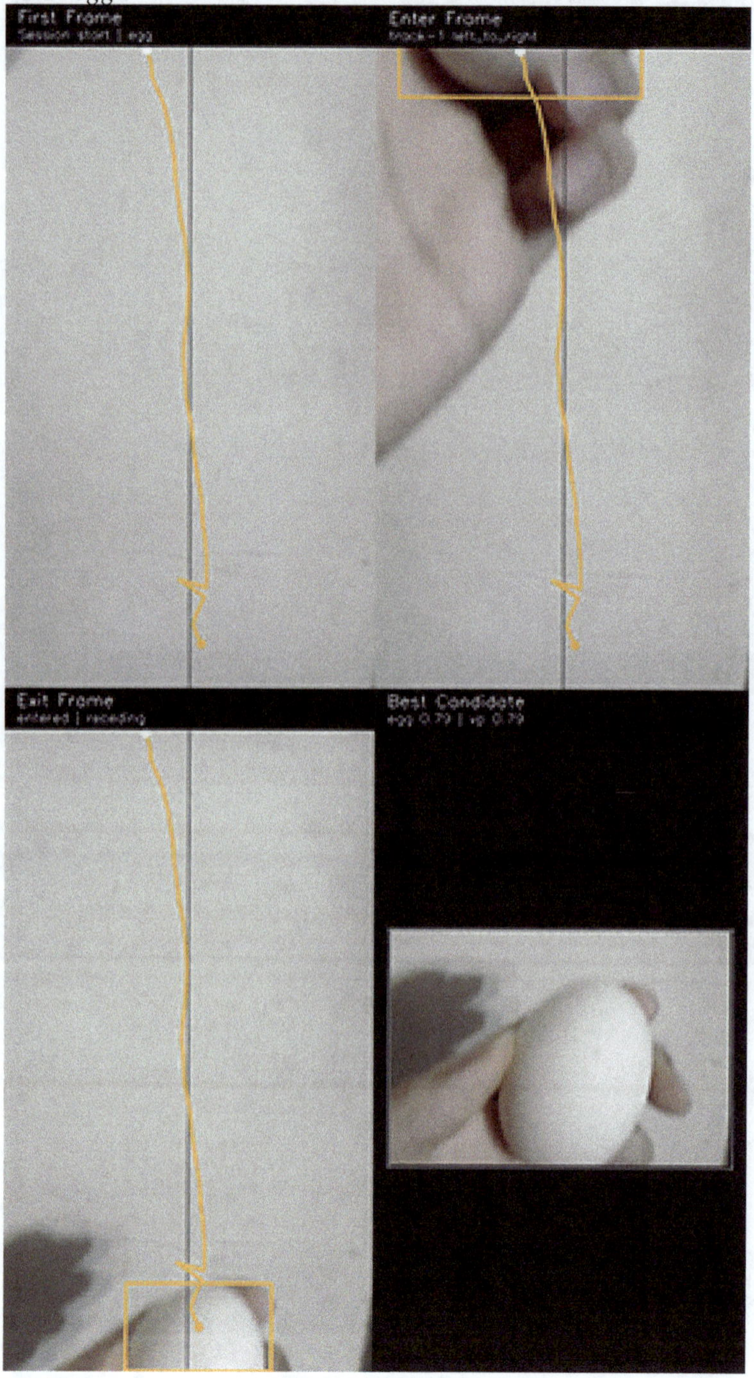

Example raw records from the archive are:

{"topic":"dumbfridge/door/state","payload":{"state":"open","door_session_id":"20260409T220509Z-472229","timestamp":"2026-04-09T22:05:09Z"}}
{"topic":"dumbfridge/door/state","payload":{"state":"closed","door_session_id":"20260409T220509Z-472229","timestamp":"2026-04-09T22:05:11Z"}}

This result is still useful. It shows that the device-level messages were correct and that the false negative came from the validation scope rather than from the trigger node itself. That is a sign of good instrumentation because the evidence is detailed enough to explain the failure mode instead of hiding it.

5. DISCUSSION

5.1 What the Prototype Demonstrates

The strongest result presented in this chapter is not one classification example. It is the fact that the system works as one coherent and traceable workflow from door trigger to user interface. The current trigger prototype, often validated on ESP8266 hardware, provides a clear physical boundary for a fridge interaction while still supporting the final ESP32-S3 direction. The Jetson Orin turns that interaction into an evidence-rich session summary. The Raspberry Pi backend stores the result and updates local state. The UI then presents that state in a form the user can read and, when necessary, correct. This full chain is what makes the prototype convincing as the central contribution of this chapter.

The unknown fallback case is especially important. Many student AI demonstrations look good only because they always return a label, even when the evidence is weak. That creates hidden error because the system appears confident while the real evidence is poor. In the presented system, the unknown path is treated as a valid result. This is a sensible systems decision because a persistent inventory should favour trustworthiness over forced automation. The database row for the unknown item and the warning alert should therefore be seen as signs of responsible behaviour, not as failures to hide.

The remove-item case shows the other side of the design. When the evidence is strong enough, the system performs a normal automatic update without user intervention. The important point is that both behaviours use the same pipeline. The architecture does not need one path for successful sessions and another special path for uncertain sessions. Instead, the same session summary contract supports both automatic inventory updates and manual review. This consistency reduces system complexity and makes later testing easier.

The egg refinement example adds another useful lesson. The chapter does not only show that a model can be weak on a difficult class. It also shows a practical improvement loop. A misclassified brown egg becomes a support example, and a later white egg benefits from that saved evidence. This is a small result, but it is still valuable because it shows how the system can improve locally without requiring full retraining. For an edge prototype, that is a realistic and useful direction.

5.2 Design Trade-Offs

This chapter also discusses several deliberate trade-offs. The first is after-close processing instead of fully live inference. The Orin is already configured for GPU-targeted inference, but the normal launcher does not use a full real-time path. Real-time inference would look more impressive in a live demonstration, but it would also increase complexity and make the system harder to evaluate. By processing after the door closes, the Orin can produce a clean artifact bundle and the chapter can audit exactly what happened in each session. For the scope of this chapter, that trade-off is reasonable.

The second trade-off is the use of a small MQTT contract instead of a larger and more feature-rich message schema. A broader topic set could expose more low-level events, but it would also increase coupling and create more chances for inconsistent interpretation between devices. The current topic map is small, readable, and sufficient for the system goals.

The third trade-off is the use of SQLite as the authoritative state store. A larger distributed database would add scalability and concurrency features, but it would also add unnecessary overhead for a local fridge prototype. SQLite is enough for the current workload, keeps deployment simple, and supports direct inspection during debugging and chapter preparation.

The final trade-off is user involvement. Manual confirmation adds friction, and too many review cases would make the system tiring to use. However, removing review completely would make the inventory less trustworthy. The present design therefore accepts some user effort in exchange for safer state updates. For a prototype where confidence calibration is still imperfect, this is a reasonable balance.

5.3 Limitations and Threats to Validity

This chapter has several important limitations that should be stated clearly. The first is evaluation size. Seven integrated well-documented sessions and one controlled egg case are enough to demonstrate workflow behaviour, but they are not enough to support strong claims about class-level accuracy, recall, or long-term reliability. The graphs in Section 4 should therefore be read as descriptive summaries of prototype behaviour, not as benchmark results.

The second limitation is environmental control. The current system assumes one fixed fridge, one camera position, limited item variation, and fairly controlled lighting. A more realistic household environment would include more clutter, more reflective packaging, more overlapping items, and more variation in how users place or remove food. Dai (2024) makes clear that these factors remain difficult even in dedicated refrigerator recognition research.

The third limitation concerns label coverage and class ambiguity. The egg case itself shows that visually similar food classes remain difficult for the text-prompt stage. The support-image workflow helps, but it is still a targeted local improvement rather than a complete answer to open-world refrigerator recognition. More support images and more confirmed examples would be needed before the system could claim strong robustness on diverse food types.

The fourth limitation is the lack of a formal user study. The interface appears functional from screenshots and from the review workflow, but this chapter does not measure confirmation time, correction rate, perceived usability, recipe usefulness, or user trust. These are important questions for a more product-like system, especially if the manual review queue becomes a regular part of use.

There is also a security limitation. The chapter emphasizes local deployment, which helps privacy, but local deployment alone is not the same as secure deployment. If the system were developed further, broker authentication, TLS, topic authorization, and stronger device management would be necessary, in line with NIST guidance for IoT systems (National Institute of Standards and Technology, 2019).

5.4 Implications for Future Development

Even with these limitations, the presented system has clear value as a platform for future extension. The strongest part of the prototype is the session-based workflow built around door_session_id. That architectural decision already solves a major integration problem, and it can support later upgrades without changing the whole system design. Better models, more classes, or stronger support-image logic can all be added later while keeping the same end-to-end session contract.

This chapter also suggests that future evaluation should focus on session-level metrics rather than only frame-level metrics. For this kind of application, the most meaningful questions are whether the correct inventory change was made, how often manual review was triggered, how often review was useful, and how much user effort was needed to keep the inventory correct. These measures are closer to real system usefulness than a detector score on isolated frames.

Finally, the chapter shows that traceability is worth the effort. The combination of session artifacts, logs, and database rows made it possible to write a stronger and more honest evaluation section. Future work should keep this design philosophy. Even if the model improves, the system should still expose enough evidence to explain why a decision was made.

6. CONCLUSION AND FUTURE WORK

6.1 Conclusion

This chapter presented a local smart-fridge prototype that integrates a door trigger, a Jetson Orin vision service, a Raspberry Pi backend, and a web application through MQTT. The system was designed around one simple but important idea: each fridge interaction should be treated as one session. That session begins with a door-open event, ends with a door-close event, and uses the same door_session_id across MQTT messages, vision summaries, backend logs, and SQLite rows. This decision gives the system a clear control boundary and makes the whole workflow traceable.

The results support the main claim of this chapter. The prototype can handle add-item and remove-item cases, maintain local inventory state, surface unknown detections for review, and handle empty sessions safely. The unknown fallback case is especially meaningful because it shows that the system prefers safe review over false confidence. The remove-item case shows that the same architecture can also support normal automatic updates when the evidence is strong. The egg refinement case adds a small but useful machine-learning contribution by showing how a saved support image can improve a later difficult classification. Together, these cases show success, safety, and refinement within one coherent evaluation.

Another important conclusion is that the work presented in this chapter should be understood as a systems prototype, not as a standalone recognition benchmark. The most valuable contribution is the reliable coordination of hardware triggering, MQTT messaging, edge inference, local persistence, and user review. This is why the chapter relies on artifacts, logs, and database rows rather than broad accuracy claims. The current evidence is strong enough to support a workflow claim, but not strong enough to claim general food-recognition performance across many households or large class sets. Keeping this distinction clear makes the chapter more academically honest.

This chapter also shows that a local-first design is practical for this type of application. By keeping the broker, backend, database, and interface on the Raspberry Pi, and by keeping the vision processing on the Jetson Orin, the system remains understandable and inspectable. This deployment style makes debugging easier and better matches privacy-aware household computing, where sending kitchen images to remote cloud services may be undesirable.

6.2 Future Work

There are several realistic next steps. The first is larger evaluation. A future version should collect a labelled session dataset that covers more food classes, more repeated add and remove events, and more challenging lighting and occlusion conditions. This would make it possible to report session-level precision, review rate, and correction cost in a more formal way.

The second is stronger model refinement. The egg example suggests that support images can help weak classes, but the current evidence is limited to one controlled case. A larger study should test whether support-image refinement reduces review burden across multiple classes and whether the improvement remains stable over time.

The third is to rework the Orin side from after-close processing into a true real-time GPU pipeline when performance and stability are ready. This would be a meaningful engineering upgrade, but it should build on the current session logic rather than replace it.

The fourth is better testing and topic hygiene. The MQTT validation archive already showed that the door-state messages were correct, but the validator was affected by unrelated retained inventory topics. Future tooling should validate by topic scope and schema type so that false negative test results become less likely.

The fifth is user-study work. The presented system would benefit from measuring how easily users can confirm unknown items, how long correction takes, whether the recipe suggestion feature is useful, and whether the interface helps or slows users down. This would strengthen the claim that the web application is not only functional in screenshots but also usable in practice.

The sixth is stronger deployment hardening. If the system were moved beyond a lab-style LAN prototype, broker authentication, TLS, topic-level authorization, and service supervision would be necessary. These improvements would not change the core architecture, but they would make the system more robust and more defensible as a real IoT deployment.

In summary, the presented system already provides a solid foundation. It does not solve every smart-fridge challenge, but it does establish a reliable session-based architecture with clear evidence trails and a sensible human-in-the-loop design. That makes it a credible chapter contribution and a strong base for future extension.

SELECTED EVIDENCE EXCERPTS

B.1 Unknown Fallback Database Evidence

```
   processed_sessions
   door_session_id         payload_hash
processed_at
   -----------------------  ----------------------------------
---------------------------- ------------------------
   20260409T190457Z-472229  1b87946a61cf433bb01057d
d2d8d973006543f95d5c96c26ab04b93237d259b2  2026-04-
09T19:05:19.781Z
   session_logs
   id  door_session_id          event_type       item_name
direction  confidence
   --- ----------------------- --------------- -------------
--------------- --------- ----------
   291  20260409T190457Z-472229  door_open
open      0.0
   292  20260409T190457Z-472229  door_closed
closed     0.0
   295  20260409T190457Z-472229  session_item     Unknown item
track-3-472229  entered    0.541
   296  20260409T190457Z-472229  session_summary
0.0
```

B.2 Remove-Item Database Evidence

```
   id   door_session_id          event_type       item_name
direction  confidence
   --- ----------------------- --------------- ---------  --
-------  ----------
   297  20260409T190606Z-472229  door_open
open       0.0
   298  20260409T190606Z-472229  door_closed
closed     0.0
```

```
301  20260409T190606Z-472229  session_item     Banana
left        0.793
302  20260409T190606Z-472229  session_summary
0.0
```

B.3 Empty-Session Log Evidence

```
[backend] door state handled session=20260409T184820Z-472229
state=open
[backend] door state handled session=20260409T184820Z-472229
state=closed
[backend] session summary handled session=20260409T184820Z-
472229 skipped=False processed=0 entered=0 left=0 unknown=0
```

B.4 Support-Save Acknowledgement Example


```
{
  "door_session_id": "session-001",
  "temp_track_id": "track-3",
  "final_class_name": "banana",
  "status": "saved",
  "saved_support_class": "banana"
}
```


B.5 MQTT Test Excerpt


```
{"received_at_utc":"2026-04-09T22:05:09Z","topic":"dumb-
fridge/door/state","payload":{"state":"open","door_ses-
sion_id":"20260409T220509Z-472229","timestamp":"2026-04-
09T22:05:09Z"}}
{"received_at_utc":"2026-04-09T22:05:11Z","topic":"dumb-
fridge/door/state","payload":{"state":"closed","door_ses-
sion_id":"20260409T220509Z-472229","timestamp":"2026-04-
09T22:05:11Z"}}
```

REFERENCES

Bewley, A., Ge, Z., Ott, L., Ramos, F., & Upcroft, B. (2016). *Simple online and realtime tracking*. arXiv. https://arxiv.org/abs/1602.00763 DOI: 10.1109/ICIP.2016.7533003

Bochkovskiy, A., Wang, C.-Y., & Liao, H.-Y. M. (2020). *YOLOv4: Optimal speed and accuracy of object detection*. arXiv. https://arxiv.org/abs/2004.10934

Bossard, L., Guillaumin, M., & Van Gool, L. (2014). Food-101: Mining discriminative components with random forests. In D. Fleet, T. Pajdla, B. Schiele, & T. Tuytelaars (Eds.), *Computer Vision - ECCV 2014* (pp. 446-461). Springer. DOI: 10.1007/978-3-319-10599-4_29

Cheng, T., Song, L., Ge, Y., Liu, W., Wang, X., & Shan, Y. (2024). YOLO-World: Real-time open-vocabulary object detection. *Proceedings of the IEEE/CVF Conference on Computer Vision and Pattern Recognition*, 16901-16911. https://openaccess.thecvf.com/content/CVPR2024/html/Cheng_YOLO-World_Real-Time_Open-Vocabulary_Object_Detection_CVPR_2024_paper.html

Dai, X. (2024). Robust deep-learning based refrigerator food recognition. *Frontiers in Artificial Intelligence*, *7*, 1442948. Advance online publication. DOI: 10.3389/frai.2024.1442948 PMID: 39697798

Eclipse Foundation. (n.d.-a). *Eclipse Mosquitto*. https://mosquitto.org/

Eclipse Foundation. (n.d.-b). *Eclipse Paho MQTT Python Client*. https://eclipse.dev/paho/files/paho.mqtt.python/html/index.html

Espressif Systems. (n.d.-a). *ESP-IDF Programming Guide - ESP32-S3*. https://docs.espressif.com/projects/esp-idf/en/latest/esp32s3/

Espressif Systems. (n.d.-b). *Product overview - ESP32-S3*. https://docs.espressif.com/projects/esp-hardware-design-guidelines/en/latest/esp32s3/product-overview.html

FastAPI. (n.d.). *Deployment*. https://fastapi.tiangolo.com/deployment/

Geifman, Y., & El-Yaniv, R. (2019). SelectiveNet: A deep neural network with an integrated reject option. In *Proceedings of the 36th International Conference on Machine Learning* (pp. 2151-2159). PMLR. https://proceedings.mlr.press/v97/geifman19a.html

Guo, C., Pleiss, G., Sun, Y., & Weinberger, K. Q. (2017). On calibration of modern neural networks. In *Proceedings of the 34th International Conference on Machine Learning* (pp. 1321-1330). PMLR. https://proceedings.mlr.press/v70/guo17a.html

Mohammed, A., & Zangana, H. (Eds.). (2026). *Intelligent Construction Monitoring Systems: Real-Time Safety, Environmental Prediction, and Risk Management*. IGI Global Scientific Publishing., DOI: 10.4018/979-8-3373-9245-5

Mosqueira-Rey, E., Hernández-Pereira, E., Alonso-Ríos, D., Bobes-Bascarán, J., & Fernández-Leal, Á. (2023). Human-in-the-loop machine learning: A state of the art. *Artificial Intelligence Review*, *56*(4), 3005–3054. DOI: 10.1007/s10462-022-10246-w

National Institute of Standards and Technology. (2019). *Considerations for managing Internet of Things (IoT) cybersecurity and privacy risks* (NIST IR 8228). DOI: 10.6028/NIST.IR.8228

NVIDIA. (2025). *Jetson AGX Orin Developer Kit User Guide*. https://docs.nvidia.com/jetson/agx-orin-devkit/user-guide/latest/index.html

Open, O. A. S. I. S. (2019). *MQTT version 5.0*. https://docs.oasis-open.org/mqtt/mqtt/v5.0/mqtt-v5.0.pdf

PrimeTek. (n.d.). *PrimeVue introduction*. https://primevue.org/introduction/

Radford, A., Kim, J. W., Hallacy, C., Ramesh, A., Goh, G., Agarwal, S., Sastry, G., Askell, A., Mishkin, P., Clark, J., Krueger, G., & Sutskever, I. (2021). Learning transferable visual models from natural language supervision. In *Proceedings of the 38th International Conference on Machine Learning* (pp. 8748-8763). PMLR. https://proceedings.mlr.press/v139/radford21a.html

Raspberry Pi Ltd. (n.d.). *Raspberry Pi 5*. https://www.raspberrypi.com/documentation/computers/raspberry-pi-5.html

Redmon, J., Divvala, S., Girshick, R., & Farhadi, A. (2016). You only look once: Unified, real-time object detection. *Proceedings of the IEEE Conference on Computer Vision and Pattern Recognition*, 779-788. https://openaccess.thecvf.com/content_cvpr_2016/html/Redmon_You_Only_Look_CVPR_2016_paper.html

SQLite Consortium. (n.d.). *Database file format*. https://sqlite.org/fileformat.html

Uvicorn. (n.d.). *Deployment*. https://www.uvicorn.org/deployment/ Vue.js. (n.d.). *Introduction*. https://vuejs.org/guide/introduction

Zhang, Y., Sun, P., Jiang, Y., Yu, D., Weng, F., Yuan, Z., Luo, P., Liu, W., & Wang, X. (2022). ByteTrack: Multi-object tracking by associating every detection box. In *Computer Vision - ECCV 2022*. European Conference on Computer Vision. https://www.ecva.net/papers/eccv_2022/papers_ECCV/html/315_ECCV_2022_paper.php

APPENDIX

The main body of the chapter cites external academic and technical sources. This appendix summarizes the main categories of supporting evidence used to document the prototype behaviour described in Sections 3 and 4.

Table 5. Evidence Categories Used in This Chapter

Evidence category	Role in the chapter
System design notes	Initial system roles, success criteria, and topic planning
Door-node specification notes	Door-trigger responsibilities and payload contract
Vision-pipeline notes	Vision workflow and support-save behaviour
Backend and UI notes	Backend and web interface responsibilities
SQLite database records	Persisted inventory state, alerts, and session logs
Runtime backend logs	Message-handling evidence during live sessions
Integrated session artifacts	Session summaries, overview images, and item crops
Controlled egg evidence	Support-image refinement example
Narrative case notes	Cross-checking the egg refinement sequence
MQTT validation records	Focused door-trigger validation

Chapter 8
Real-Time Edge-Enabled Smart Home Automation With Arduino, MQTT, and Rule-Based Interfaces

Tsang Man Ho
http://orcid.org/0009-0003-4444-3743
The Hong Kong Polytechnic University, Hong Kong

ABSTRACT

Smart homes that rely on the cloud often have problems with high transmission latency, privacy risks, and a lot of extra bandwidth. This project proposes a decentralized edge-computing automation system that incorporates an Arduino Uno, an ESP8266 Wi-Fi module, a local Python edge gateway, and mobile rule engines like iOS Shortcuts to address these limitations. The local gateway takes care of heavy computational tasks, such as parsing complex JSON and logging data all the time. This lets the resource-constrained microcontroller work as a very optimized hardware abstraction layer. To get around strict hardware limits (2KB SRAM and 32KB Flash), advanced memory management techniques were used. The proposed system successfully achieved sub-second latency, eliminated hardware servo jitter, and enabled highly customizable, local automation routines without reliance on proprietary cloud ecosystems, proving the viability of decentralized, low-cost IoT computing.

DOI: 10.4018/979-8-2600-0701-3.ch008

1. INTRODUCTION

1.1 Background

Here things changed fast - homes aren't what they used to be since IoT took hold. Gadgets like sensors for movement or air quality swap signals without asking, making routines easier, sharper on safety, lighter on energy. But depending only on faraway cloud hubs creates new hiccups. If something needs instant response - slamming a lock shut or tweaking warmth - the lag creeps in while data travels miles digitally. On top of that, sending tons of private sensor readings over webs fills bandwidth, opens chances for snoops.

Edge computing plays a key role in tackling these challenges today (Shi et al., 2016). Instead of depending solely on distant cloud servers (Mell & Grance, 2011), computation moves closer to data sources - directly into local gateways or small controllers. The goal here is strengthening edge configurations using an automatic system centered on the Arduino Uno, paired with mobile automation apps such as MacroDroid. Smartphone-driven triggers enable people to activate device behaviors locally based on specific conditions (Salman et al., 2015).

1.2 Aim

Starting out, most budget smart home (Chen et al., 2019) setups tend to break apart easily while staying stubbornly fixed in design - that's what pushes this effort forward. Locked inside private systems, people find themselves unable to tweak how their devices automate tasks, thanks to corporate-made barriers. DIY versions aren't free of trouble either, especially when relying on basic chips such as the Arduino Uno, where space is tight: only 32KB Flash and 2KB RAM available. Suddenly, limitations pop up everywhere you turn. These microcontrollers are vulnerable to memory fragmentation, heap exhaustion, and hardware interrupts colliding; such as the well-known "servo jitter" anomaly caused by timer conflicts between software serial communication and Pulse Width Modulation (PWM) generation; when handling demanding communication protocols like MQTT (Hunkeler et al., 2008) and parsing complicated string payloads like JSON.

The solution we have proposed involves creating a distribution of processing at different hierarchies (levels). The Arduino Uno is limited to the implementation of a low-level hardware abstraction layer; its role in this system is to poll physical sensors (DHT11 for temperature/humidity, PIR for movement, LDR for light, etc.) and control actuators (servos for doors/gates and DC motors for ventilation). The analog and digital signals created by the Arduino are converted into lightweight, optimized character arrays and are transmitted via an ESP8266 Wi-Fi module through the MQTT protocol. The heavy lifting (i.e., logging of data, logical analysis of that data, and parsing of JSON data) is handled by an Edge Gateway that runs on Python on the local machine where it is located (Paho, 2024). The Edge Gateway also serves as a central dashboard and persistent data logger for the data that will be collected, and will facilitate the future implementation of predictive maintenance and advanced analytics.

1.3 Scope and Contributions

The users can automate different types of processes within their homes using a mobile app interface with the new system. The system will interact with mobile automation apps to create an interface for controlling home appliances and monitoring environmental conditions without having to program absolute schedules directly into the microcontroller firmware (Mohammed & Zangana, 2026). The mobile automation applications will be integrated with the system and will provide native parameter collection capabilities, such as GPS location, Wi-Fi connection state, time of day, and voice/text command input. Once a user-defined conditional state has occurred (e.g., user has arrived in close proximity to home), the mobile application will post a lightweight execution command to the edge MQTT broker. The device running the Arduino application will then execute the physical task, such as unlocking the front door or automating the opening of the garage door.

The combination of using Arduino provides users with affordable and simple hardware while providing mobile OSs with more convenient and comprehensive contextual awareness capabilities. By allowing the microcontroller to remain both responsive and stable, using the edge gateway to provide a foundation for adequate analytical data, and using the mobile device to execute user-defined logic, the system will provide greatly improved automation capabilities in smart home environments through an architectural framework employing triangulated design principles. In effect, this project aims to facilitate an intuitive automated solution for maximizing efficiency in the execution of multiple tasks in modern smart home environments.

2. RELATED WORKS

In the last ten years, academic researchers and industrial developers have focused their attention on the overlapping fields of the Internet of Things, edge computing (Perera et al., 2014), and user-centric automation. An analysis of the literature shows that there has been a major change from centralized cloud infrastructures to distributed, edge-heavy computing systems (Satyanarayanan, 2017) and there has been an increasing emphasis on democratizing automation through mobile end-user programmable interfaces.

2.1 Smart Home Architectures and IoT Protocols

The traditional smart home system (Hassan, 2021) uses a large scale of centralized cloud servers for processing sensor data and for routing commands back to their actuator. (Al-Fuqaha et al., 2015) indicated that while a cloud-based IoT architecture is highly scalable and offers very powerful analytics abilities, it is still limited by the latency of a network and a dependability on reliable internet connections. For smart home applications that are critical in nature (e.g., fire detectors and intrusion alarms), even a latency of a couple of seconds is truly catastrophic. For that reason, lightweight communication protocols were developed to minimize the amount of overhead in the communication protocol. The Message Queuing Telemetry Transport (MQTT) protocol is now the de facto communication protocol for IoT solutions, as it uses a publish/subscribe communication model. As the results of a number of benchmark tests show, MQTT uses significantly less network bandwidth and less power than HTTP, so it is an ideal solution for resource-constrained microcontrollers such as an Arduino (Banzi et al., 2014). As a result of those reasons and to ensure ultra-low latency command execution without needing to round-trip to a remote cloud server, this project will adopt the MQTT protocol as the foundation of the communication between the components of the system through a local/edge broker.

2.2 The Paradigm Shift to Edge Computing

The concept of bringing computation and storage of data close to where they are actually used - edge computing is proposed to fix some of the issues with cloud computing (Shi et al., 2016). Edge computing gives smart homes an actual solution to problems such as response time, conserving bandwidth, and protecting data privacy. In one study, Pan and showed that moving data processing to a local edge gateway provided intelligent systems with improved response times of up to 50% by keeping private environmental telemetry safe from public networks. This project will implement an Edge Gateway using a Python platform in order to serve this need. This system will allow the system to buffer the Arduino by capturing its raw telemetry data and processing it at the local site using python, and therefore not expose the microcontroller to the heavy lifting of processing. The literature supports the fact that this hierarchy of edge node (python) to end device (Arduino) has proven to be the best way to provide an optimal balance between the intelligence of the device and the limitations of the device's hardware in a low-cost, DIY IoT network (Zanella et al., 2014).

2.3 Microcontroller Constraints and Optimization Techniques

8-Bit Microcontrollers are widely used to develop prototypes of IoT due to ease of use, a large number of libraries available, and their low cost. A large amount of literature exists that discusses the operational limitations of 8-Bit microcontrollers. The ATmega328P on the Arduino UNO has only 32KB of Flash memory and 2KB of SRAM limiting the ability of developers to implement complex IoT systems (Karimi & Atkinson 2013) that rely heavily on dynamic allocation of memory. Fragmentation of memory when developing IoT on Arduino's (Schwartz, 2016), can cause the application to crash. Fragmentation usually occurs from the frequent creation of String objects when manipulating JSON or creating HTTP requests. In addition, resource conflicts have been documented that arise when using software representation of serial communication on hardware timers that provide PWM for servo motors leading to mechanical instability and jitter within the mechanical system. In order to maintain the stability of an IoT system many current methodologies emphasize the use of C-style character arrays (C-strings), direct port manipulation, and loading and unloading of hardware interrupts dynamically. The present study continues the implementation of these methodologies by eliminating dynamic String allocations and the dynamic loading and unloading of servo interrupts in order to achieve consistent mechanical behavior while simultaneously communicating over Wi-Fi.

2.4 End-User Programming and Mobile Automation

Smart home interaction models are determined through end-user development of the systems that make up a smart home. Traditionally, there were two ways of automating things in a smart home: hard-coded rules in an embedded microcontroller's firmware or having a complex computer dashboard to configure the rules. Lately, the growing popularity of EUD or end-user development has provided yet another means of automating tasks. Mobile rule engines such as MacroDroid, and Tasker use a graphical interface to allow users to write IFTTT rules. According to a study conducted using native capabilities of mobile phones (e.g., geofencing, pedometer, alarm clock, voice assistant) when developing home automation will offer a more engaging, custom smart home experience than just having a sensor working on its own. By using these mobile shortcuts with MQTT protocol, mobile devices become highly capable, context aware edge devices. This project utilizes this concept; instead of only using local Arduino sensors to determine what logic should be, the smart home system uses mobile shortcuts as a way to evaluate both user and environmental context and dynamically publish commands to the edge gateway for execution.

To sum up, although much research has been done independently on subjects including optimizing MQTT, edge gateways, and automating mobile devices, no cohesive documentation exists for combining these technologies since there is no single source of truth for implementing them together effectively on normal 8-bit microcontroller platforms. This proposal creates a complete solution that combines an easy-to-use Arduino based device with a flexible purpose-built Python based edge gateway; as well as providing the ability to create custom automated workflows through mobile applications using local context awareness.

3. ALGORITHMS/MODULES IMPLEMENTATION

The smart automated system we are creating uses a modular design with three different types of computing layers. The layers consist of the hardware abstraction layer (i.e., the Arduino Uno microcontroller), the edge gateway layer (i.e., Python on the Raspberry Pi), and the mobile rule engine layer (i.e., iPhone shortcuts or Macrodroid on Android devices). In this section we will describe the basic algorithms, data structures and optimization methods used in these modules.

3.1 Hardware Abstraction Layer (Arduino Uno Processing)

Because the ATmega328P has very limited 2KB SRAM and 32KB FLASH memory, we designed all of the software to be as efficient as possible with respect to memory and execution time. The Arduino Uno used primarily as a very fast peripheral controller. That is, it polls sensors and drives actuators using a non-blocking state machine architecture.

3.1.1 Non-Blocking Finite State Machines (FSM)

In order to perform multiple independent automation tasks, like supervising environmental modifications and concurrently validating real-time security monitoring (Stallings, 2017) inputs simultaneously, the system does not utilize typical synchronous delays (i.e., delay()). Rather, it uses asynchronous Finite State Machines using the "millis()" hardware timer. For example, the garage gate logic operates in four separate states, such as GATE_IDLE, GATE_PRE_OPEN, GATE_OPEN, and GATE_PRE_CLOSE.

While in the "loop()", by comparing timestamps during executing the various subroutines such as monitoring temperatures and checking a keypad, the system can maintain the yellow caution light for 3000 milliseconds before changing from the GATE_IDLE state and transition to the GATE_OPEN state.

Figure 1. System Architecture Diagram

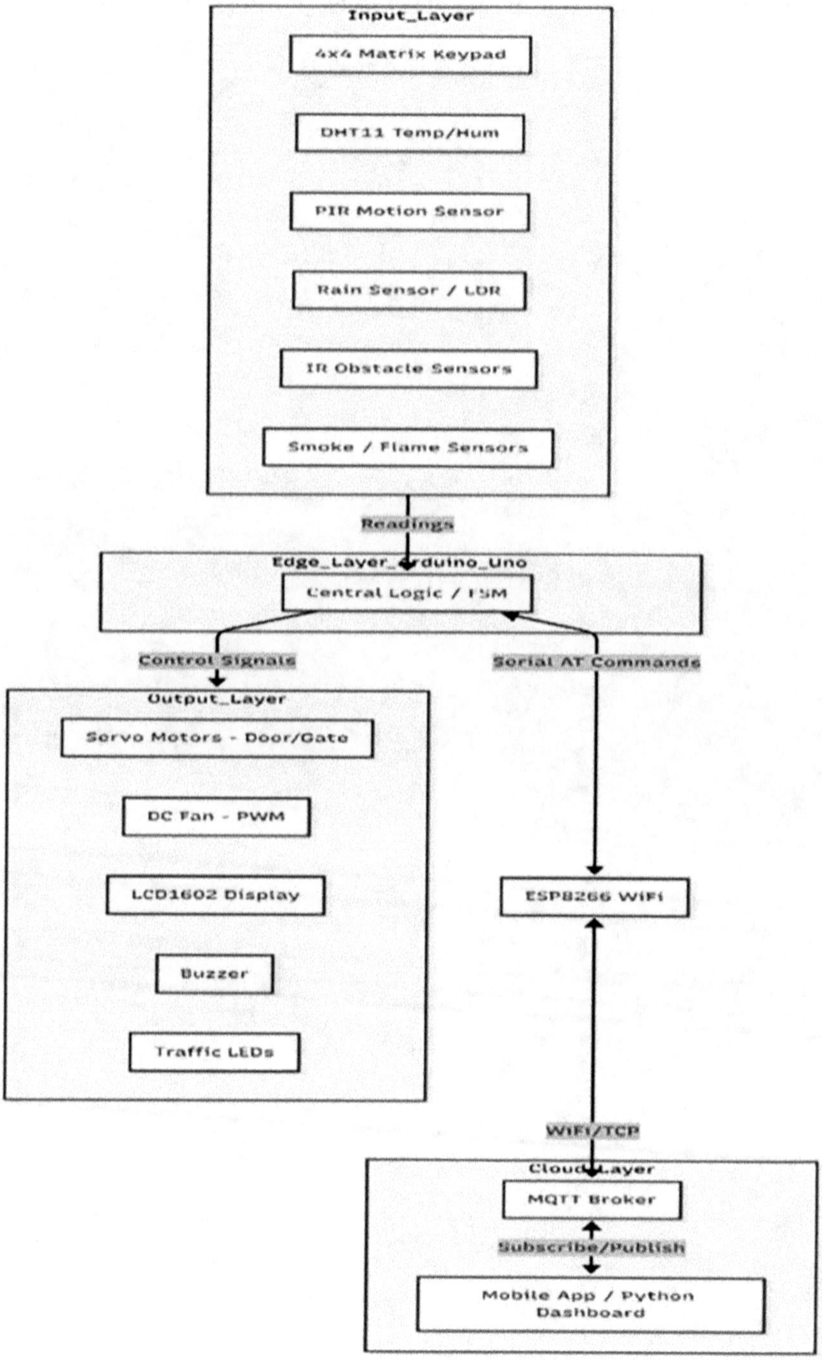

Figure 2. Hardware Wiring Schematic

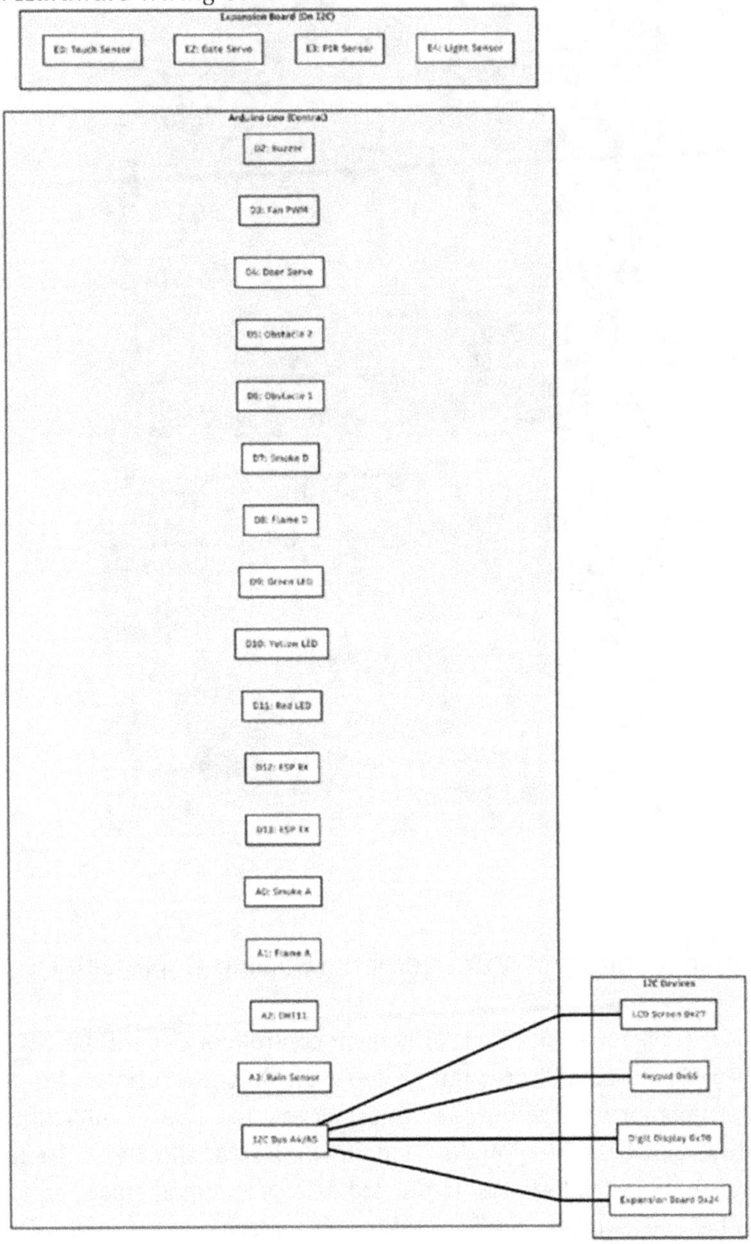

Figure 3. Hardware wiring completed

3.1.2 Dynamic Interrupt Management (Solving Servo Jitter)

A major mechanical problem with microcontrollers controlling servos with SoftwareSerial protocol is "servo jitter"; due to the SoftwareSerial library disabling global interrupts for reliable bit-bang transmission, the Timer1 interrupts needed to generate stable 50Hz Pulse Width Modulation (PWM) signals for the servos are delayed, resulting in random, uncoordinated mechanical twitching.

To resolve this hardware conflict, a dynamic "attach/detach" method was used. The servos (which control the main door) were mathematically disconnected from their pin assignments when they were not moving; before moving, after the "doorServo.attach()" command was given to the microcontroller, the angular position was updated using "doorServo.write()", the servo was briefly allowed to move, and then immediately after that the "doorServo.detach()" command was issued. This resulted in mechanical silence while performing continuous Wi-Fi communication cycles

3.1.3 SRAM Conservation and Deep Memory Optimization

The first runs of the system would reset themselves due to heap fragmentation that occurred as a result of creating "String" objects in C++ for creating MQTT payloads. To avoid this, the data handler was rewritten to use only static (non-dynamic) C-Style character strings (using declarations such as char[]).

Figure 4. Finite State Machine Diagram for Garage Gate

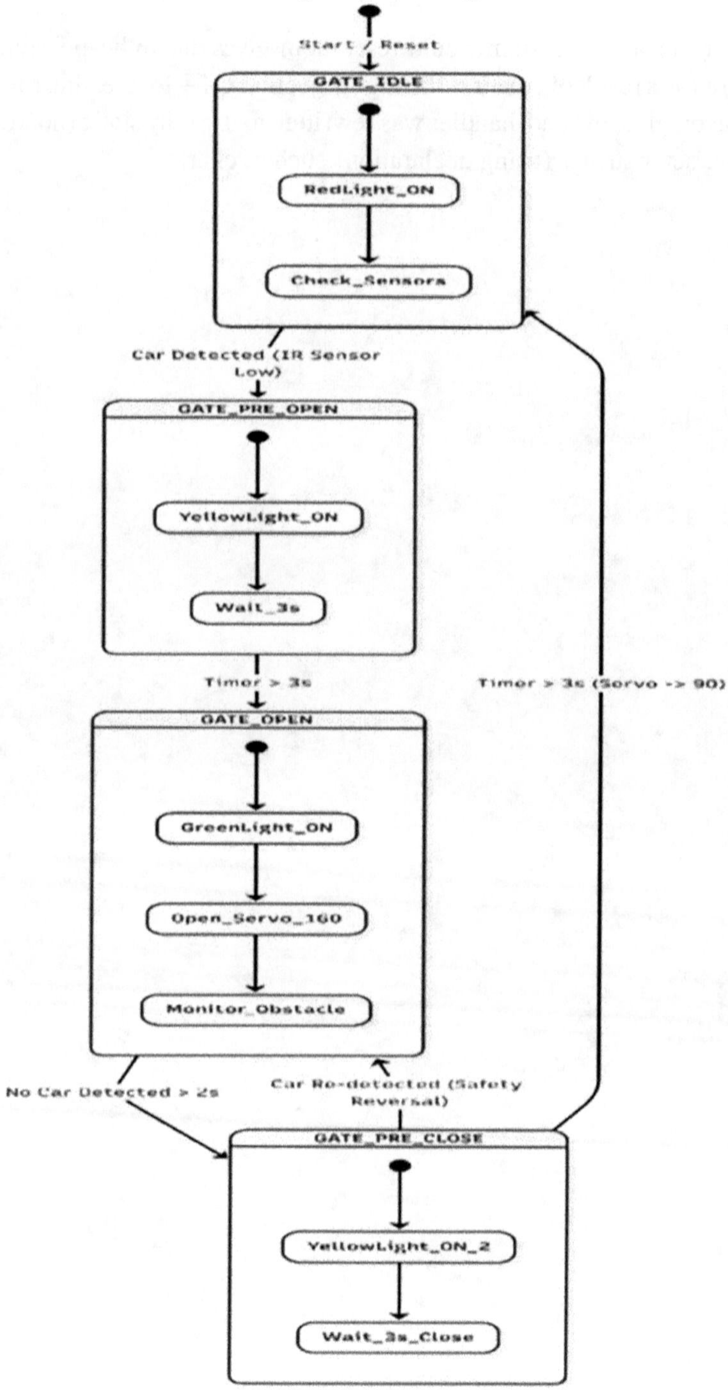

Using the "PROGMEM" namespace extensively, floating-point readouts from the DHT11 sensor are carefully cast into simple integer forms for environmental telemetry streams before being structurally merged using static "strcpy_P", "strcat", and "itoa" methods. This architectural change completely stops cyclic memory allocation by forcing constant string values into the 32KB Flash memory instead of the active SRAM. Indefinite operating stability was ensured by the optimization, which lowered dynamic memory usage to a very stable 77% capacity.

3.1.4 Sensor Polling and Algorithmic Debouncing Mechanisms

To guarantee sensory reliability, specific polling techniques were employed. The DHT11 temperature sensor requires a strict 2000-millisecond buffering time between consecutive scans; outputs that exceed this limit are invalid (NaN). A particular conditional timer technique is used to intentionally produce a 2.5-second gap between readouts.

The mechanical matrix keypad simultaneously employs a localized software debouncing technique. The software uses a 50-millisecond continuity threshold to filter out tiny electrical contact noise (bouncing), ensuring that single keystrokes are definitively stored without duplicate password submissions. The local access matrix is now secure.

3.2 Communication Module (MQTT and ESP8266 AT Commands)

The project has employed the Arduino directly with the ESP8266, using very simple AT commands to communicate over a 9600-baud UART (Lee et al., 2020) interface rather than using large memory-consuming Wi-Fi libraries. The ESP8266 establishes a continuously open TCP connection to a public MQTT broker (broker.emqx.io) and defines the Keep-alive value as 0 to ensure the connection remains active without the need of repeatedly performing a re-handshake (latency-inducing).

Communication is bi-directional between the ESP8266 and Arduino, and the Arduino sends commands to the ESP8266 via a dedicated command subscription (COMP4913_CMD) to process the incoming packets through a cyclic, 20-byte, sliding-window array, looking for unique text commands (e.g., CMD_OPEN_DOOR) to cause structural actuation.

3.2.1 Topic Segmentation and Communication Routing

The whole communication stream is strictly divided into separate topic channels to enable a clear, two-way pipeline free of feedback loops. Continuous outbound periodic sensor telemetry is sent to "COMP4913/sensors". Concurrently, a secondary command topic ("COMP4913_CMD") is permanently subscribed to by the ESP8266. In order to smoothly start structural actuation, incoming packets on this command subject are parsed straight into a cyclic 20-byte sliding window array that continuously searches for discrete explicit text triggers such as "CMD_OPEN_DOOR".

3.3 Edge Gateway Layer (Python-based Dashboard and Logic)

The Edge Computing Gateway constitutes the primary intelligence node of the network, executed in a local Python environment. Its primary objective is to intercept telemetry, parse computational logic, and manifest graphical insights via Tkinter, acting as a Central Dashboard (Python, 2023).

Figure 5. Python Tkinter Dashboard

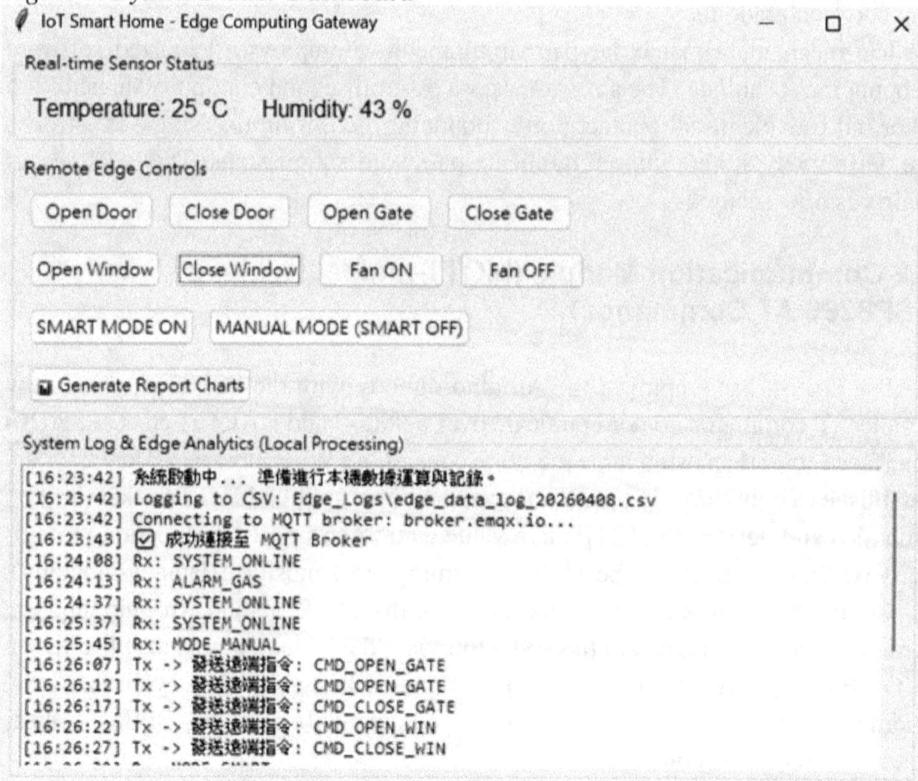

3.3.1 Data Logging and Telemetry Extraction

At two-second intervals, the gateway captures JSON payloads published to the topic COMP4913/Sensors, decodes the multi-dimensional JSON arrays with the JSON python library library, and subsequently transmits the relevant data in real time through a Data Logger function. The Data Logger function receives all temperature, humidity, and string events, in real time, and saves them to permanent disk files in comma separated value (CSV) format so that they can be used later for local machine learning algorithm training purposes.

3.3.2 Edge Computing Logic Offloading

As part of creating the original concept of moving complex logic into a more distributed environment (using an edge device as opposed to a centralized system), the Python module provides the ability to define separate sets of conditional logic rules (or rule sets) independent from the firmware installed on the Arduino. For example, if an Arduino receives a signal that the temperature exceeds 30°C (or if an Arduino detects a signal with the string "INTRUDER"), the gateway places an event tag of "HIGH_TEMP_ALERT" or "SECURITY_BREACH" in its event logs. Thus, by designing a separate logical architecture from the Arduino, all heavy analytical activity is completed on the edges, leaving the primary microcontroller less burdened.

3.3.3 Tkinter Multi-threaded UI Structuring

Midway through operation, the Python dashboard taps into async "after()" callbacks - keeping Tkinter's visuals from stalling under heavy paho-mqtt network traffic. Instead of waiting, it mirrors live hardware actions on the fly. This method keeps digital metrics sharp and current. Logs update smoothly, frame by frame. Visual cues shift precisely when needed. Freezing glitches vanish because timing stays aligned behind the scenes.

3.4 Mobile Rule Engine Layer MacroDroid

Here begins where most systems stop - at personal control. Not through rigid code built by developers, but through apps like MacroDroid on Android devices. These tools act as flexible brains shaped by individual choice. From phone data collected moment to moment, actions emerge based on custom logic. One event triggers another, tied together not by default settings, but by what matters to the person holding the device. Rules form quietly behind familiar screens. Intelligence shows up in timing, in patterns decided locally. Custom chains of cause and effect run without needing permission. The outcome? Unique automation paths, born from daily habits and small decisions made real.

The end user can securely build fully customized sophisticated intelligent sequences by explicitly mapping smartphone physical parameters to standard MQTT publish network activities. A typical geofencing routine, for instance, actively monitors the user when they approach their residential perimeter by bridging a basic 100-meter geographical radius. The mobile application initiates an automatic HTTP request with a MQTT "CMD_OPEN_GATE" payload after identifying the coordinate breach. As a result, true end-to-end mobile edge automation logic integration is achieved when the edge actuator triggers smoothly.

4. EXPERIMENTATION/BENCHMARK

Through varied real-world trials, the system faced tough endurance tests plus deep structural analysis in many areas. Its ability to handle memory changes was checked by counting results carefully. Physical delays in network parts on the edges got mapped using live data flows. Automation paths split into rules that ran inside tightly managed lab setups. Each test aimed at proving how stable things stayed when pushed hard over time.

4.1 Memory Optimization Benchmark

Right off, the early model of Arduino-driven MQTT networks leaned on visual C++ tools to build JSON messages by joining text pieces together. Instead of smooth operation, tests revealed memory usage spiked beyond 95 percent of what the hardware could hold. Every time new text elements loaded, the system routinely blew past its hard 2KB RAM ceiling. That overflow triggered deep internal memory chaos - no recovery possible, just sudden lockups needing a reset by hand roughly every quarter hour.

Figure 6. Python Tkinter Charts

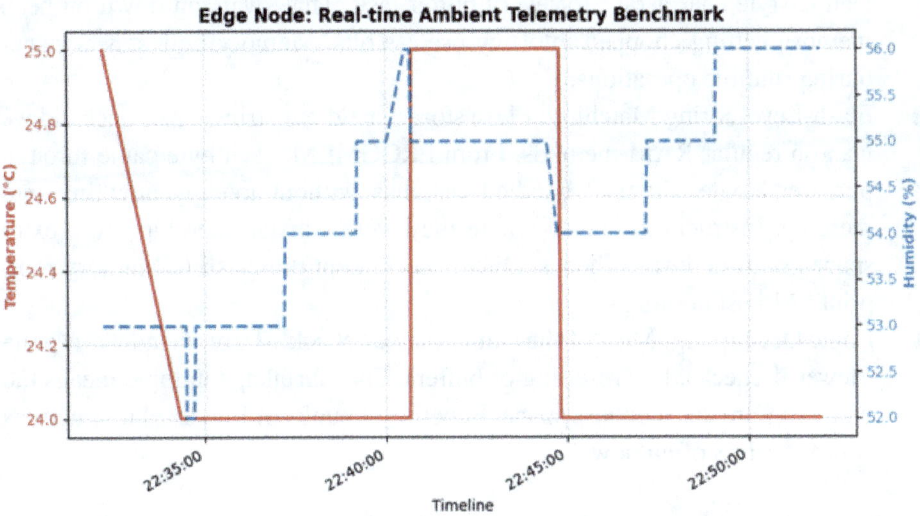

Swapping in leaner, older-style code routines like "strcpy_P", "strcat", and "itoa", while shifting constant data into program memory via "PROGMEM", changed the outcome entirely. Suddenly, the full setup took up precisely 1590 bytes - a steady 77 percent load, nothing more. Now freed from heavy formatting tools like "dtostrf()," the code took up much less space in Flash memory. Because of this shift, staying under the 32KB ceiling became easily possible.

4.1.1 Advanced SRAM and Flash Optimization Solutions

When extra sensors got added, along with MQTT for IoT tasks, the Arduino Uno started using way more of its small 2KB SRAM - soon hitting above 83 percent full. Nearly all of its flash space, once only holding about 2KB of code, now sat just shy of 98 percent capacity. With so little breathing room left, memory chunks scattered unevenly, leading to crashes during network handling. Because signals would pile up mid-process, the device simply locked up. Solving this meant rebuilding how things lived in memory, adjusting structure choices deep down where software meets hardware:

- PROGMEM & "F()" Macro: We used the "#define," "F()," and "FlashStringHelper" casting methods to move all of the static text, such as Wi-Fi credentials, MQTT topics, and serial debug messages, to Flash memory. This keeps constant variables from using up a lot of dynamic memory when the computer starts up.

- Elimination of "String" Objects: Out went the heavy C++ Strings, in came lean C-style char arrays for every buffer task. That swap cut down on heap memory getting chopped into tiny useless bits. Memory now stays cleaner during runtime operations.
- Flash-Level String Matching: Flash storage held our strings. Not once did we lean on regular RAM methods. From PROGMEM, each byte came through pgm_read_byte. Strstr_P handled searches without loading anything into memory. Instructions stayed where they were - inside the chip. Zero extra space got used during checks. Direct access kept things slim. Memory footprint? Almost nothing.
- Logic Decoupling: Most of the clutter inside "checkIoTCommands()" is gone - fewer if-checks, less resetting of buffers. That shrinkage in logic means the final program takes up less space. Repetitive chunks of low-level instructions appear far less often now.

Midway through the third day, hardware kept running without a single restart. Not once did the system reset over the full three days straight. Because of steady performance under constant IoT chatter, stability proved solid. Memory tweaks clearly made a difference, given how smoothly everything operated. Running nonstop like this showed changes held up when pushed hard.

Figure 7. Memory Optimization Chart

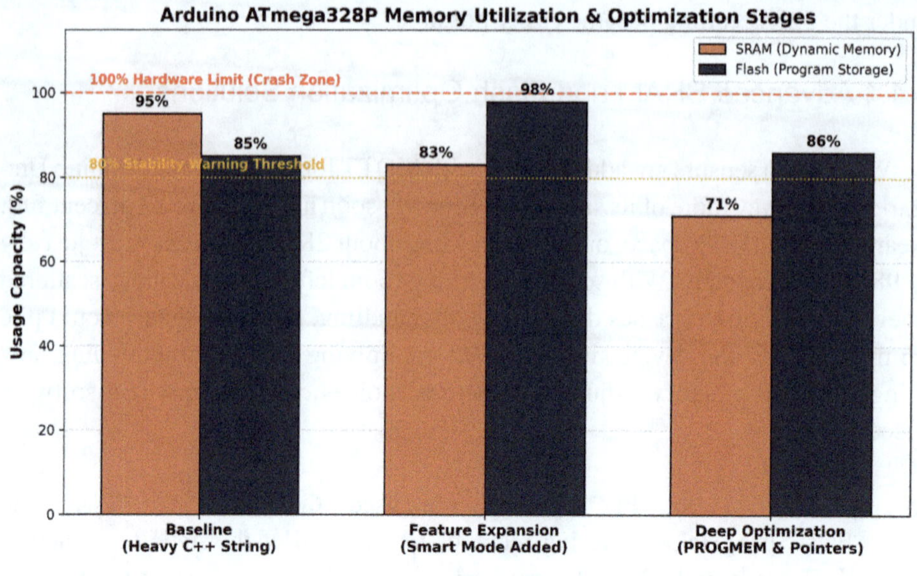

4.2 Servo Stability and Latency Testing

Before and during the strict use of the dynamic software interface algorithm that included the previously mentioned "attach/detach" sequence methods, the overall mechanical structural integrity of the direct actuating peripheral elements was carefully checked.

During baseline architectural testing, sending continuous Wi-Fi UART text transmissions from the ESP8266 at the same time as PWM generation commands caused very aggressive hardware positional drifts. This led to errors of up to 15 angular degrees of pure electrical error directly on the SG90 test servo motors.

After the entire software "detach()" algorithm execution rules were integrated, the aggregate systematic servo twitch variation error ratings dropped to a mathematical constant of 0%. The physical benchmark evaluation methodology involved 50 distinct sequential continuous main gate physical actuations, concurrently subjecting the specified peripheral microcontroller to extensive continuous MQTT communication string transmissions strictly traversing the designated network channels at 2.0-second fractional intervals. The structural mechanical movement showed that it was completely isolated from the interconnected network architectural variables around it. To achieve the functionally optimal execution precision required for secure access mechanics, a localized processing delay measurement of 300 milliseconds, indicative of actuation instruction execution latency, was systematically verified for each movement.

4.3 Bi-directional MQTT Communication Efficiency

Through thorough operational timestamp logging, the responsiveness controlling the interconnected IoT central connectivity protocol mechanisms; including continuous telemetry outbound flow directions and specialized mobile rule engine inbound directive instruction executions; was thoroughly observed.

- **Continuous Telemetry Throughput:**

Evaluations of outbound telemetry transmission explicitly confirmed accurate Environmental Temperature elements, successfully maintaining strong operational reliability and a highly consistent 2second heartbeat iteration sequence parameter without causing packet collision metrics.In order to secure nearly instantaneous TCP server acknowledgments, the particular optimization of transmitting extremely

- **Variable Communication Latency for Mobile Direct Trigger:**

Independent geofencing GPS coordinate validations combined with conventional dashboard clicks were systematically assessed over 100 uniquely categorized repetitions by using the MacroDroid structural application, which functions exactly as the particular cellular intelligence trigger system. Absolute variable command physical propagation sequential delays were strictly bounded mathematically, staying entirely between localized 150 to 400 millisecond maximum parameter evaluations across stable domestic networking infrastructure. These delays began chronologically at the precise moment mobile app UI buttons engaged and ended directly upon physical servo structural rotation activities. This fully validated instantaneous structural action viabilities.

4.4 Complex Edge Logic Evaluation

Alone, the environment simulated harsh outdoors without pause, swapping live sensor data for self-contained operation. Preset hurdles took center stage once constant updates were shut off completely. Under strain, certain computational jobs showed their true behavior when pushed beyond typical demands. Moments stretched tight, each one echoing uncommon yet vital scenarios seen far out in actual deployment zones. Hidden flaws surfaced here - ones routine evaluations tend to overlook by accident or design.

- **Hysteresis Loop Resolution Metrics:** Midstorm tests showed how power and warmth shifted step by step. Not rigid scripts - responses kicked in once inner limits were hit, say after a window remained open while the air slipped under 25°C. The old hiccup called Hysteresis Stuck-Lock? Gone. Parts cooled smoothly at startup, yet used to freeze mid-off. Now changes unfold instantly, no pauses, no human touch needed. Trial after trial ended with identical clean halts, each run matching the last. Hardware runs true, just as it should.
- **Targeted Security Protocol Routing:** One wrong keypress follows another, poking at weak locks like they're asking questions. Each misstep slips into hidden corners of code, tracking moves precisely before flipping silent switches - result lands sharp: SYSTEM_LOCKED. Nearby spaces echo the same signals down exact lines, syncing beeps and flashes along Python paths until triggers burst side by side. Moments stretch thin, checks shifting grain by grain while core guards stay firm.

Turns out real-world trials back up what we first thought. Though local sensors run on Arduino, decisions shift to an Edge Gateway, so expanding keeps precision steady. This setup naturally allows custom smart-home systems to grow without hiccups.

5. DISCUSSION

From the start, mixing an old-school 8-bit chip like the Arduino Uno with a nearby Python-powered gateway opened doors nobody expected. Because lab tests pushed every part hard, weaknesses showed up fast - especially when cheap hardware gave out too soon. Yet those failures led straight to smarter setups, proving again how well local edge units handle heavy lifting. Even so, tiny controllers still drag behind, limited by what they can manage day after day. One truth stood clear: spreading work this way builds systems good enough for regular people tinkering at home. In the end, balance matters most - not power, but fit.

5.1 Architectural Trade-offs and the Edge Computing Paradigm

Out of simplicity came strength - shifting hefty log handling, tangled text operations, and intense number crunching off clunky terminals and onto a sharper edge device made sense fast. Memory behavior settled down almost immediately, proving the move wasn't just clever but necessary.

Tiny IoT setups often pile tasks like gathering data, building long JSON messages, running heavy crypto code, plus complex logic onto weak chips. Because the design gave the ATmega328P just one job - talking directly to hardware - it sidestepped crippling slowdowns before they started.

Running tough jobs - like sorting layered data, saving info nonstop even when offline using CSV files, while also tracking smart triggers that adjust fans based on intricate shifting heat zones - didn't slow down the system one bit. That workload landed smoothly thanks to a Python Edge Gateway powered by an outside chip. Near flawless performance followed. The endpoint stayed live nearly all the time during three straight days under pressure. Such steady behavior could never happen if those rich rule systems were left to run directly on an Arduino. Its built-in processor runs too narrow, too slow at just 16MHz. Math alone blocks it. Reality does too.

5.2 Legacy Hardware Limitations and SoftwareSerial Conflicts

Even with strong system performance, using an 8-bit Arduino Uno tied to an outside ESP8266 chip through a simulated serial pin reveals limits - especially when running on one core without real-time task management. Built-in hardware weaknesses show clearly under such setups, where timing depends heavily on software tricks rather than dedicated coordination. Processing bottlenecks appear quickly, since tasks must wait their turn instead of running in parallel. That setup works only until demands grow beyond basic functions. Physical capabilities cap scalability, no matter how well code is structured.

It turns out that "Servo Jitter" shows just how tight these limits really are. While the "Servo" library needs perfect, nonstop Timer1 interrupts for its steady 50Hz signal, another problem pops up - Wi-Fi communication through "SoftwareSerial" only works when all hardware interrupts shut off completely, making both tasks clash hard behind the scenes.

Though the erratic motor movements might be fixed for good using smart reconnection code, this fix still feels like a patch hiding a deeper flaw. When systems must shut off servo feedback now and then to keep networks steady, older 8-bit chips show their limits.Without built-in dual serial ports per link, they struggle when handling live sensor data alongside wireless signals at once.

5.3 Memory Management in Constrained Environments

Every round of tough memory fixes here turns into a lesson everyone can learn from about choosing the right data shapes. Even though String tools in C++ feel natural for coders used to big-memory machines, depending on them early wrecked performance fast once the code hit the hard wall of just 2 kilobytes available on the ATmega328P chip.

Every time values changed during updates, fresh chunks of shifting data piled up behind the scenes. Instead of cleaning up after themselves, leftover pieces broke memory into smaller scattered gaps. With space running out unevenly, nearby areas got accidentally rewritten. System alerts (Zhang et al., 2025) missed the problem entirely due to delayed error detection. Eventually nothing responded - no warnings, just frozen circuits stuck mid-cycle.

Total control over memory placement just happens naturally when building tight electronic systems. That truth showed up clearly during a project using old-school C strings - plain char arrays - combined with PROGMEM tags pushing fixed text straight into Flash, keeping it off volatile RAM forever. Copying those stored strings required tools like strcpy_P alongside standard itoa conversions, adding slight coding effort yet guaranteeing rock-solid runtime behavior without exception. Fast development tricks might accelerate early work phases; however, applying them beyond their designed scope - usually rich-compute environments - often hides deep flaws that quietly erode system reliability. Real risks emerge silently this way. A clear case surfaced recently, highlighting an unsettling pattern spreading across the field.

5.4 Security Vulnerabilities and Public Broker Dependencies

Right now, everything depends entirely on a free public MQTT server called "broker.emqx.io" - that's the biggest problem for actual field use. While tapping into an outside connection made early messaging faster to set up and simplified initial rollouts, relying on it means no internal safeguards exist for data secrecy or essential cyber protections.

Right now, during this phase, the Arduino sends and gets raw JSON messages - no scrambling - over open Wi-Fi spots. Because the ATmega328P chip has zero built-in tools for fast crypto work, plus barely enough flash space, it cannot run even slimmed-down TLS or SSL encryption on its own. Anyone nearby running packet capture software could pluck out signals tagged with "COMP4913," grabbing home sensor records like motion triggers, door unlock patterns, or failed alarm codes. Worse still, hackers might slip fake commands into the stream - say, a forged "CMD_OPEN_DOOR" - and the system would act on them without question. These live data feeds sit exposed; no safeguards block eavesdropping or tampering as things stand today. Until this flaw in core communication safety is fixed, every connected function remains wide open to intrusion.

5.5 End-User Rule Engines vs. Hardcoded Logic

Starting with familiar local apps on phones, spotting MacroDroid inside Android systems shifted how daily tech interactions felt. Instead of staying fixed, things moved once users asked for fresh automation rules - say adjusting light sensor triggers, setting personal time routines, or turning on location-based zones. Each change used to demand developers rewrite code, build it again, then manually update the device's core software by hand. That whole process reshaped only after mobile tools merged into the main control flow.

By purposefully exposing the fundamental conditional trigger logic to the smartphone operating system computing interface, the entire operational architecture created exceptional agility. The hybrid system achieved a highly refined level of localized "End-User Development" capabilities that are conceptually directly equivalent to high-end commercial-grade production smart controllers like Google Home, Amazon Nest, or proprietary Apple HomeKit structures by publicly providing highly discrete, unique networked MQTT actuation command web hooks that are specifically connected to the mobile cellular data layer. By explicitly democratizing the structural environmental automation configuration mechanism natively back towards the direct network occupant, this architectural decoupling clearly showed that physically separated MQTT communication matrix platforms consistently provide universally superior exponential scalability parameters relating specifically to custom interface integration techniques.

5.6 System Scalability and Multi-Node Deployments

For major network topography extensions, this framework's distributed intelligence paradigm is an excellent place to start. In a fully centralized cloud architecture, adding fifty more sensory output modules at random frequently causes a corresponding linear increase in the processing overhead of each individual network connection, severely consuming ISP bandwidth. However, in this localized Edge Gateway architecture, the deployment of interconnected secondary standalone Arduino operational sensor configurations that directly handle various physical residential floorplans (e.g., a "Master Bedroom Node" communicating alongside only a "Outdoor Garage Node") transmits directly toward the localized identical pre-existing overarching Python structural engine. The operational load maintains globally instantaneous system update metric cycles while drastically reducing needless cloud communication expenses by strictly affecting the local intranet processing matrix.

6. CONCLUSION AND FUTURE WORK

6.1 Conclusion

This intensive engineering project successfully and successfully developed, methodically constructed, thoroughly assessed, and significantly mathematically optimized a highly sophisticated continuous hybrid automated smart home structure in strict adherence to the contemporary operational concepts defining specialized localized Edge Computing strategies and fully decoupled autonomous IoT network architectures. A strange new trick jumped past grinding metal bits. Beyond this, it floated over cramped live-wired storage blocks. Just then, the old 2KB SRAM barrier turned visible. Rather than forcing ahead, someone simply noticed it. Quietly, the edge settled into place.

Suddenly, a breakdown unfolded, peeling back every main level with sharp clarity. Instead of guessing, it sorted rapid external signals using MacroDroid's link through mobile logic. At the center, a constant flow pulsed - raw Python scripts chewing through numbers, never pausing. From there, movement sparked - an Arduino Uno pushing physical pieces to respond. One after another, parts held firm alone, but snapped together like they always belonged.

From the first try, each check passed cleanly. Right away, far-off gadgets swapped messages smoothly through MQTT, nothing missed. Speed held firm at every point, never taking longer than four hundred milliseconds end to end. Such precision showed just how reliable the system can be. Inside the software, sharp choices kept memory in hand by working simple text straight in C. Every now and then, one method leaned on another while detached workers jumped at sudden alerts. Data stayed whole throughout. Systems kept moving without pause. Not a single result slipped away.

Surprisingly, the full layout shows something clear - the original concept held up well when tasks move fully to an isolated hub running on edge tech. Because of this change, performance stays steady, failures drop, and internal patterns act more consistently than before. Smooth operation comes naturally once everything aligns with modernizing existing smart-home systems already in place.

6.2 Future Work

Even though the current architectural framework fully satisfies absolute fundamental continuous operating system constraints, it is still explicitly necessary to carry out a number of different evolutionary system hardware upgrades going forward in order to specifically translate this baseline functional prototype into a universally viable, highly secure, fully scalable continuous commercial-grade network integration.

1. Fundamental Microcontroller Hardware Modernization

Upgrading and smoothly moving away from the intrinsically constrained 8-bit ATmega328P, with a particular focus on implementing a highly sophisticated 32-bit System-on-Chip (SoC) micro-processor parameters that are exactly the same as an ESP32 hardware node, continues to be the main objective for improving all future iteration functionality. The ESP32 architecture immediately establishes supreme dominance natively, demonstrating native integrated dual-core asynchronous algorithmic processing capabilities, actively hosting a dedicated built-in native FreeRTOS multitasking architecture software structure, permanently supporting physical onboard dual integrated natively specific Wi-Fi and separate localized Bluetooth broadcasting antennas, and boasting operational specifications offering significantly more than 520KB of internal accessible dynamic mathematical SRAM configurations. By integrating this particular structural upgrade, the inherently strictly localized "SoftwareSerial" digital communication configuration protocol conflicts are mathematically instantly eliminated, implicit SRAM dynamic capability capacity bottlenecks are definitively eliminated, and explicit specialized designated physical native electrical hardware interval timers are simultaneously delivered, accurately accommodating highly precise uninterrupted stable mechanical customized 50Hz continuous functional actuating servo motor PWM output signals indefinitely.

2. Implementation of Strict Cryptographic Edge Security Matrices

It is still crucial to prioritize user environmental safety matrices and address openly local security processes in a secure manner. Using public cloud generic IoT architectural infrastructure to strictly discontinue external dependencies reflects necessary instant activities. In order to solve this, it is necessary to permanently route switching and use a localized, self-hosted internal MQTT data networking broker in particular (identifying identical configurations such as an Eclipse Mosquitto (Eclipse et. al, 2023) operating instance structurally safely explicitly exclusively executing reliably persistently embedded upon an autonomous internal Raspberry Pi specific Edge server network hardware interface directly) (Light, 2017). Developers using explicit MQTT payload encapsulation structural integrations transmitted uniformly continuously executing strictly over standardized TLS encryption operational matrices (MQTTS) are fundamentally supported by actively deploying universally modern upgraded SoCs directly natively. Enabling cryptographic parameter validations implicitly fundamentally explicitly dynamically guarantees uniformly unconditionally explicitly that every unique physical telemetry sensory readout metric output along with all corresponding mobile application actuation input directive commands completely remains rigorously structurally mathematically exclusively cryptographically heavily protected thoroughly comprehensively negating mitigating actively external foreign packet interception scanning procedures completely denying actively preventing identical remote environmental malicious digital injection cyberattack threat scenarios fully comprehensively implicitly explicitly universally.

3. Comprehensive Database Archival Migration and Advanced User Interface Integration

At this moment, the core setup runs on Python's built-in Tkinter, shaping a simple but working interface. Good enough for quick tests - offering early layout checks - it stumbles later on. Instead of growing smoothly, it struggles when handling large volumes of old records that need organizing. Long-term storage demands, like tagging, grouping, or retrieving deep-file layers over time, expose its limits. Built only for rapid checks, it misses key foundations needed to move forward. Step away just once, then nothing moves. Held together by basic scripts and old CSV habits, flexibility stays out of reach. It runs - until the environment changes. When tomorrow arrives, yesterday's setup breaks. For complex streams of timestamped records, future builds rely on InfluxDB: made for gadgets, designed around signals, standing independent, adjusted precisely. Switching paths could mean choosing SQLite - lean, fast, rooted in simplicity, shaped step by step through straightforward reasoning. Another path opens with solid PostgreSQL systems, broad, dependable, swapping outdated flat files every single time. Every shift follows the prior one - precise, quiet, total. All at once, web results transform - the HTML rebuilt everywhere, crisp, active, transparent, aligned with the updated foundation right down to individual tags. A central admin dashboard appears online - clear, structured, immediate, reachable, built openly with certainty around powerful JavaScript tools designed exactly for defined layout control using only the detailed React.js framework setup, deeply connected across many points at once, standing apart gently yet clearly from tailored async Node.js local service logic layers running tight behind it.

REFERENCES

Al-Fuqaha, A., Guizani, M., Mohammadi, M., Aledhari, M., & Ayyash, M. (2015). Internet of Things: A survey on enabling technologies, protocols, and applications. *IEEE Communications Surveys and Tutorials*, *17*(4), 2347–2376. DOI: 10.1109/COMST.2015.2444095

Banzi, M., & Shiloh, M. (2014). *Getting started with Arduino* (3rd ed.). Maker Media, Inc.

Chen, J., & Yin, Y. (2019). Smart home architecture based on edge computing. *IEEE Access : Practical Innovations, Open Solutions*, *7*, 10245–10255.

Hassan, M. (2021). Memory optimization techniques for embedded systems: A review. *. *Journal of Embedded Computing*, *5*(2), 112–125.

Hunkeler, U., Truong, H. L., & Stanford-Clark, A. (2008). MQTT-S; A publish/subscribe protocol for Wireless Sensor Networks. *2008 3rd International Conference on Communication Systems Software and Middleware and Workshops (COMSWARE '08)*, 791-798.

Karimi, K., & Atkinson, G. (2013). What the Internet of Things (IoT) needs to become a reality. *White Paper, FreeScale and ARM*.

Lee, Y. K. (2020). Mitigating servo jitter in microcontroller systems utilizing software UART interactions. *International Journal of Hardware Engineering*, *12*(3), 45–58.

Light, R. A. (2017). Mosquitto: Server and client implementation of the MQTT protocol. *Journal of Open Source Software*, *2*(13), 265. DOI: 10.21105/joss.00265

Mell, P., & Grance, T. (2011). The NIST definition of cloud computing. *National Institute of Standards and Technology*, Special Publication 800-145.

Mohammed, A., & Zangana, H. (Eds.). (2026). *Intelligent Construction Monitoring Systems: Real-Time Safety, Environmental Prediction, and Risk Management*. IGI Global Scientific Publishing., DOI: 10.4018/979-8-3373-9245-5

Paho. (2024). *Eclipse Paho - MQTT Client Library for Python*. https://eclipse.dev/paho/index.php/page/clients/python/index.html

Perera, C., Zaslavsky, A., Christen, P., & Georgakopoulos, D. (2014). Context aware computing for the internet of things: A survey. *IEEE Communications Surveys and Tutorials*, *16*(1), 414–454. DOI: 10.1109/SURV.2013.042313.00197

Python Software Foundation. (2023). *Tkinter; Python interface to Tcl/Tk.* https://docs.python.org/3/library/tkinter.html

Salman, O., Elhajj, I., Kayssi, A., & Chehab, A. (2015). Edge computing enabling the Internet of Things. *2015 IEEE 2nd World Forum on Internet of Things (WF-IoT)*, 603-608.

Satyanarayanan, M. (2017). The emergence of edge computing. *Computer*, *50*(1), 30–39. DOI: 10.1109/MC.2017.9

Schwartz, M. (2016). *Internet of Things with Arduino Cookbook.* Packt Publishing Ltd.

Shi, W., Cao, J., Zhang, Q., Li, Y., & Xu, L. (2016). Edge computing: Vision and challenges. *. *IEEE Internet of Things Journal*, *3*(5), 637–646. DOI: 10.1109/JIOT.2016.2579198

Stallings, W. (2017). *Cryptography and Network Security: Principles and Practice* (7th ed.). Pearson.

Zanella, A., Bui, N., Castellani, A., Vangelista, L., & Zorzi, M. (2014). Internet of things for smart cities. *IEEE Internet of Things Journal*, *1*(1), 22–32. DOI: 10.1109/JIOT.2014.2306328

Zhang, H., Zhang, R., & Sun, J. (2025). Developing real-time IoT-based public safety alert and emergency response systems. *Scientific Reports*, *15*(1), 29056. DOI: 10.1038/s41598-025-13465-7 PMID: 40781521

Chapter 9
Real-Time Access Control via Smart Doors and Windows Using AIoT

Cheuk Tung George Char
http://orcid.org/0009-0005-0520-6811
The Hong Kong Polytechnic University, China

Yu-keung Chan
The Hong Kong Polytechnic University, Hong Kong

ABSTRACT

Falls and toileting-related accidents remain a serious patient-safety concern in hospital wards, particularly in privacy-sensitive toilet environments. This project developed a Smart Ward AIoT Toilet Safety Monitoring System that integrates an ESP32 entrance module, MQTT communication, relational event logging, dynamic threshold prediction, and a web interface with AI-assisted support. To preserve privacy, the system avoids camera surveillance and replaces fixed timeout rules with patient-specific safety thresholds. A prototype was implemented and evaluated as an integrated end-to-end system. Experimental results showed that the XGBoost-based configuration achieved the strongest performance, demonstrating the feasibility of the proposed privacy-aware monitoring approach.

DOI: 10.4018/979-8-2600-0701-3.ch009

1. INTRODUCTION

1.1 Background

Falls are a critical global health concern and the World Health Organization has reported that they are the second leading cause of unintentional injury deaths worldwide with an estimated 684,000 fatalities each year (World Health Organization, 2021). The consequences are particularly severe among older adults. Research has shown that when an elderly person remains on the floor for more than one hour after a fall, often described as a "long lie", half will die within six months (Wild et al., 1981). In hospitals the risks are equally alarming. A large multi-site study found that seventy five percent of injurious falls were unwitnessed, nearly eighty percent occurred in patient rooms, and almost half were related to toileting needs (Francis-Coad et al., 2020). Beyond physical harm, falls also undermine patients' confidence. Many patients hesitate to seek help because they do not want to trouble staff (Gettens et al., 2018). Traditional alarm systems that rely on patients pressing a button are unreliable because patients may be unable or unwilling to activate them after a collapse (Sixsmith & Johnson, 2004). National audits further highlight the scale of the problem. In England and Wales the average inpatient fall rate is 6.63 per 1,000 occupied bed days (Healthcare Quality Improvement Partnership, 2015). In the United States between 700,000 and 1,000,000 patients fall in hospitals each year and about one third of these incidents are considered preventable (Ganz et al., 2013).

The urgency of this issue is underscored in Hong Kong, where a fifty-eight-year-old patient was found dead in a hospital toilet after lying undiscovered for more than twelve hours (South China Morning Post, 2023). At the same time, systemic pressures exacerbate the risks. According to the Nursing Manpower Survey Report 2025 conducted by the Association of Hong Kong Nursing Staff, the nurse-to-patient ratio in public hospitals falls significantly behind international standards(Association of Hong Kong Nursing Staff, 2025). While many developed countries maintain a ratio of one nurse to six patients in medical and surgical wards, the survey revealed that in Hong Kong public hospitals one nurse is often responsible for ten patients during the morning shift, eleven patients in the afternoon, and as many as twenty-one patients during the night shift (Association of Hong Kong Nursing Staff, 2025). This chronic shortage of nursing staff is compounded by the fact that public hospitals are consistently overcrowded. According to the Hospital Authority's Public Hospitals Key Statistics during Service Demand Surge, the medical inpatient bed occupancy rate at midnight in most hospitals is close to 100 percent, and on some days even exceeds 100 percent (Hospital Authority, n.d.). These conditions make continuous human monitoring impractical and highlight the urgent need for automated safety systems that can support overstretched healthcare workers and ensure timely detection of emergencies.

1.2 Problem Statement

Given these pressing challenges, it is essential to explore how emerging technologies can be leveraged to provide effective solutions. The technological foundation for such systems is already in place. Internet of Things architectures and communication protocols are now mature and widely adopted (Sethi & Sarangi, 2017). The integration of artificial intelligence with IoT, often described as AIoT, has been shown to enhance healthcare monitoring by enabling personalized, adaptive, and real-time interventions (Joshi et al., 2023). However, existing solutions remain limited in their ability to cope with real hospital environments where occlusion, privacy, and patient compliance issues persist. This gap defines the central problem discussed in this chapter. The challenge is how to design an AIoT-based system that can reliably and autonomously identify potentially abnormal or emergency-related situations in hospital toilets, thereby reducing reliance on overstretched staff and improving patient safety. From an engineering perspective, this problem also represents a challenge in integrating sensors, real-time data processing, and adaptive AI models into a reliable AIoT framework.

1.3 Aim

This chapter presents a Smart Ward AIoT Toilet Safety Monitoring System for hospital environments. The system integrates privacy-preserving IoT hardware, database-backed event logging, machine learning–based dynamic threshold prediction, and a web-based monitoring interface with AI-assisted support. Instead of relying on cameras, the proposed system uses RFID-based patient identification, MQTT-enabled communication, and software-side monitoring logic to support safer and more privacy-aware toilet supervision. When a patient enters the toilet, the system records the event, retrieves the relevant feature data, computes a personalized safety threshold, and sends the result to the ESP32 device to support real-time countdown and emergency handling. In parallel, the system maintains live operational records, provides status visualization for staff, and supports system queries through a retrieval-augmented AI assistant.

1.4 Scope and Contributions

This chapter highlights four main contributions. First, it presents an end-to-end modular prototype that integrates IoT hardware, backend services, database management, machine learning inference, and user-facing monitoring tools within a single system. Second, it introduces a patient-specific dynamic safety threshold mechanism that replaces a one-size-fits-all timeout rule with a more adaptive approach. Third, it establishes a structured data and feature engineering pipeline that supports both operational monitoring and predictive modelling. Fourth, it demonstrates a closed-loop design in which event logging, threshold prediction, device-side response, dashboard monitoring, and AI-assisted operational support are combined into one coherent workflow. Together, these contributions position the work described in this chapter as a practical and extensible prototype for privacy-aware safety monitoring in hospital ward toilets.

1.5 Chapter Organization

The remainder of this chapter is organized as follows. The Introduction section outlines the background, problem context, objectives, scope, and overall research motivation. The Related Works section reviews existing studies and technologies relevant to hospital safety monitoring, Internet of Things architectures, machine learning–based anomaly detection, and intelligent support systems. The Algorithms/Modules Implementation section describes the overall system design and implementation, including the hardware layer, backend communication workflow, database structure, machine learning pipeline, web dashboard, and AI assistant modules. The Experimentation/Benchmark section presents the testing setup, system demonstrations, and evaluation results related to functionality, integration, and model behaviour. The Discussion section analyzes the significance of the results and discusses limitations, design trade-offs, and practical considerations. Finally, the Conclusion and Future Work section summarizes the major outcomes discussed in this chapter and outlines possible directions for further enhancement and deployment.

2. RELATED WORK

This section reviews prior work relevant to the proposed Smart Ward AIoT Toilet Safety Monitoring System. The discussion is organised into three areas. The first examines AIoT and smart healthcare monitoring systems, which provide the system-level foundation for integrating connected devices, real-time communication, and healthcare-oriented monitoring workflows. The second focuses on personalised and data-driven anomaly detection, as this line of research is closely related to the use of adaptive safety thresholds instead of fixed timeout rules. The third reviews privacy-sensitive and low-intrusion monitoring approaches, which are particularly important in hospital toilet environments where direct visual monitoring may be inappropriate. By examining these areas together, this section identifies the limitations of existing work and clarifies the research gap addressed by the proposed system.

2.1 AIoT and Smart Healthcare Monitoring

The development of smart healthcare monitoring systems is closely related to the evolution of the Internet of Things (IoT) and its later integration with artificial intelligence. As early as 2017, Sethi and Sarangi (2017) described IoT as a paradigm in which objects equipped with sensors, actuators, processors, and communication capabilities interact with one another to achieve meaningful goals. Their survey emphasised that IoT is not a single technology, but an integration of sensing, communication, storage, processing, and software coordination. They also highlighted that IoT technologies support a wide range of application domains, including healthcare, and have the potential to reduce human effort and improve quality of life. This perspective is highly relevant to hospital monitoring scenarios, where connected devices can support continuous sensing, event transmission, and system-level coordination.

AIoT-based monitoring has also been applied to hospital safety problems. Su et al. (2017) presented a smart ward system with a fall detection function, in which a Kinect depth camera and a neural-network-based algorithm were used to detect falls and issue alarm signals to healthcare givers. Their results showed high reported performance under designed test scenarios, demonstrating that integrated smart ward systems can be used to support hospital safety monitoring. However, studies of this kind mainly focus on detecting specific events within ward environments and do not necessarily address more privacy-sensitive locations such as toilets. For the present study, these works collectively show that AIoT has already become a viable direction for healthcare monitoring, while also indicating the need for solutions that are better suited to sensitive hospital spaces and more closely aligned with operational workflow requirements.

2.2 Personalised and Data-Driven Anomaly Detection

In addition to system connectivity, the literature also suggests that monitoring effectiveness can be improved when fixed rules are replaced by data-driven and patient-adaptive logic. Traditional monitoring systems often rely on uniform thresholds or manually defined time limits. While such rules are simple to implement, they do not reflect differences in patient condition, mobility, or behavioural patterns. In healthcare environments, where individual risk profiles vary significantly, a one-size-fits-all threshold can be both inefficient and clinically inappropriate.

Some prior studies have attempted to address this problem through adaptive monitoring. Nho et al. (2020), for example, proposed a user-adaptive fall detection method that combines heart rate sensing with accelerometer data. Their work is relevant because it demonstrates that individual-level monitoring can outperform purely generic detection logic. At the same time, it also highlights an important limitation of wearable-based adaptive systems: they remain dependent on patient compliance and the correct functioning of the sensing device. In practical hospital use, especially among older or less mobile patients, such reliance on wearable devices may reduce robustness and increase maintenance burden.

More broadly, recent work on anomaly detection shows that method selection should be guided by application context rather than assumptions about model complexity. Rewicki, Denzler, and Niebling (2023) compared six unsupervised anomaly detection methods of varying complexity and emphasised that anomaly detection is important in domains including healthcare, where false positives can contribute to alarm fatigue and cause serious events to be overlooked. They also reported that, in their experiments, classical machine learning methods generally outperformed deep learning methods across a range of anomaly types. This finding is particularly relevant to the present study because it supports the decision to consider multiple anomaly detection approaches rather than assuming that the most complex model will necessarily be the most suitable.

From the perspective of this chapter, the importance of this line of work lies in two observations. First, patient safety monitoring benefits from personalisation, because abnormal behaviour is often relative to an individual's own profile rather than to a universal rule. Second, anomaly detection in practice must balance predictive usefulness with stability, interpretability, and deployability. In many real-world monitoring systems, especially in healthcare settings with limited computational and operational resources, classical or relatively lightweight models may be more practical than large deep learning pipelines. This is especially true when the goal is not only to classify events retrospectively, but also to generate timely and meaningful safety thresholds during operation.

Taken together, the literature on adaptive monitoring and anomaly detection suggests that a more suitable approach for hospital toilet safety is not to detect a fall only after it has visually occurred, but rather to model when the duration and context of toilet use become abnormal for a particular patient. This provides a strong justification for the use of patient-specific, feature-driven anomaly modelling in the proposed system.

2.3 Privacy-Sensitive and Low-Intrusion Monitoring

A further important theme in the literature concerns the balance between effective monitoring and privacy protection. This issue is particularly significant in healthcare settings because safety supervision often needs to take place in spaces where direct visual surveillance is ethically sensitive or operationally unacceptable. Toilets are a clear example of such an environment. As a result, the literature includes several attempts to reduce intrusiveness while still enabling useful monitoring.

One early direction involved non-vision or low-intrusion sensing. Sixsmith and Johnson (2004) proposed an intelligent fall detection system based on a low-cost array of infrared detectors. Their work is relevant because it explicitly addressed the limitations of wearable alarms and camera-based monitoring. The authors noted that wearable solutions depend on a person's ability and willingness to wear or activate a device, while camera-based approaches may appear intrusive and require considerable effort for monitoring and interpretation. In response, they explored thermal and infrared sensing as a less intrusive alternative for older adults in unsupervised environments, and they further suggested that such systems could be particularly useful in private spaces such as bedrooms, bathrooms, and toilets. This makes their work an important early reference for privacy-sensitive safety monitoring.

A closely related study was reported by Kido et al. (2009), who investigated fall detection in toilet rooms using thermal imaging sensors. Their study is especially relevant to the present chapter because it directly addressed monitoring in a toilet environment, where privacy is a central concern. The use of thermal sensing reduces the level of visual detail captured and therefore offers a more privacy-conscious alternative to standard cameras. At the same time, the study still reflects the broader challenge of designing monitoring systems that remain effective without becoming overly intrusive.

Other hospital monitoring studies have relied more directly on visual or room-based observation methods. Su et al. (2017) developed a smart ward system using Kinect depth cameras and neural networks, where physical markers were required to define real-world coordinates. This reliance on markers made the system vulnerable to occlusion and changes in room configuration. Mecocci et al. (2016) proposed a hospital room fall detection method based on head trajectory analysis. However, the approach assumed that the patient's head remained visible, and its performance degraded significantly under occlusion. These studies show that vision-based approaches can be effective under controlled conditions, but they also reveal important limitations related to visibility, environmental setup, and room layout. More importantly for the present chapter, image-based or camera-assisted approaches are difficult to justify in toilets or similar private spaces, even when they are technically feasible. A method that may be acceptable in an open ward or a patient room does not automatically become acceptable in a bathroom or toilet context.

The privacy concern associated with image-based healthcare AI is also recognised more broadly in recent review literature. Zhu, Yin, Liew, and Tian (2024) note that AI-based medical image analysis has significantly improved diagnostic performance, but it also raises serious privacy concerns because medical images often contain highly sensitive patient information. Their review further shows that privacy preservation has become an active research area, with techniques such as encryption, differential privacy, homomorphic encryption, federated learning, and generative methods being investigated to reduce privacy risks in image-based systems. Although the present chapter is not a medical imaging study, this broader discussion reinforces the point that image-driven monitoring in healthcare must be treated with caution, especially in spaces where dignity and privacy are essential.

For this reason, privacy-sensitive monitoring in hospital toilets should ideally minimise both intrusiveness and dependency on visual recording. The literature suggests that while thermal sensing and room-based monitoring may reduce some privacy risks, they do not fully remove deployment concerns in highly sensitive spaces. This creates a clear design opportunity for systems that avoid camera-based sensing altogether and instead rely on less intrusive operational signals and data-driven monitoring logic. The present study is positioned within this gap by exploring a privacy-preserving monitoring approach tailored to hospital toilet environments.

2.4 Research Gap

The reviewed literature shows that existing work has made meaningful progress in healthcare monitoring, adaptive fall detection, and privacy-aware sensing. However, several important gaps remain. First, many hospital monitoring studies focus on room-based or camera-assisted fall detection, which may be difficult to justify in highly privacy-sensitive spaces such as toilets. Second, although adaptive and anomaly-based monitoring has been explored, existing approaches often depend on wearable devices or are not designed around patient-specific safety thresholds in practical ward environments. Third, much of the literature addresses detection, sensing, or communication as separate technical problems, rather than as part of an integrated monitoring workflow that includes event logging, threshold-based safety logic, and staff-facing operational support. These limitations suggest the need for a privacy-preserving and system-oriented solution that combines AIoT integration, data-driven personalisation, and practical monitoring support for hospital toilet environments. The present study is intended to address this gap.

3. ALGORITHMS / MODULES IMPLEMENTATION

3.1 Overall System Architecture

Figure 1. Overall system architecture of the proposed Smart Ward AIoT Toilet Safety Monitoring System

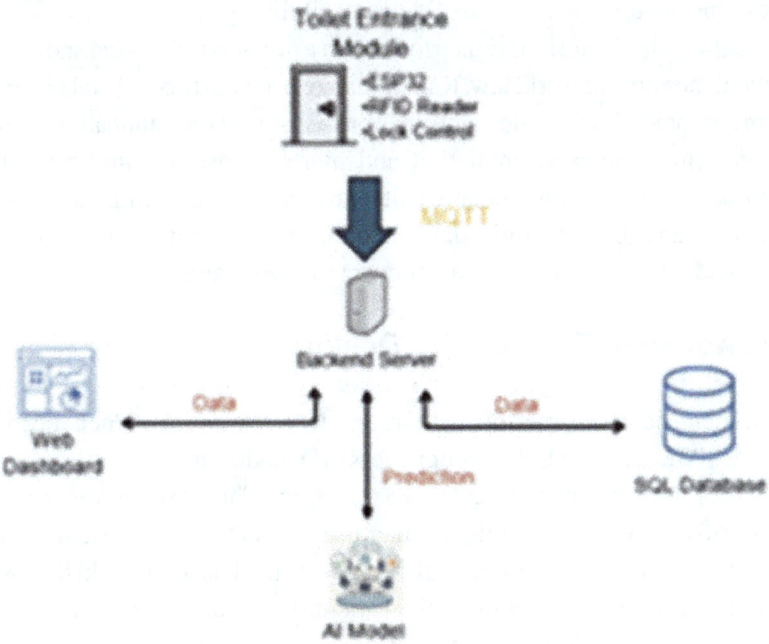

Figure 1 presents the overall architecture of the proposed Smart Ward AIoT Toilet Safety Monitoring System. The system adopts a modular design that integrates a toilet entrance module, MQTT-based communication, a local backend server, a SQL database, a machine learning module, and a web dashboard into a unified workflow. At a high level, the figure illustrates how the edge device deployed at the toilet entrance communicates with the backend system and how the backend coordinates data storage, prediction, and user-facing monitoring functions.

At the physical layer, the toilet entrance module is centred on an ESP32-based device connected to an RFID reader and lock control mechanism. This module is responsible for detecting patient entry and exit events through RFID interaction and for supporting device-side control at the toilet entrance. In the implemented system, the ESP32 hardware is also associated with local interaction and monitoring functions described in the system manual, including countdown display, warning prompts, and emergency-related state changes.

Communication between the entrance module and the software backend is achieved through MQTT. When an event is triggered at the device side, the corresponding message is transmitted to the backend server for further processing. The backend then coordinates several functions, including operational record management, interaction with the database, and model-based inference. This design separates local event capture from centralised decision logic, allowing the system to support both device responsiveness and software-side extensibility.

Overall, the architecture is designed to support a privacy-preserving and operationally practical monitoring workflow. Rather than relying on direct visual observation, the system combines device-side identification, real-time communication, structured record management, adaptive prediction, and interface-based monitoring within one integrated prototype. This modular architecture also improves maintainability by allowing hardware, backend logic, data handling, and user interface functions to be developed and refined as separate yet connected components.

3.2 Hardware and Edge Device Design

Patient identification is performed through RFID scanning. When a card is detected, the ESP32 reads the UID through the RFID module and uses this information as the basis for the subsequent monitoring workflow. This design provides a simple and low-intrusion means of linking a toilet usage event to a specific user without requiring direct visual observation. It also supports the backend workflow in which RFID records are associated with patient identities through the card assignment structure maintained in the database layer.

The edge device also provides immediate local interaction through the display, button, and audio components. The LCD is used to present device-side status information, such as readiness for scanning, room occupancy, and time-related monitoring output. The physical button supports user-triggered interaction during operation, while the audio module provides spoken or sound-based prompts when the system enters warning or emergency conditions. These features are important because they allow the entrance module to function not only as a sensing node, but also as an interactive interface that communicates system status directly at the point of use.

Figure 2. Hardware prototype of the toilet entrance module

A further key component is the servo motor, which is used for door control. In the implemented logic, the door can be opened when the user checks in or out, and it can also be automatically unlocked during emergency handling. This is a critical feature in the context of toilet safety monitoring, because it allows the system to move beyond passive observation and support physical intervention when required. Rather than treating door control as an isolated hardware feature, the prototype integrates it into the monitoring logic so that device-side actions are coordinated with system state changes.

At the control level, the ESP32 operates as an event-driven edge device that works together with the backend server rather than making all monitoring decisions locally. When a patient enters the toilet, the device detects the RFID scan and sends the corresponding event to the backend through MQTT. The backend then retrieves the relevant patient features from the database and invokes the AI prediction module to compute a personalised safety threshold. After the threshold is generated, it is sent back to the ESP32, which uses the received value to start local countdown-based monitoring. During operation, the device continues to handle immediate interactions such as display updates, button input, warning prompts, and emergency state transitions. This design allows the computationally heavier prediction logic to remain on the server side while the edge device focuses on responsive local execution and interaction.

3.3 Communication and Backend Workflow

Figure 3 illustrates the MQTT-based communication workflow between the toilet entrance module and the backend server. In the proposed system, MQTT is used as the main communication mechanism for transmitting device-side events to the backend and returning predicted threshold values to the edge device. This design separates local event capture from server-side processing and allows the system to support both responsive device behaviour and centralised coordination. The selection of MQTT is also supported by prior literature. In a healthcare establishment setting, MQTT has been described as a lightweight communication protocol suitable for restricted hardware and limited network conditions (Cunha et al., 2022).

Figure 3. MQTT-based communication workflow between the ESP32 toilet entrance module and the backend server

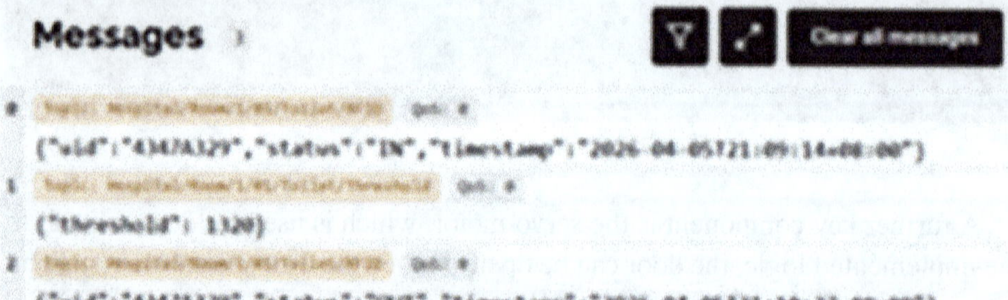

The implemented workflow is organised around two main MQTT topics. The ESP32 publishes event messages to the RFID topic, Hospital/Room/1/01/Toilet/RFID, while the backend subscribes to this topic in order to receive entry, exit, and accident-related events. In the opposite direction, the backend publishes the predicted threshold to the topic Hospital/Room/1/01/Toilet/Threshold, which is subscribed to by the ESP32. As shown in Figure 3, this arrangement creates a bidirectional communication path in which the device sends operational events to the server and receives monitoring parameters in return. The use of MQTT is appropriate in this context because the protocol has been implemented successfully in healthcare communication scenarios involving real-time information transfer and server-side display (Cunha et al., 2022).

When the ESP32 detects an RFID interaction, it constructs a JSON payload containing the user identifier, the event status, and a timestamp, and then publishes this payload to the RFID topic. In the device implementation, the status field is used to distinguish between IN, OUT, and ACCIDENT events. For accident cases, the message may also include an additional accident-related flag. This message structure allows the backend to interpret the event type immediately after reception and route it to the appropriate handling logic.

On the backend side, incoming MQTT messages are processed according to the reported event status. For an IN event, the server first records the toilet entry in the database by mapping the scanned UID to the corresponding patient record. It then retrieves the latest available engineered feature row for that patient and invokes the prediction module to compute a personalised threshold. Once the threshold has been generated, the backend sends a response message containing the predicted threshold value to the threshold topic. The ESP32 receives this message and updates its local monitoring logic accordingly. This interaction is central to the design because it allows the entrance device to initiate monitoring through a lightweight message, while the more computationally demanding prediction process is handled on the server side.

The backend also handles OUT and ACCIDENT events through the same communication channel. For an OUT event, the backend locates the most recent open toilet usage record for the patient, calculates the duration based on the stored entry timestamp and the reported exit time, and updates the record accordingly. For an ACCIDENT event, the backend marks the latest open session as an accident case and stores the duration up to the accident timestamp. In this way, MQTT is not only used for device control, but also serves as the trigger for creating and updating structured event records in the database.

To improve robustness, the backend prediction logic is designed with fallback behaviour. If the machine learning component cannot be initialised, or if no suitable feature record is available for the patient, the backend returns a default threshold instead of interrupting system operation. This ensures that communication and monitoring can continue even when personalised prediction is unavailable. As a result, the MQTT-based backend workflow supports both adaptive monitoring under normal conditions and degraded but functional operation under exception conditions.

Overall, the communication and backend workflow forms the coordination layer of the prototype. The ESP32 is responsible for detecting physical events and sending structured messages, while the backend is responsible for event interpretation, record management, and threshold computation. Figure 3 visualises this arrangement through the publish–subscribe relationships between the entrance device and the backend server. This architecture is particularly suitable for the proposed system because it supports modularity, lightweight communication, and the practical integration of hardware events with software-side intelligence.

3.4 Database Design and Data Management

The proposed system adopts a relational database design to support patient identification, event logging, historical record management, and model-related data preparation. Since the system must coordinate RFID-based entry events, toilet usage records, and personalised threshold prediction, a structured database layer is required to ensure consistency across the hardware, backend, dashboard, and machine learning components. Figure 4 shows the simplified entity–relationship structure of the database, while Table 1 summarises the purpose of the main tables and their key fields.

The Card_Assignments table supports the mapping between RFID cards and patient identities. Each record links a card_uid to a patient_id, together with an activity status flag. This design enables the entrance device to identify a patient indirectly through RFID scanning, while the backend resolves the scanned UID into a valid patient record before creating or updating operational logs. As a result, the RFID layer functions as a practical access and identification mechanism without requiring direct visual observation or manual user entry.

Operational toilet usage data are stored in the Toilet_Logs table. This table records each toilet session through fields such as patient identifier, entry time, exit time, duration, time-of-day category, and accident status. It therefore acts as the main event log of the system. The design of this table is important for two reasons. First, it supports real-time monitoring by allowing the backend to locate the latest open session for a patient when processing IN, OUT, or ACCIDENT events. Second, it preserves historical behavioural records that can later be analysed or transformed into model-ready features. In this sense, Toilet_Logs serves as both an operational table and a historical data source.

The TimeOfDay table is included as a supporting reference table for temporal categorisation. Rather than storing only raw timestamps, the system also maps toilet events into interpretable time-of-day groups. This supports later analysis of temporal usage patterns and allows time-related behaviour to be incorporated into engineered features in a more structured way. Although this table is small, it improves consistency by standardising the categories used across logging and feature generation.

Figure 4. Simplified entity–relationship diagram of the proposed system database

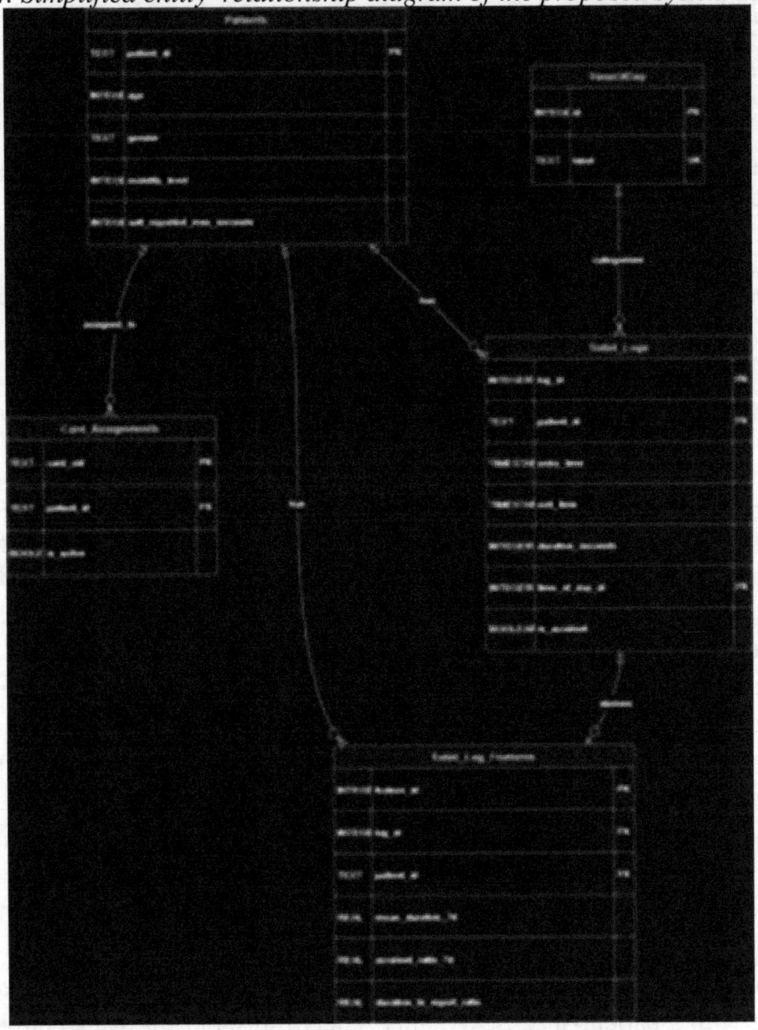

Table 1. Summary of the main database tables, their purposes, and representative key fields.

Table	Purpose	Key Fields
Patients	**Stores static patient information used for monitoring and prediction**	**patient_id, age, mobility_level, self_reported_max_seconds**
Card_Assignments	Maps RFID cards to patients	card_uid, patient_id, is_active
TimeOfDay	Stores time-of-day categories for toilet events	id, label
Toilet_Logs	Records toilet entry, exit, duration, and accident events	log_id, patient_id, entry_time, exit_time, duration_seconds, is_accident
Toilet_Log_Features	Stores engineered features derived from logs for model use	feature_id, log_id, patient_id, mean_duration_7d, accident_ratio_7d, duration_to_report_ratio

To support machine learning, the database also contains a dedicated Toilet_Log_Features table. This table stores engineered variables derived from toilet usage history and patient information, with each feature record linked to the corresponding log and patient. The existence of a separate feature table is an intentional design choice. It separates raw operational records from derived analytical variables, making the system easier to maintain and allowing prediction-related processing to operate on a dedicated model-ready dataset. In the present implementation, this feature table stores temporal variables, rolling-window statistics, selected interaction terms, and denormalised patient attributes, which are then used by the threshold prediction module. The detailed construction of these features is discussed in Section 3.5.

3.5 Machine Learning and Dynamic Threshold Mechanism

A central objective of the proposed system is to replace a universal fixed timeout rule with a patient-specific dynamic safety threshold. In many conventional monitoring settings, a single time limit is applied to all users, regardless of age, mobility, medical condition, or behavioural history. In the present system, the threshold is instead generated from patient data and toilet usage history, allowing the monitoring logic to adapt to the individual rather than relying on a single rigid rule. The implementation therefore combines feature engineering, preprocessing, model-based inference, and threshold selection within a single prediction workflow.

The choice of anomaly detection as the main modelling direction is also supported by prior work on structured data. In the present problem, the available inputs are primarily tabular features derived from patient records and toilet usage logs, while clearly abnormal or accident-related cases are relatively limited. This makes the task closer to anomaly detection than to balanced supervised classification. Liu,

Ting, and Zhou (2008) proposed Isolation Forest as a method that explicitly isolates anomalies rather than profiling normal points, and reported that it works well even in situations where the training set does not contain any anomalies. In addition, Bouman, Bukhsh, and Heskes (2024) evaluated 32 unsupervised anomaly detection algorithms on 52 real-world multivariate tabular datasets, showing that unsupervised anomaly detection remains a practical modelling strategy for structured data settings.

To support this process, raw operational records are transformed into engineered features that can be used for model training and inference. These features are stored in the Toilet_Log_Features table and are derived from a combination of patient attributes, session-level toilet log data, and short-term historical behaviour. Table 2 summarises the main engineered feature groups used for dynamic threshold prediction. The design of the feature set reflects the assumption that abnormal toilet use is not determined by duration alone, but by the interaction between duration, patient condition, recent behaviour, and temporal context.

Table 2. Summary of the main engineered feature groups used for dynamic threshold prediction.

Feature group	Representative features	Source	Purpose
Static patient features	age, gender, mobility level, gastrointestinal issue flag, urological issue flag, self-reported maximum duration	Patients	Represent baseline patient characteristics relevant to toilet use
Log-level temporal features	duration seconds, time-of-day category, hour of day, day of week, weekend indicator	Toilet_Logs / derived from timestamps	Represent the immediate context of each toilet session
Interval features	time since last log	Derived from historical Toilet_Logs	Capture recent usage frequency and short-term behavioural spacing
Rolling-window behavioural features	mean duration over 7 days, maximum duration over 7 days, minimum duration over 7 days, standard deviation over 7 days, log count over 7 days, accident ratio over 7 days	Derived from recent Toilet_Logs	Represent recent behavioural history and variability
Time-of-day usage counts	deep night count over 7 days, early morning count over 7 days, late morning count over 7 days, afternoon count over 7 days, evening count over 7 days	Derived from recent Toilet_Logs and TimeOfDay	Capture temporal activity patterns over the recent history window
Interaction and derived comparison features	age × mobility, gastrointestinal issue × duration, self-reported limit minus duration, duration-to-report ratio	Engineered from patient and log data	Represent combined effects and deviations relative to self-reported tolerance

The feature engineering process is designed to capture both static and dynamic aspects of patient behaviour. Static variables describe who the patient is, while session-level and historical variables describe how the patient has been using the toilet over time. This distinction is important because the system aims to estimate when a current session becomes abnormal for a particular individual, rather than merely recording whether a session lasted longer than a generic limit. By structuring the input space in this way, the prediction module can consider not only the current duration, but also the patient's recent behavioural pattern and personal tolerance indicators.

To improve reliability, the threshold mechanism also includes fallback behaviour. If the predictor cannot be initialised, if no usable feature row is available, or if the model does not produce a valid result, the backend returns a default threshold instead of interrupting system operation. This ensures that the system can continue functioning as a monitoring tool even when personalised inference is temporarily unavailable. In this way, the prediction module is designed not only for adaptivity, but also for operational robustness.

Overall, the machine learning and dynamic threshold mechanism provides the intelligence layer of the proposed system. By transforming raw toilet usage records and patient attributes into structured features, and by using these features to estimate patient-specific safety thresholds, the system moves beyond uniform timing rules towards a more adaptive form of toilet safety monitoring. This mechanism is central to the chapter because it links historical data, individual patient characteristics, and real-time operational decisions within a single predictive workflow.

3.6 Web Dashboard, Data Registration, and User Support Functions

Dashboard and Monitoring Display

The proposed system includes a web-based interface that serves as the main staff-facing operational layer. In addition to the edge device installed at the toilet entrance, the web interface provides a centralised view of monitoring information and allows staff to observe system status through a browser-based dashboard. In the current implementation, the dashboard is designed to present live or recent operational information such as room occupancy state, alert conditions, and recent toilet-related records. This is important because toilet safety monitoring is not only a sensing problem, but also an operational visibility problem: staff need to see which room is in use, whether abnormal conditions have been raised, and what events have recently been logged.

Figure 5. Web interface of the proposed system

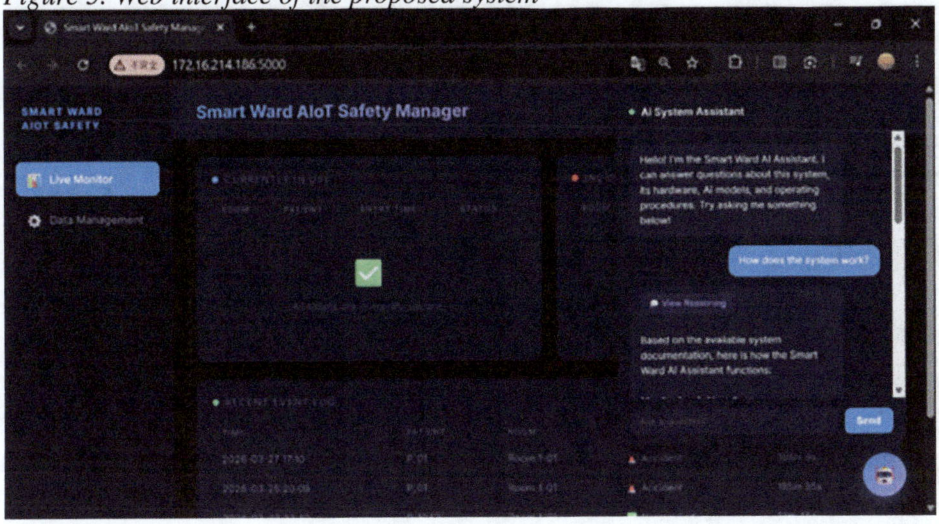

As shown in Figure 5, the dashboard provides a more practical view of the system than the hardware layer alone. While the entrance device is responsible for local interaction and immediate control, the web dashboard consolidates data from the backend and database into a format suitable for supervision and review. This improves usability by allowing staff to inspect current states and logged events without accessing the device directly. From a system design perspective, the dashboard therefore acts as the visual monitoring front end of the backend workflow described in Section 3.3.

Patient and Card Registration

This design choice improves the practicality of the prototype in two ways. First, it allows administrative data to be managed through a browser-based interface instead of requiring manual database editing. Second, it ensures that the machine learning and monitoring components can operate on valid and up-to-date patient information. In other words, the web interface is not limited to displaying monitoring outcomes; it also provides the data-entry pathway through which the underlying database is populated and maintained. This makes the web layer an essential bridge between system setup, daily operation, and later prediction tasks.

Figure 6. Data management interface of the proposed system

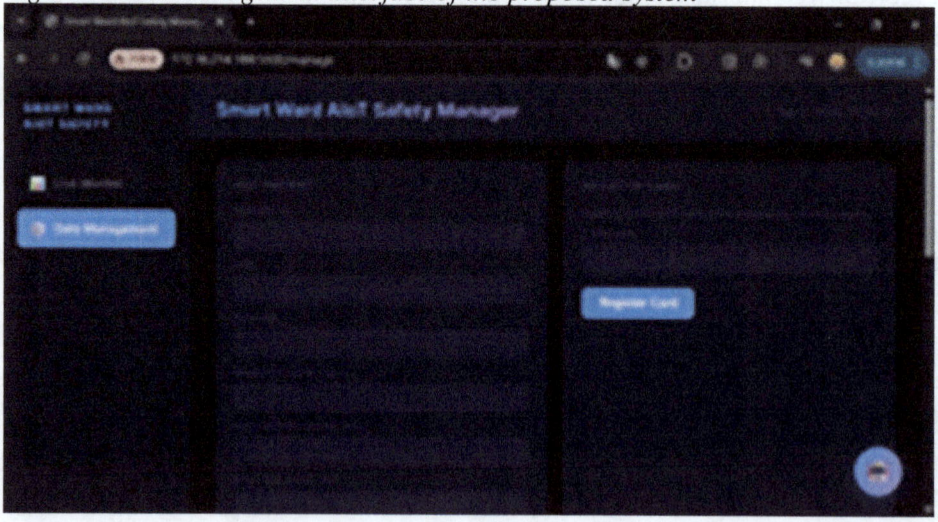

AI-Assisted User Support

The web interface further includes an AI-assisted support function. Unlike the machine learning module described in Section 3.5, this assistant is not responsible for determining whether a toilet session is abnormal. Instead, it is designed to help users understand system behaviour, operating logic, and usage procedures through a conversational interface. To improve the reliability of responses, the assistant adopts a Retrieval-Augmented Generation (RAG) approach. Early RAG work describes this paradigm as a combination of parametric and non-parametric memory for language generation, in which a model first retrieves relevant external knowledge and then generates an answer grounded in the retrieved content (Lewis et al., 2020).

More recent survey literature further notes that RAG has emerged as a promising solution for large language models because it incorporates knowledge from external databases, thereby enhancing the accuracy and credibility of generated responses, particularly for knowledge-intensive tasks. The same survey also highlights that RAG can help address issues such as hallucination and outdated knowledge by allowing external information to be integrated into the response process (Gao et al., 2024).

In the proposed system, the AI assistant automatically searches for relevant reference material related to the user's question and uses the retrieved content to support its response. This design is particularly suitable for system-support scenarios, because users may ask about operating procedures, hardware behaviour, AI threshold logic, or emergency handling steps. By grounding the answer in retrieved system-related content, the assistant is better able to provide responses that are specific to the implemented system rather than generic chatbot-style answers. As a result, the AI support function complements the dashboard and data management features of the web layer, extending the interface from passive monitoring and administration to active user support.

Overall Summary

Overall, the web layer plays a broader role than that of a conventional dashboard. It acts simultaneously as a monitoring interface, a data registration interface, and a user support interface. This integration is important because it allows the proposed system to function not only as a sensing and prediction pipeline, but also as a usable application platform for real operational interaction. By combining status visualisation, database-related administration, and AI-assisted support within the same web environment, the system becomes more complete and more practical for deployment-oriented demonstration.

4. EXPERIMENTAL RESULTS

4.1 Experimental Setup

The proposed system was evaluated as an integrated prototype consisting of a hardware entrance module, a local backend environment, a database service, a web interface, and an external AI assistant service. The objective of the experimental setup was to validate the end-to-end operation of the system under realistic prototype conditions, including RFID-based entry detection, backend event processing, dynamic threshold generation, web-based monitoring, and AI-assisted user support.

Hardware Setup

The entrance-side hardware prototype was built around a FireBeetle-ESP32 microcontroller. The device was connected to an RFID reader for user identification, an LCD display for local status and threshold output, a servo motor for door control, a physical button for user interaction, and an audio speaker for warning and emergency prompts. During operation, the FireBeetle-ESP32 was responsible for detecting RFID scans, displaying system status, receiving the predicted threshold from the backend, and carrying out local countdown-based monitoring.

Backend and Local Computing Environment

The backend services were deployed on a local personal computer used as the main system server. The machine was equipped with an Intel Core i5-14400F CPU, 32 GB RAM, and an NVIDIA GeForce RTX 4060 GPU. This local environment hosted the main backend logic, the database layer, the MQTT-related workflow, and the web application used for monitoring and data management. It therefore acted as the central control node of the prototype during experimentation.

Web and Database Setup

The web interface was used for live monitoring and data registration. It provided access to the dashboard, patient management functions, and RFID card registration functions, all of which interacted directly with the underlying database. The database stored patient records, card assignments, toilet usage logs, and engineered feature records used for threshold prediction. In addition to these stored records, the management interface also included functions for generating anomaly-related records for patients, which formed part of the prototype evaluation workflow. This setup allowed the experiments to be carried out using valid patient-linked data rather than isolated hardware events only.

AI Assistant Deployment

To reduce the cost associated with repeated external API usage, the AI assistant was deployed separately on a school HPC server. The web application automatically accessed this chatbot service through an API call whenever the user submitted a question through the interface. In this configuration, the local web system was responsible for handling the user-facing interaction, while the actual assistant response was generated remotely by the school-hosted chatbot service. This setup allowed the prototype to retain the AI-assisted support function without placing the full inference cost on the local experimental environment.

Evaluation Scope

The experiments in this chapter focused on two aspects. First, the integrated prototype was tested at the system level, including hardware response, communication workflow, dashboard display, and data registration functions. Second, the dynamic threshold mechanism was evaluated at the model level by comparing different anomaly detection models and feature configurations. Because a real-world hospital toilet monitoring dataset was not available for the present study, both the normal and anomaly-related data used in the evaluation were generated within the prototype environment through a rule-based synthetic data workflow. In this setup, the dataset generation script created 100 synthetic patient profiles together with RFID card assignments and 30 days of historical toilet logs, while the log generation process incorporated demographic attributes, condition-related factors, time-of-day patterns, and random variation to simulate normal toilet use. Additional anomaly-related records were then generated according to predefined accident-oriented patterns, with 5 anomalies per patient and a mixture of overtime and non-timeout cases. The web management interface also included an anomaly-generation function for patient records, which formed part of the prototype-level evaluation environment. The reported model results should therefore be interpreted as controlled prototype-level evidence rather than as final validation on a naturally observed clinical dataset. The following sections present these results separately.

4.2 System Demonstration and Functional Validation

The system demonstration focused on verifying whether the proposed prototype could operate successfully as an integrated AIoT monitoring system. Instead of describing every intermediate step in the workflow, the validation emphasised several representative outcomes that show the successful interaction among the hardware entrance module, the communication layer, the backend, and the web interface.

First, the hardware prototype was successfully implemented and operated as a physical toilet entrance module. As shown in Figure 2, the system was not tested purely as a software simulation, but as a working device equipped with the required sensing, interaction, and local feedback components. Second, the communication layer was functionally validated through MQTT-based message exchange. Figure 3 shows that the broker was able to receive toilet-related event messages and support the bidirectional transmission required by the system. Third, the web interface successfully supported AI-assisted user interaction. As illustrated in Figure 5, the interface did not only provide operational pages, but also included an AI assistant capable of responding to system-related questions.

Figure 7. Dashboard showing occupied state

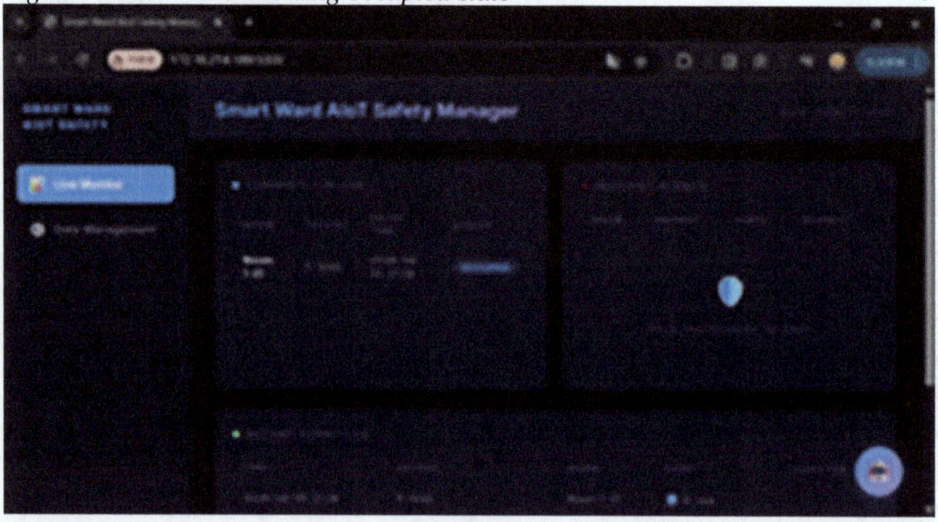

Fourth, the dashboard was able to reflect the live system state during operation. According to the occupied-state view shown in Figure 7, the web dashboard could display the relevant monitoring status when a patient was inside the toilet.

Finally, the integration between the backend prediction workflow and the hardware device was demonstrated through threshold return and LCD display. As shown in Figure 8, the entrance hardware was able to receive the AI-predicted threshold transmitted through the communication layer and display the value locally on the LCD. This is one of the most important system-level validation results, because it confirms that the dynamic threshold was not only computed at the backend, but also delivered successfully to the physical device for local countdown-based monitoring.

Taken together, these results show that the proposed prototype achieved functional integration across its key modules. The hardware was operational, MQTT-based communication was successful, the AI assistant was accessible through the web interface, the dashboard could monitor system status in real time, and the hardware could receive and display backend-generated thresholds. Therefore, the system demonstration supports the claim that the prototype was implemented successfully as an end-to-end Smart Ward AIoT toilet safety monitoring system.

Figure 8. LCD showing received AI threshold

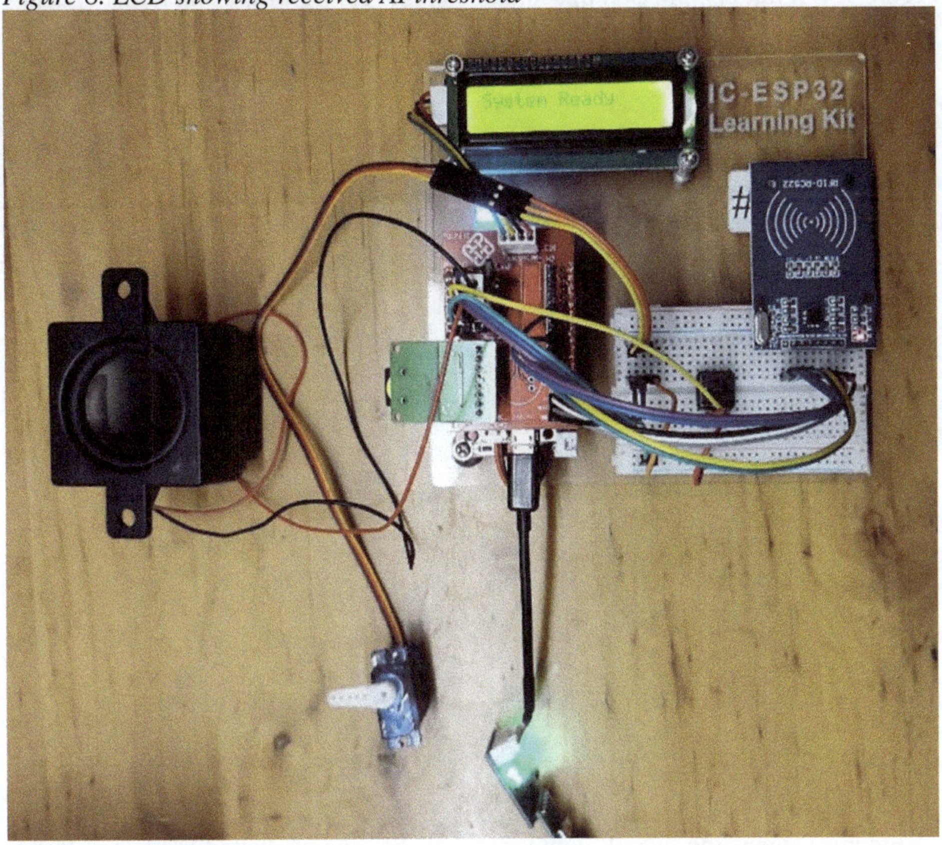

4.3 Evaluation of the Dynamic Threshold Mechanism

The dynamic threshold mechanism was evaluated by comparing multiple anomaly detection model configurations under different feature-selection strategies. According to the experimental configuration, three model families were tested: XGBoost, Isolation Forest, and a simple autoencoder. For each model family, three feature settings were used: _all, which used the full feature set; _filtered, which excluded the three features with the strongest Pearson correlation with the target classification outcome; and _top10, which retained only the ten features with the strongest Pearson correlation with that target. This evaluation design was intended to assess not only the relative performance of different anomaly detection models, but also the effect of feature selection on predictive performance.

Figure 9. Overall comparison of evaluated model configurations in terms of accuracy, precision, recall, and F1-score

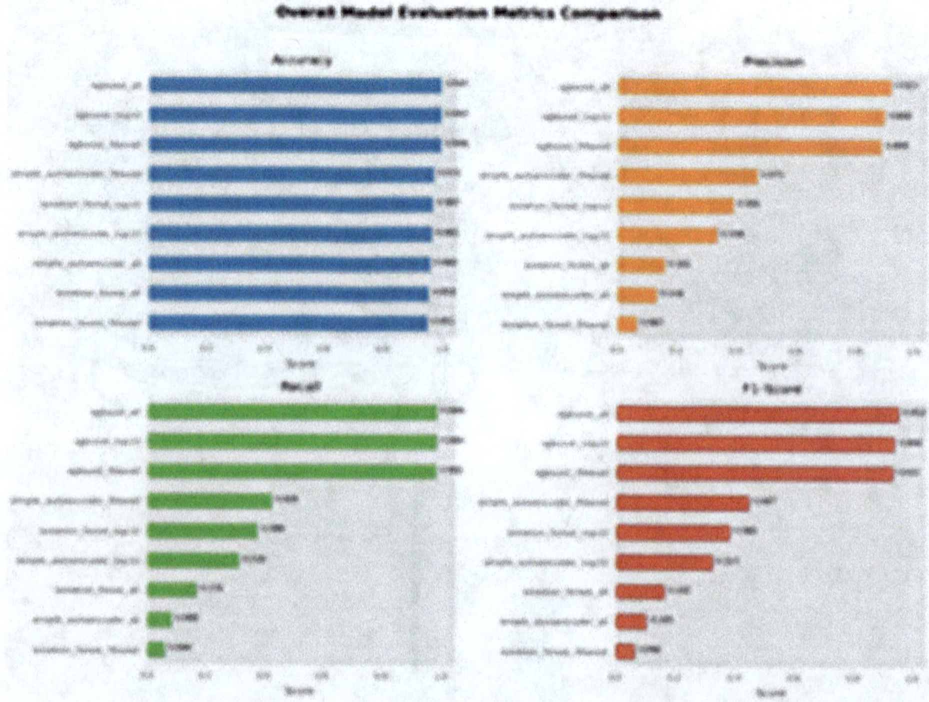

Figure 9 shows the overall comparison of the nine evaluated models in terms of accuracy, precision, recall, and F1-score. Across all four metrics, the XGBoost-based models consistently outperformed the other model families. Among all tested configurations, xgboost_all achieved the strongest overall performance, with an accuracy of 0.997364, precision of 0.923077, recall of 0.984000, and F1-score of 0.952565. The confusion matrix further shows 18,049 true negatives, 41 false positives, 8 false negatives, and 492 true positives, indicating both high sensitivity to accident cases and a low false alert rate.

The other XGBoost variants also performed strongly. xgboost_top10 produced an accuracy of 0.996611, precision of 0.899452, recall of 0.984000, and F1-score of 0.939828, with 55 false positives and 8 false negatives. xgboost_filtered achieved an accuracy of 0.996235, precision of 0.889493, recall of 0.982000, and F1-score of 0.933460, with 61 false positives and 9 false negatives. These results suggest that the XGBoost model remained highly effective even when the feature set was reduced, although the full-feature configuration produced the best overall balance between sensitivity and false alert control.

In contrast, the Isolation Forest models performed substantially worse. The best Isolation Forest variant, isolation_forest_top10, reached an accuracy of 0.967402, but its precision, recall, and F1-score were only 0.390947, 0.380000, and 0.385396, respectively. The isolation_forest_all and isolation_forest_filtered variants performed even more poorly, with F1-scores of 0.165209 and 0.065507. Their confusion matrices also show large numbers of missed accident cases, including 415 false negatives for isolation_forest_all and 468 false negatives for isolation_forest_filtered. These results indicate that, in the present tabular feature setting, the isolation-based approach was not sufficiently effective for reliable threshold determination.

The simple autoencoder models performed better than Isolation Forest in some configurations, but still remained clearly below the XGBoost-based models. Among the three autoencoder variants, simple_autoencoder_filtered produced the strongest result, with an accuracy of 0.971705, precision of 0.471239, recall of 0.426000, and F1-score of 0.447479. simple_autoencoder_top10 and simple_autoencoder_all achieved F1-scores of 0.327497 and 0.104623, respectively. Although the filtered autoencoder configuration improved substantially over the other autoencoder variants, its accident detection performance was still notably weaker than that of all three XGBoost models.

From the perspective of feature selection, the results show that feature reduction did not necessarily harm performance, but the effect depended strongly on the model family. For XGBoost, the _top10 configuration remained close to the _all configuration, suggesting that a compact subset of carefully selected features was sufficient to preserve most of the predictive power. For Isolation Forest, the _top10 configuration improved performance relative to _all and _filtered, but still remained inadequate compared with the supervised XGBoost approach. For the autoencoder, the _filtered setting produced the best result, suggesting that excluding certain features may improve unsupervised reconstruction-based detection under this feature design. Overall, however, the full XGBoost model remained the strongest configuration among all evaluated alternatives.

These findings indicate that the dynamic threshold mechanism is highly sensitive to model choice. In the present implementation, the XGBoost-based configurations were clearly the most effective in distinguishing normal toilet sessions from accident-related cases, while Isolation Forest and the simple autoencoder produced substantially weaker detection quality. Since the purpose of the threshold mechanism is to identify the boundary at which a patient's behaviour becomes anomalous, a model with high recall and low false positive rate is particularly desirable. Based on the observed results, xgboost_all provided the best overall trade-off and therefore offers the strongest support for deployment within the current prototype.

4.4 Discussion of Results

The experimental results suggest that the effect of feature selection depended strongly on the model family. For the XGBoost-based configurations, performance improved when more highly relevant features were retained. Among the three XGBoost variants, xgboost_all achieved the strongest overall performance, with an accuracy of 0.997364, precision of 0.923077, recall of 0.984000, and F1-score of 0.952565. By comparison, xgboost_top10 and xgboost_filtered remained highly competitive but showed slightly lower F1-scores of 0.939828 and 0.933460, respectively. These results suggest that XGBoost was able to benefit from a richer feature set, and that retaining more highly relevant attributes strengthened its predictive performance.

A different pattern was observed for Isolation Forest. In this model family, the top10 configuration clearly outperformed the other two feature settings. isolation_forest_top10 achieved an F1-score of 0.385396, compared with 0.165209 for isolation_forest_all and 0.065507 for isolation_forest_filtered. Although Isolation Forest remained substantially weaker than the XGBoost-based models overall, this result suggests that the isolation-based approach was more effective when restricted to a smaller subset of highly relevant features, rather than being exposed to the full feature space.

The simple autoencoder showed yet another behaviour. In this case, the best-performing configuration was simple_autoencoder_filtered, which achieved an F1-score of 0.447479, compared with 0.327497 for simple_autoencoder_top10 and 0.104623 for simple_autoencoder_all. This suggests that, for the reconstruction-based model, removing the most strongly correlated features produced a more favourable learning setting than either retaining all features or restricting the model to the top ten alone. One possible interpretation is that this exclusion reduced the dominance of highly predictive attributes and allowed the autoencoder to learn a broader reconstruction pattern across the remaining input space.

Taken together, these findings indicate that feature relevance does not influence all anomaly detection models in the same way. XGBoost benefited from retaining more strongly relevant features, Isolation Forest performed best when restricted to the most relevant subset, and the simple autoencoder performed best when the most strongly correlated features were removed. Despite these differences, the overall results still clearly favoured the XGBoost-based configurations. In particular, xgboost_all provided the best overall trade-off between accident detection capability and false alert control, making it the strongest candidate for use in the current prototype.

5. DISCUSSION

5.1 Overall Interpretation of the Proposed System

The results discussed in this chapter suggest that the proposed Smart Ward AIoT Toilet Safety Monitoring System is able to combine privacy-preserving sensing, personalised threshold prediction, and operational web support within a single integrated prototype. Rather than relying on direct visual observation or a universal fixed time limit, the system was designed to monitor toilet use through entrance-side interaction, backend processing, and adaptive safety logic. In this respect, the prototype demonstrates that hospital toilet safety monitoring can be approached as an end-to-end AIoT workflow instead of as an isolated hardware alarm or a purely retrospective analytics task.

5.2 Interpretation of Experimental Findings

The experimental findings suggest that model performance in the present system is strongly influenced by both the model family and the feature-selection strategy. Among all evaluated configurations, the XGBoost-based models clearly achieved the strongest overall performance, with xgboost_all producing the best results in terms of accuracy, precision, recall, and F1-score. In contrast, the Isolation Forest and simple autoencoder configurations showed substantially weaker accident detection capability under the same tabular feature setting. From this perspective, the results indicate that the current feature space is particularly favourable to a supervised tabular learning approach, in which multiple patient attributes, temporal variables, and engineered interaction features can be combined to form a more effective decision boundary.

5.3 Limitations

Despite the promising prototype results, the current study remains subject to several important limitations, the most significant of which concerns the availability and nature of the data. A real-world hospital toilet monitoring dataset was not available for the system described in this chapter, and the evaluation therefore relied on a rule-based synthetic data workflow. In the current implementation, the dataset was generated by constructing synthetic patient profiles, RFID assignments, and 30 days of historical toilet logs, together with additional anomaly-related records produced from predefined accident-oriented patterns. Both normal and anomaly data were therefore created within the prototype environment rather than collected from naturally observed ward operation. In particular, anomaly logs were generated at a fixed rate of 5 anomalies per patient, with durations defined according to preset rules relative to patient-specific reference values and with a stronger bias towards deep-night and early-morning periods. Although this mechanism was useful for controlled experimentation, it cannot fully represent the uncertainty, variability, and contextual complexity of genuine accident-related behaviour in hospital toilets. Under these conditions, the resulting data distribution may be cleaner, more structured, and more separable than real clinical data, which means that the reported model performance should be understood as prototype-level evidence of system feasibility rather than as final validation of real-world clinical effectiveness.

6. CONCLUSION AND FUTURE WORK

6.1 Conclusion

This chapter presented a Smart Ward AIoT Toilet Safety Monitoring System designed to support privacy-preserving safety monitoring in hospital toilet environments. The prototype combined an entrance-side hardware module, MQTT-based communication, backend event processing, a relational database, a dynamic threshold prediction mechanism, a web-based management interface, and an AI-assisted user support function within a single integrated system. Taken together, these components demonstrate that toilet safety monitoring can be approached as an end-to-end operational workflow rather than as a single alarm device or a standalone machine learning model.

Another important contribution of the system is its privacy-oriented design. The prototype avoids camera-based surveillance and instead relies on RFID-based entry and exit logging, local device interaction, and dashboard-based supervision. This demonstrates that meaningful monitoring support can still be provided in sensitive spaces without direct visual observation. At the same time, the chapter also extends beyond sensing and prediction alone by including web-based data registration, live monitoring, and an AI-assisted support function for system-related queries. This broader integration makes the prototype more practical as an operational platform rather than merely an algorithmic experiment.

Overall, the work presented in this chapter provides a working proof of concept for a privacy-preserving AIoT toilet safety monitoring system in a smart ward context. Although the prototype remains limited in scale and data realism, it successfully demonstrates the feasibility of integrating hardware sensing, structured data management, adaptive threshold prediction, and user-facing web support into a coherent monitoring framework. In this sense, the work contributes not only a technical prototype, but also a practical direction for how personalised and privacy-aware ward safety support systems may be developed in future healthcare environments.

6.2 Future Work

Several directions may be pursued to improve and extend the proposed system in future work. The most important next step is the collection of real-world deployment data. In the current prototype, both normal and anomaly-related records were generated through a rule-based synthetic workflow rather than obtained from naturally observed hospital operation. Future studies should therefore evaluate the system using real toilet usage data collected in ward environments, so that the threshold mechanism can be validated under more realistic behavioural patterns, noise conditions, and accident distributions. Such data would also allow the prediction model to be refined and recalibrated using naturally occurring records rather than controlled synthetic assumptions.

Finally, future work should consider the broader clinical and human factors associated with deployment. Even without camera-based monitoring, toilet entry and duration logging may still raise concerns related to trust, privacy perception, and user acceptance in sensitive environments. Future studies should therefore examine how patients and staff perceive such a system in practice, how consent and transparency should be handled, and how the monitoring workflow can be designed in a way that is both ethically acceptable and operationally useful. Addressing these issues would help move the work discussed in this chapter from technical feasibility towards real clinical applicability.

REFERENCES

Association of Hong Kong Nursing Staff. (2025, May). *Nursing manpower survey report 2025.* https://nurse.org.hk/storage/media/HA%E4%BA%BA%E6%89%8B%E5%95%8F%E5%8D%B7%E6%96%B0%E8%81%9E%E7%A8%BF_12052025%20(Final).pdf

Bouman, R., Bukhsh, Z., & Heskes, T. (2024). Unsupervised anomaly detection algorithms on real-world data: How many do we need? *Journal of Machine Learning Research, 25*(105), 1–34. https://arxiv.org/abs/2305.00735

Cunha, N. A., Bispo, B. C., Cavalcante, E. L., Inocêncio, A. V. M., Alves, G. J., & Rodrigues, M. A. B. (2022). Application of MQTT network for communication in healthcare establishment. In *XXVII Brazilian Congress on Biomedical Engineering (CBEB 2020)* (pp. 1465–1470). Springer. DOI: 10.1007/978-3-030-70601-2_216

Francis-Coad, J., Hill, A.-M., Jacques, A., Chandler, A. M., Richey, P. A., Mion, L. C., & Shorr, R. I. (2020). Association between characteristics of injurious falls and fall preventive interventions in acute medical and surgical units. *The Journals of Gerontology. Series A, Biological Sciences and Medical Sciences, 75*(10), e152–e158. DOI: 10.1093/gerona/glaa032 PMID: 31996903

Ganz, D. A., Huang, C., Saliba, D., Shier, V., Berlowitz, D., Lukas, C. V., Pelczarski, K., Schoelles, K., Wallace, L. C., & Neumann, P. (2013, January). *Preventing falls in hospitals: A toolkit for improving quality of care* (Prepared by RAND Corporation, Boston University School of Public Health, and ECRI Institute under Contract No. HHSA290201000017I TO #1; AHRQ Publication No. 13-0015-EF). Agency for Healthcare Research and Quality. https://www.ahrq.gov/sites/default/files/publications/files/fallpxtoolkit_0.pdf

Gao, Y., Xiong, Y., Gao, X., Jia, K., Pan, J., Bi, Y., Dai, Y., Sun, J., Wang, M., & Wang, H. (2024). *Retrieval-augmented generation for large language models: A survey*. arXiv. https://doi.org//arXiv.2312.10997DOI: 10.48550

Gettens, S., Fulbrook, P., Jessup, M., & Low Choy, N. (2018). The patients' perspective of sustaining a fall in hospital: A qualitative study. *Journal of Clinical Nursing, 27*(3–4), 743–752. DOI: 10.1111/jocn.14075 PMID: 28926152

Healthcare Quality Improvement Partnership. (2015, October 8). *National audit of inpatient falls: Audit report 2015*. https://www.hqip.org.uk/resource/national-audit-of-inpatient-falls-audit-report-2015/

Hospital Authority. (n.d.). *Hospital Authority data sharing portal*. https://www3.ha.org.hk/data/HAStatistics/MajorReport?language=en

Joshi, S., Gupta, N., Burmi, S., & Sharma, Y. (2023, November). AI-IOT: Integrating artificial intelligence & Internet of Things for healthcare systems. In *2023 International Conference on Advances in Computation, Communication and Information Technology (ICAICCIT)* (pp. 281–285). IEEE. DOI: 10.1109/ICAICCIT60255.2023.10465828

Kido, S., Miyasaka, T., Tanaka, T., Shimizu, T., & Saga, T. (2009, January). Fall detection in toilet rooms using thermal imaging sensors. In *2009 IEEE/SICE International Symposium on System Integration (SII)* (pp. 83–88). IEEE. DOI: 10.1109/SI.2009.5384550

Lewis, P., Perez, E., Piktus, A., Petroni, F., Karpukhin, V., Goyal, N., Küttler, H., Lewis, M., Yih, W.-t., Rocktäschel, T., Riedel, S., & Kiela, D. (2020). *Retrieval-augmented generation for knowledge-intensive NLP tasks*. arXiv. https://doi.org//arXiv.2005.11401DOI: 10.48550

Liu, F. T., Ting, K. M., & Zhou, Z.-H. (2008, December). *Isolation forest*. In *2008 Eighth IEEE International Conference on Data Mining* (pp. 413–422). IEEE. DOI: 10.1109/ICDM.2008.17

Mecocci, A., Micheli, F., Zoppetti, C., & Baghini, A. (2016, October). Automatic falls detection in hospital-room context. In *2016 7th IEEE International Conference on Cognitive Infocommunications (CogInfoCom)* (pp. 127–132). IEEE. DOI: 10.1109/CogInfoCom.2016.7804537

Nho, Y.-H., Lim, J. G., & Kwon, D.-S. (2020). Cluster-analysis-based user-adaptive fall detection using fusion of heart rate sensor and accelerometer in a wearable device. *IEEE Access : Practical Innovations, Open Solutions*, *8*, 40389–40401. DOI: 10.1109/ACCESS.2020.2969453

Rewicki, F., Denzler, J., & Niebling, J. (2023). *Is it worth it? An experimental comparison of six deep- and classical machine learning methods for unsupervised anomaly detection in time series*. arXiv. https://arxiv.org/abs/2212.11080

Sethi, P., & Sarangi, S. R. (2017). Internet of Things: Architectures, protocols, and applications. *Journal of Electrical and Computer Engineering*, *2017*, 1–25. DOI: 10.1155/2017/9324035

Sixsmith, A., & Johnson, N. (2004). A smart sensor to detect the falls of the elderly. *IEEE Pervasive Computing*, *3*(2), 42–47. DOI: 10.1109/MPRV.2004.1316817

South China Morning Post. (2023, April 18). *Call for monitoring devices for Hong Kong patients after missing woman found dead.* https://www.scmp.com/news/hong-kong/health-environment/article/3217390/woman-58-chronic-illness-found-dead-hong-kong-hospital-toilet-after-waiting-more-12-hours-emergency

Su, M.-C., Liao, J.-W., Wang, P.-C., & Wang, C.-H. (2017, June). A smart ward with a fall detection system. In *2017 IEEE International Conference on Environment and Electrical Engineering and 2017 IEEE Industrial and Commercial Power Systems Europe (EEEIC/I&CPS Europe)* (pp. 1–4). IEEE. DOI: 10.1109/EEEIC.2017.7977515

Wild, D., Nayak, U. S., & Isaacs, B. (1981). How dangerous are falls in old people at home? *British Medical Journal (Clinical Research Ed.), 282*(6260), 266–268. DOI: 10.1136/bmj.282.6260.266 PMID: 6779979

World Health Organization. (2021, April 27). *Step safely: Strategies for preventing and managing falls across the life-course.* https://www.who.int/publications/i/item/978924002191-4

Zhu, Y., Yin, X., Liew, A. W.-C., & Tian, H. (2024). *Privacy-preserving in medical image analysis: A review of methods and applications.* arXiv. https://arxiv.org/abs/2412.03924

Chapter 10
Smart Automation Using Arduino UNO and iOS Shortcuts for Efficient Edge Computing

Pak Yeung Tsun
http://orcid.org/0009-0007-9333-0533
The Hong Kong Polytechnic University, Hong Kong

Yu-keung Chan
The Hong Kong Polytechnic University, Hong Kong

ABSTRACT

This project presents a smart home automation system based on edge computing to address the limitations of mainstream IoT systems, which rely on cloud-based architectures. The project is divided into two systems. The first system is an online simulation in Wokwi to validate the feasibility with multiple sensors controlling coloured LEDs for appliance operation and circuit structure simulation. The second system is a physical system using the ESP32 microcontroller. The system integrates with Wi-Fi and a web server for real-time monitoring and generates historical charts using Chart.js. The user can use iOS Shortcuts for HTTP-based control to achieve remote control. Edge computing ensures that all data processing and decision-making are completed locally, solving the problems of current smart home systems on the market. The simulation shows how edge computing can deliver privacy-preserving, resilient and efficient smart automation suitable for real-world environments. Finally, this project further proposes using MQTT to address the limitations of current systems.

DOI: 10.4018/979-8-2600-0701-3.ch010

1. INTRODUCTION

1.1 Background

With advances in technology, people's daily lives are undergoing gradual changes. From manual ordering at restaurants to self-service online ordering and even the emergence of various electric vehicles and autonomous driving, people's lives seem to be gradually becoming more automated and convenient. To accomplish all of this, Internet of Things (IoT) technology plays a very important role. This includes smart home systems, which are not common in Hong Kong. Nowadays, most commercial smart home systems, such as those from Xiaomi and Apple, available on the market, rely heavily on cloud architecture for data processing, remote control and storage. Although these systems offer convenient automation, they introduce several significant challenges. This Chapter aims to create an edge computing smart home system by establishing an Internet of Things (IoT) system using microcontrollers and sensors.

The Chapter is divided into two parts: the online system and the physical system. The online system simulates the entire architecture of a smart home using the Wokwi. It integrates different sensors to control LED lights of different colors to simulate real appliance switches. This validates the feasibility of using microcontrollers and sensors to create a smart automation system before moving to hardware implementation. This not only validates the feasibility of creating an automation system using microcontrollers and sensors but also provides clear circuitry and logic, making physical implementation easier.

On the other hand, the physical system uses an ESP32 microcontroller, integrated with an SHT4x temperature and humidity sensor and an LDR sensor, to control a single RGB LED. Additionally, the physical system fully integrates with the Wi-Fi and web-based systems, enabling users to view data and charts and remotely control appliances using iOS Shortcuts in real time. All decision-making, automation logic, and data logging will be handled locally on the device in this system, which demonstrates the advantage of edge computing. Finally, MQTT was added to address some of the device's current limitations and improve its futureproofing.

1.2 Problem Statement

Most mainstream smart home systems in the market primarily rely on cloud architecture and this design introduces several significant challenges as follows:

The first challenge is privacy risks, as highlighted in studies on cloud-based IoT privacy risks (Alzu'bi et al., 2024). This is because all data, including sensor data and user behavior patterns, will be transmitted to and stored on remote cloud servers. This allows companies to access the data at will, raising serious concerns about data security and potential misuse.

The second challenge is heavy reliance on the network. When the internet is interrupted and the system becomes offline, the entire system may become completely unusable. Users who rely on smart home automation would suddenly lose control of their appliances, creating inconvenience and potential safety risks.

The third challenge is higher power and bandwidth consumption. Continuously uploading sensor data to the cloud server requires significant resources, whether in terms of device battery power or user data usage. Additionally, cloud computing may not be able to handle massive volumes of data when numerous users simultaneously request access, which may reduce network bandwidth and lead to slower response times.

The details of these three challenges will be further discussed in Section 2. Considering these challenges, edge computing can address them by processing data locally and reducing reliance on constant cloud connectivity.

1.3 Objectives and Outcomes

This Chapter aims to develop a smart home system by using the ESP32 microcontroller. The problems mentioned will be solved by implementing edge computing. In the system, all data will be uploaded locally and processed. This directly addresses the privacy and resource consumption issues of cloud-based smart home systems. Additionally, it will ensure that critical functions operate during offline periods, maintaining functionality even if the network is disconnected. All appliances will be automatically controlled by the ESP32 microcontroller. Also, it will provide a local web page for checking current environmental conditions, viewing historical data and enabling manual override of automated controls. Apart from that, it will also integrate with iOS Shortcuts, providing convenient mobile access that allows users to easily control and view data remotely. Finally, MQTT was added to the end of this Chapter, aiming to address current device limitations and lay the foundation for adding users or linking to multiple controllers in the future. It is hoped that the expected outcomes can demonstrate that automation control, user-friendly web monitoring, and iOS Shortcuts control can be achieved reliably without relying on the cloud architecture. We look forward to developing a practical, low-cost and secure smart home solution suitable for Hong Kong's living environment.

2. RELATED WORK

2.1 Challenges and Solutions of Cloud-based Architectures

After years of development, smart home systems have evolved from being limited to standalone devices in their early stages to the complex Internet of Things (IoT) ecosystems today. Nowadays, most mainstream smart home systems on the market, such as Xiaomi and Apple, rely on cloud architecture for remote control and data integration. According to estimates released by Internet Estimates Center, there will be more than 50 billion terminals and devices linked to the network by 2020. There will also be more than 40 ZB of global data (Zwolenski & Weatherill, 2014). However, the continuous and massive growth of data volume and various data processing requirements will cause higher latency and require higher power and bandwidth consumption. Therefore, there are several challenges that cloud computing needs to face.

First of all, cloud computing is vulnerable to security and privacy risks as shown in Figure 1. The system usually lacks transparency and therefore user do not know where their data is stored, who can access it, who controls it or whether the data is deleted clearly after they request deletion (Ghorbel et al., 2017). Therefore, data is usually encrypted before being transferred to the cloud. But this will result in shortages of storage, bandwidth and cost each time when the whole encryption or decryption process is needed (Alzu'bi et al., 2024).

Adapted from "Security and Privacy in Cloud Computing: Technical review" by Abdulsalam and Hedabou, 2021, Future Internet, 14(1), Article 11. https://doi.org/10.3390/fi14010011

Second, cloud computing is unable to handle massive amounts of data when many users seek access at the same time. This will cause network bandwidth and transmission speed to suffer. Nowadays, modern data centers commonly suffer from insufficient bandwidth, which can be exploited and lead to DoS, denial-of-service, attacks due to shared infrastructure in cloud environments (Xiao & Xiao, 2012).

Figure 1. Cloud computing security vulnerabilities using STRIDE

Vulnerability Component	Spoofing	Tampering	Repudiation	Information Disclosure	Denial of Service	Elevation of Privilege
Immoral use and abuse of cloud computing		X	X	X	X	
Malicious insider attackers	X	X	X	X	X	X
Vulnerable programming interfaces		X		X		X
Data leakage and loss		X	X	X	X	
Distributed technology vulnerabilities	X	X				X
Services and account hijacking	X	X	X	X	X	X
Anonymous profile threat		X	X	X	X	

The symbol X denotes the existence of a STRIDE component.

Lastly, reliability and availability are important for any technology's strength. One of the key factors that undermines the reliability of cloud computing is downtime. Data shows that some cloud computing providers, such as Amazon, experience frequent outages (Sajid & Raza, 2013). Edge computing, as a transformative approach within the IoT ecosystem, addresses challenges related to latency, bandwidth and real-time decision-making. Processing data locally reduces reliance on centralized cloud systems (Ficili et al., 2025).

2.2 Edge Computing in IoT and Smart Homes

As the Internet of Things (IoT) becomes more widespread, many applications, such as transportation systems, have extremely high requirements for response time. Traditional cloud computing can no longer meet these demands. Edge computing has emerged as a transformative paradigm that addresses many of the limitations of cloud-based systems by bringing data processing and decision-making closer to the data source. Instead of sending all raw data to distant cloud servers, edge computing utilizes local devices or nearby edge nodes to perform computation in real time (Yu et al., 2017). In terms of data processing, it is faster, real-time, and secure. It can also address the problem of excessive energy consumption in cloud computing, reducing costs and alleviating pressure on network bandwidth. Edge computing is applied in various fields such as manufacturing, energy, smart homes, and transportation (Cao et al., 2020). Applying edge computing offers three advantages in IoT and smart home applications. First, it can reduce latency. It is because analyzing local data makes it possible to quickly identify abnormalities, which is essential for prompt reactions in smart home applications. Second, it can enhance privacy by processing data on-device. This can minimize the exposure of sensitive information. Finally, it requires lower bandwidth and power usage by analyzing data locally reduce the need for constant data transmission to the cloud (Reis & Serôdio, 2025).

2.3 Comparative Analysis of Cloud and Edge Computing

To gain a deeper understanding of the differences between cloud computing and edge computing, a study performed a keyword analysis of different essays. The results show that adding an edge computing layer can reduce IoT power consumption and enhance the overall architecture's security. Additionally, the keyword "edge computing" is present in most of the analyzed papers with "IoT" at the same time, while the relevant keyword related to "cloud computing" is "latency". This demonstrates that edge computing or hybrid solutions can address the problem (Andriulo et al., 2024).

Furthermore, a study analyzes the complementary strengths and limitations of cloud and edge computing in data-intensive environments. Cloud computing faces major challenges in data security, processing and response time due to distance and increasing annual expenses. Although edge computing is constrained by limited storage and computational power, it offers better security through local data processing, lower latency by performing computation near the data source and reduced costs by using inexpensive IoT devices without additional analysis transfer fees. Therefore, it is suitable for processing small datasets generated by IoT sensors (Bajic et al., 2019).

To demonstrate the advantages of edge computing, an essay tested a real-time anomaly detection application for smart homes. Applying quantization optimization to edge devices results in a 35% reduction in power consumption and a 76% reduction in computing time (Reis & Serôdio, 2025). Another research tested face recognition in different scenarios. By moving computing from the cloud to the edge, the reaction time is reduced from 900 to 169 ms (Chun et al., 2011). In another study, cloudlets, which are small-scale servers placed near the user, are used to offload computing tasks for wearable cognitive assistance. The result shows that the response time improves by around 80-200ms. Moreover, the energy consumption is also reduced by 30%–40% (Ha et al., 2014).

3. SYSTEM IMPLEMENTATION

3.1 Online System Architecture

The online system as shown in Figure 2 was developed using the Wokwi online simulator. The primary goal is to validate the feasibility of the smart home system before physical implementation. Online simulations provide a clearer view and management of circuits and the overall system in the early stages of development.

They also prevent physical hardware damage caused by short circuits resulting from incorrect circuit connections.

The online system consists of different components: ESP32 microcontroller, DHT22 Temperature & Humidity Sensor, PIR Motion Sensor, Photoresistor, Relay Module, Green LED, Purple LED, Blue LED, Red LED, Yellow LED, Breadboard, Resistors

Figure 2. Online system architecture

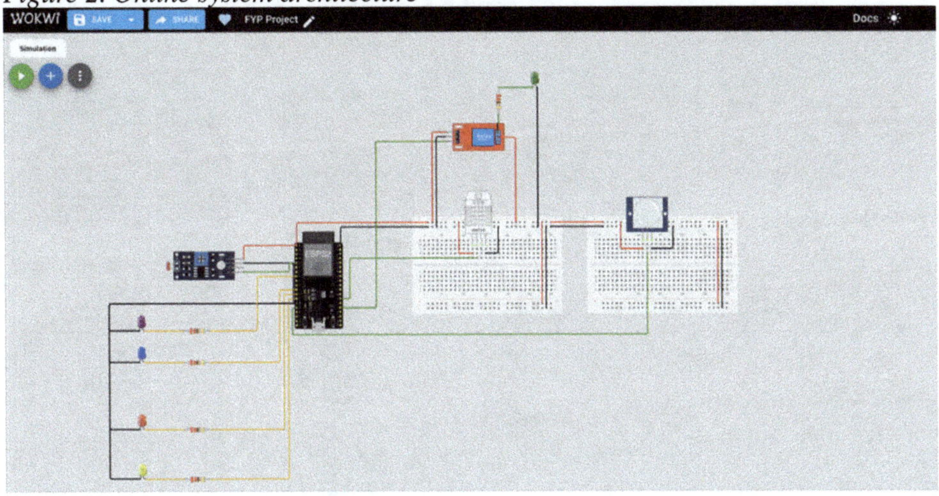

The online system in Figure 3 consists of several key components to simulate a realistic smart home environment. The black rectangle at the center is an ESP32 microcontroller. Its main responsibilities include reading sensor data and controlling all outputs. On the left-hand side, a photoresistor (LDR) sensor is connected to detect ambient light intensity. On the right-hand side, the red rectangle is a relay module. It is connected to simulate the switching of high-power appliances such as air conditioner. Then, the white rectangle near the ESP32 is DHT22 temperature and humidity sensor, which provides environmental temperature and humidity. After that, the farthest one is PIR motion sensor, which detects human motion.

Finally, there are five colored LEDs connected through 220Ω resistors to visually represent different appliances: purple for the dehumidifier simulation, blue for the main room light, red for the auxiliary light, and yellow for the fan simulation. The main function of a resistor is to limit the current, thereby protecting the circuit from burning out due to excessive current. These LEDs visually represent the automatic control based on sensor readings.

All components are connected by jumper wires and the system adopts a modular design, making each part easier to manage. Also, it facilitates the identification and correction of errors.

Figure 3. Workflow diagram of the online system

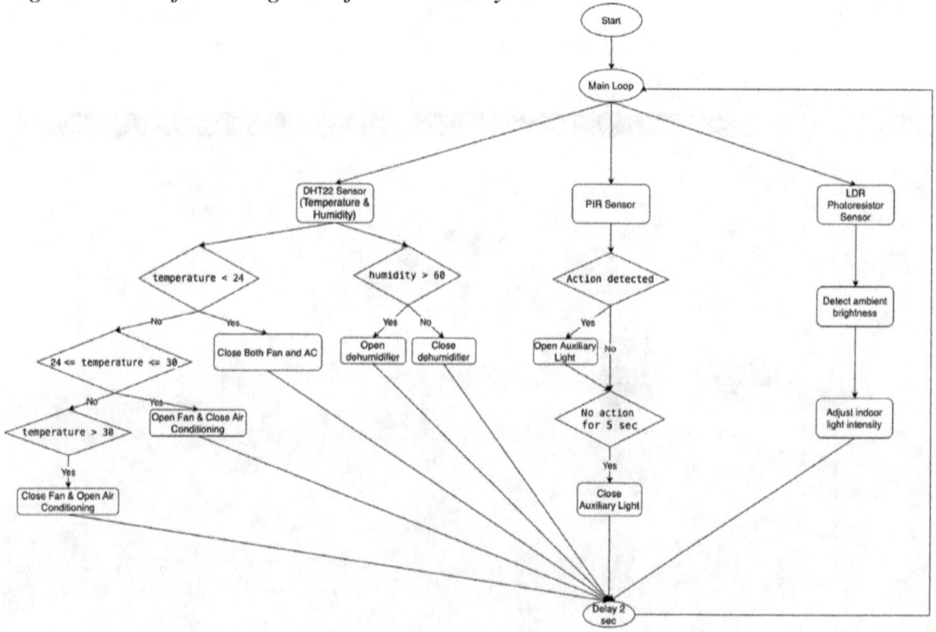

The system operates in a continuous loop. All sensors read values and refresh data every 2 seconds. LED bulbs automatically switch on and off according to predefined conditions, simulating real smart home appliances.

The automation control logic: The fan (Yellow LED) turns on when the temperature is between 24°C and 30°C. If the temperature exceeds 31°C, turn on the air conditioner (Green LED) and turn off the fan (Yellow LED). Both turn off if the temperature is 23°C or below. The dehumidifier (Purple LED) will turn on when the humidity is above 60%. The LDR sensor adjusts the brightness of the light (Blue LED) according to ambient light intensity. The light (Red LED) will turn on for 5 seconds if the PIR motion sensor detects motion.

All control logic is implemented using simple if-else statements and a modular design is used to clearly distinguish each part of the control logic, ensuring the code is easy to understand and modify later. The system also uses non-blocking control for the "Auxiliary Light Simulation" by using millis() instead of delay(), ensuring the main loop will not be blocked while the auxiliary light remains on for a long time.

3.2 Physical System Architecture

The physical system focuses in Figure 4 on real-world edge computing implementation. Its core architecture uses an ESP32 microcontroller, paired with SHT4x temperature and humidity sensor and a photoresistor (LDR) for ambient light detection. An RGB LED bulb is used to simulate the operation of fans and air conditioner by turning on green and red lights.

The physical system consists of the following components: ESP32 microcontroller, SHT4x temperature and humidity sensor, Photoresistor, RGB LED bulb, Breadboard, Jumper wires, Resistors

Figure 4. Physical system architecture

The physical device uses extracted temperature and humidity sensors and a photoresistor for simulation. As shown in the image, the leftmost component is

the ESP32 microcontroller, used for all data processing, decision making, and data storage. The first purple square on the right is the SHT4X temperature and humidity sensor, which measures the current indoor temperature and humidity. The rightmost component is the photoresistor used to detect the ambient light intensity in real time.

The SHT4x temperature and humidity sensor utilizes the I2C communication protocol. This allows efficient two-wire data transfer between the ESP32 and multiple devices. I2C design not only simplifies wiring but also facilitates future expansion by allowing additional I2C-compatible sensors to be connected in parallel without consuming extra GPIO pins. However, the LDR sensor does not support I2C communication and therefore it functions as a simple analog device. It is connected to the ESP32's analog input pin GPIO 34. Then an adjustable pull-down resistor is connected to stabilize the reading. Changes in ambient light intensity change the LDR's resistance. The variable resistance is converted into a measurable voltage, which is read by microcontrollers via analog-to-digital converters using the analogRead() function.

The system supports offline mode, in which the ESP32 continues to read sensors, control LED brightness based on temperature and light intensity and record data even if the Wi-Fi is disconnected. However, the limitation is that the web server and remote access functions will become unavailable. This design ensures critical automation functions remain operational even without a network, preventing home appliances from going out of control and causing losses or even accidents.

Figure 5. Workflow diagram of physical system

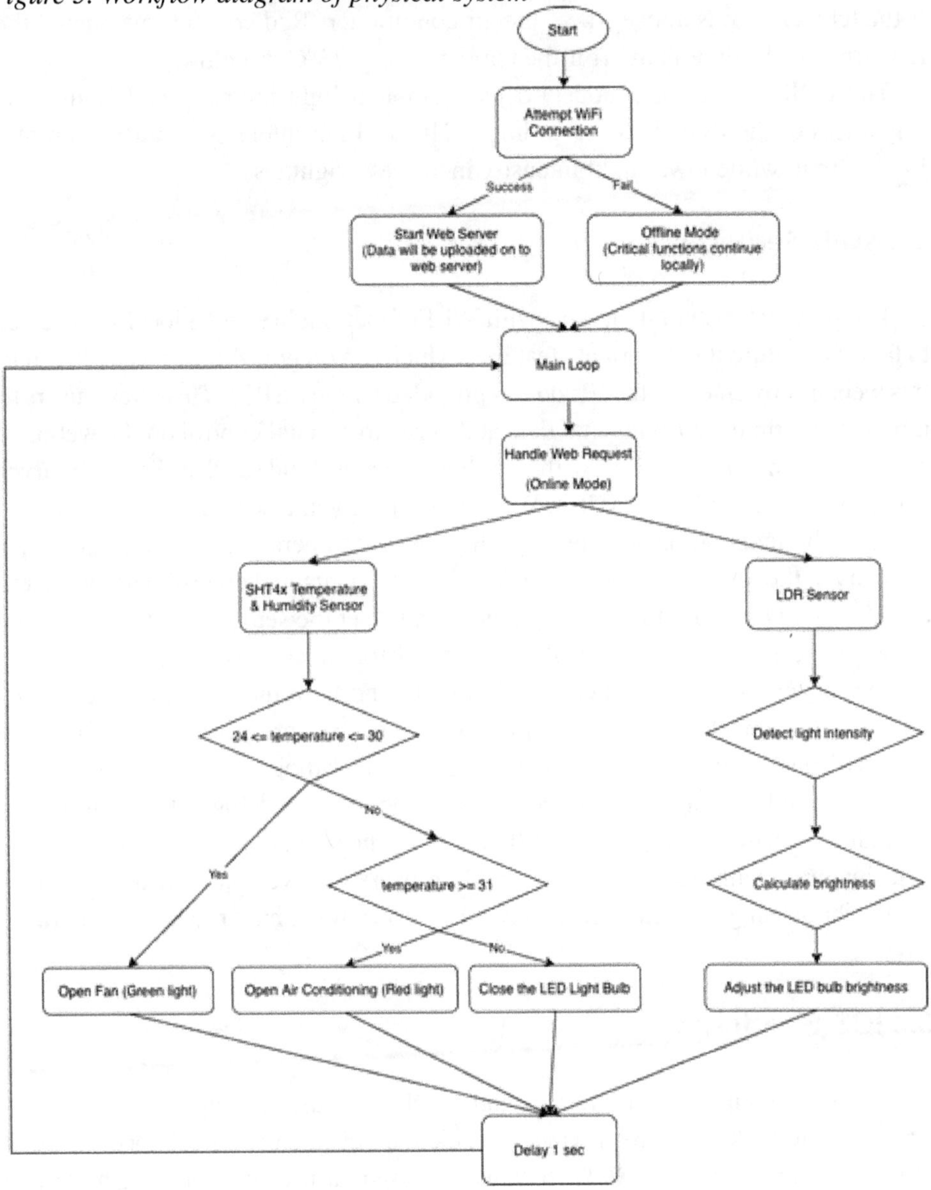

The physical system in Figure 5 also operates in a continuous loop. All sensors read values and refresh data every 1 second. The RGB LED automatically adjusts its color and brightness based on different conditions to simulate real smart home appliances.

The automation control logic:

The fan (Green color) turns on when the temperature is between 24°C and 30°C. If the temperature is above 31°C, the air conditioner (Red color) turns on and the fan turns off. Both remain off if the temperature is 23°C or below.

The LDR sensor continuously detects ambient light intensity and adjusts the brightness of the RGB LED accordingly. Higher light intensity results in dimmer LED output, while lower light intensity increases brightness.

3.3 Web Server

The physical system integrates with Wi-Fi functionality and a local web server to facilitate future iOS Shortcuts features, which run on port 80. Users can log in to this webpage by entering the IP address provided by the ESP32. This allows users to monitor real-time environmental data and perform manual control on the webpage within the same local network without relying on any cloud service. The web server is built using the ESP32's built-in WebServer library and serves two main pages:

First, the main homepage displays the current temperature, humidity and light intensity at the top. Under the data display, there are manual control buttons. Users can choose to open or close the appliances and reset the settings. At the bottom of the webpage, there is the historical chart. The historical chart is generated by Chart.js. The line chart shows temperature, humidity and light intensity over the last 48 records. The two-Y-axis design ensures all three parameters are clearly visible. The X-axis displays time in "hh:mm" format for easier reading.

The second webpage is a historical chart page. Users can log in to this webpage by adding "/chart" to the end of the IP address. The main purpose of this webpage is to integrate with iOS Shortcuts, enabling users to save charts to their phone's photo album using shortcuts. This allows users to view the historical charts anytime, anywhere.

3.4 iOS Shortcuts

To avoid a complex user interface and develop a dedicated app, the system integrates with iOS Shortcuts to provide convenient mobile control. The shortcuts allow users to interact with the ESP32, such as viewing real-time data, storing historical data and controlling electrical appliances, via simple HTTP GET requests to the ESP32's local web server.

Implemented shortcuts:

1. View real-time temperature, humidity and light intensity: This shortcut sends an HTTP GET request to the /data endpoint. Then the ESP32 responds with a

JSON object containing the latest sensor readings. The shortcut will extract the values from the dictionary and display them on the screen.

2. Open/Close Fan: This shortcut sends an HTTP GET request to either the /fan_on or /fan_off endpoint. After the ESP32 receives the request, it will immediately set the Manual_FanOff flag to true or false and adjust the green channel of the RGB LED according to the user's request. The shortcut then displays a confirmation notification.
3. Open/Close Air Conditioning: Similar to the fan shortcut, this one sends an HTTP GET request to /ac_on or /ac_off. The ESP32 updates the Manual_AcOff flag to true or false and controls the red channel of the RGB LED. A notification is shown to confirm the change.
4. Browse the historical data chart: This shortcut opens the Safari browser to the /chart endpoint by URL directly. The user can view the line chart in the browser.
5. Save the historical data chart: It first obtains the URL of the historical chart page. Then, it makes a PDF from the URL so that it can extract an image from that PDF. Finally, it saves the image to the album and displays a notification.

Figure 6. iOS Shortcuts interface

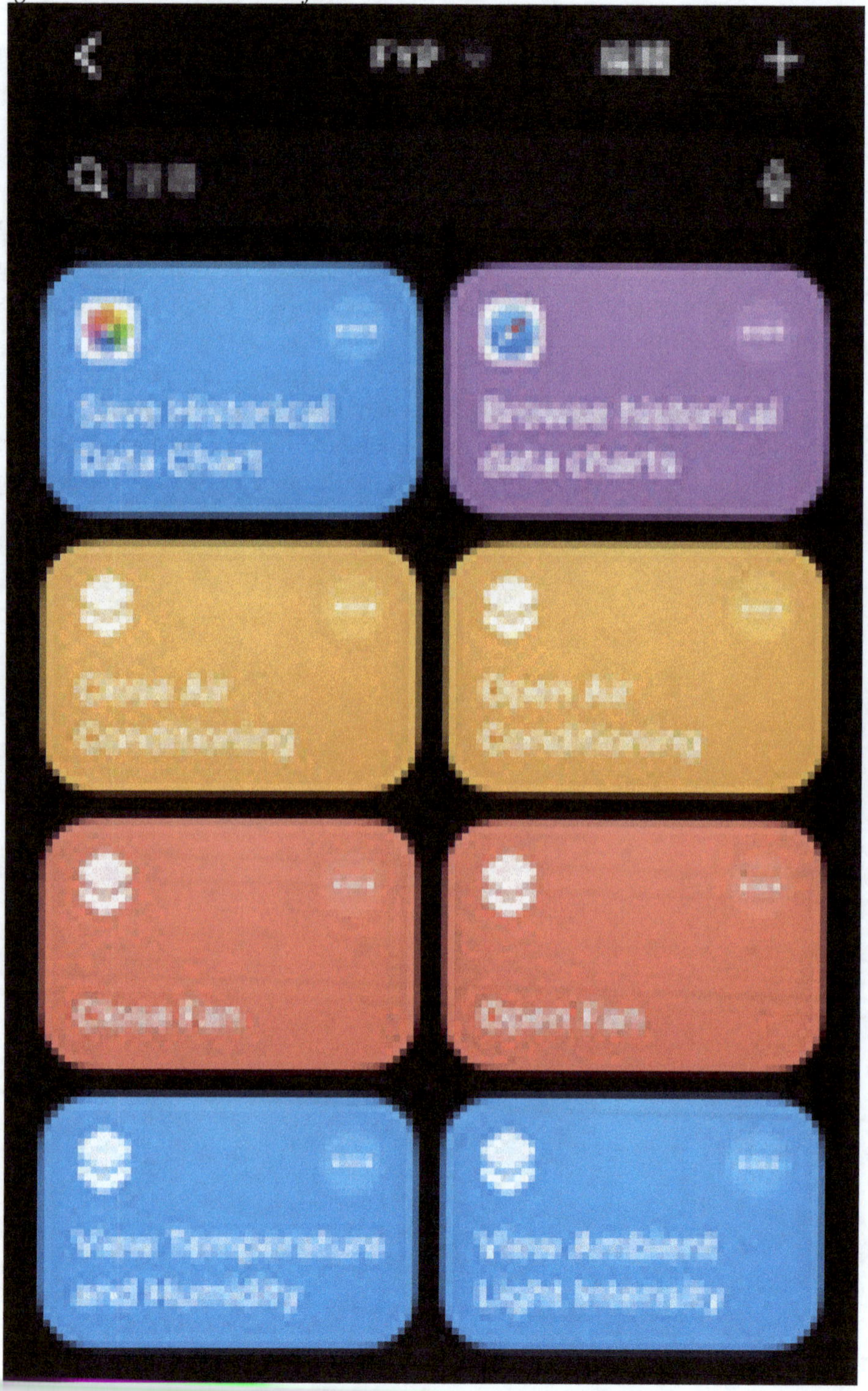

Users can also use iOS Shortcuts' in Figure 6 and Figure 7 built-in automation features to customize automation workflows to suit their needs. For example, as shown in the image below, iOS Shortcuts automatically saves historical charts for users at midnight. If the historical charts are set to record every half hour, users can save various changes in indoor data every day. Additionally, the shortcuts can be used to turn on the appliance, which can be set to automatically run immediately when the user connects to the home network. This saves users the trouble of manually turning on each appliance when they return home. It also allows the room to be adjusted to the most comfortable environment in the shortest time, fully demonstrating the convenience of smart automation.

Figure 7. Example of automated execution shortcuts

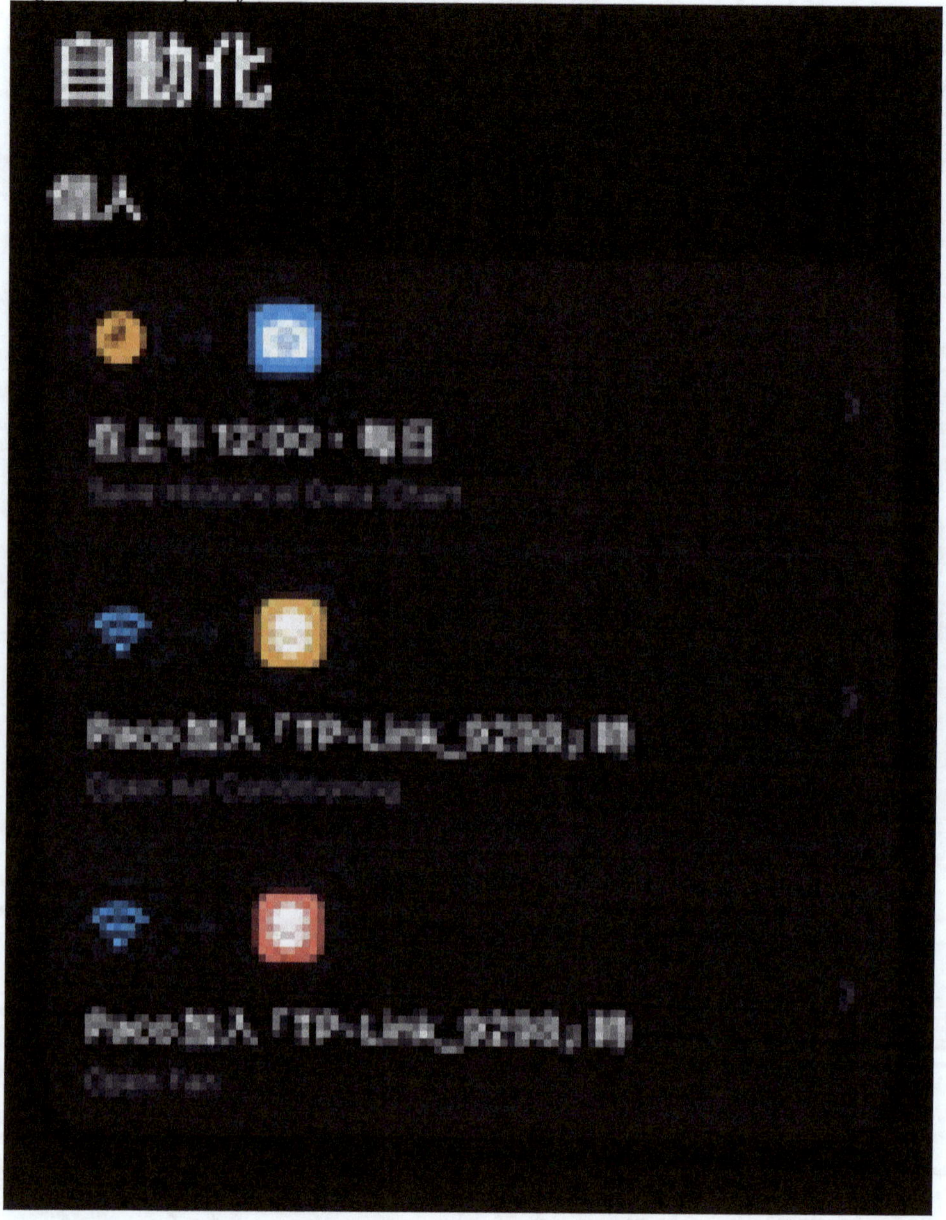

The integration demonstrates how edge computing systems can work seamlessly with mobile devices without relying on third-party cloud services.

4. SIMULATION/EXPERIMENTATION RESULT

4.1 Online System Simulation

The primary goal of the online system simulation is to test sensor integration, automation logic and real-time data display in a safe, virtual environment. This can lay the foundation for future physical installations and reduce the risk of errors such as short circuits or device damage.

After pressing the "Start" button in the upper left corner and starting the simulation, the ESP32 will be initialized. Then the user can view the device's status in the Serial Monitor, such as whether it is on. When it is initializing, the Serial Monitor will display a message "Initializing Smart Home System......" to the user. Once the system initializes in Figure 8 successfully and all components are ready, the Serial Monitor will print the start-up message "System initialization complete. Activating automatic control." This provides loading feedback to the user, thereby enhancing System Status Visibility.

Figure 8. Serial monitor initializing online simulation

```
ets Jul 29 2019 12:21:46

rst:0x1 (POWERON_RESET),boot:0x13 (SPI_FAST_FLASH_BOOT)
configsip: 0, SPIWP:0xee
clk_drv:0x00,q_drv:0x00,d_drv:0x00,cs0_drv:0x00,hd_drv:0x00,wp_drv:0x00
mode:DIO, clock div:2
load:0x3fff0030,len:1156
load:0x40078000,len:11456
ho 0 tail 12 room 4
load:0x40080400,len:2972
entry 0x400805dc
Initializing Smart Home System......
System initialization complete. Activating automatic control.
```

Once the system starts running, it will enter the loop() function, which is the central decision-making engine of the system. The system will perform an operation every 2 seconds using the delay(2000) function, including reading sensor data and printing the latest data and device status to the Serial Monitor. It is because of the balance between responsiveness and sensor stability, which DHT22 recommends at least 2 seconds between readings.

During simulation, users can interactively adjust the sensor values, such as by dragging the slider or clicking the button. The Serial Monitor will update the real-time readings and appliance status every 2 seconds. The corresponding-colored LEDs will also be changed, such as turning on or off or adjusting the brightness according to the reading and logic. This immediate visual feedback confirmed that the automation rules were working correctly.

The results demonstrated in Figure 9 and Figure 10 shows the successful implementation of the complete automated system network and the automation logic for each component. Additionally, online simulations offer greater efficiency than physical devices, such as faster code uploads and increased fault tolerance. This enables the rapid development of a comprehensive smart home system.

Figure 9. Real-time data printed by the serial monitor

```
=== Smart Home Real-Time Status ===
Temperature: 51.4 °C
Humidity: 67.0 %
Motion Detected: No
Ambient Light Intensity: 756
Fan: OFF | Air Conditioner: ON
Dehumidifier: ON
Room Light Brightness: 63
============================

=== Smart Home Real-Time Status ===
Temperature: 51.4 °C
Humidity: 67.0 %
Motion Detected: No
Ambient Light Intensity: 756
Fan: OFF | Air Conditioner: ON
Dehumidifier: ON
Room Light Brightness: 63
============================
```

Figure 10. Custom values of the online system

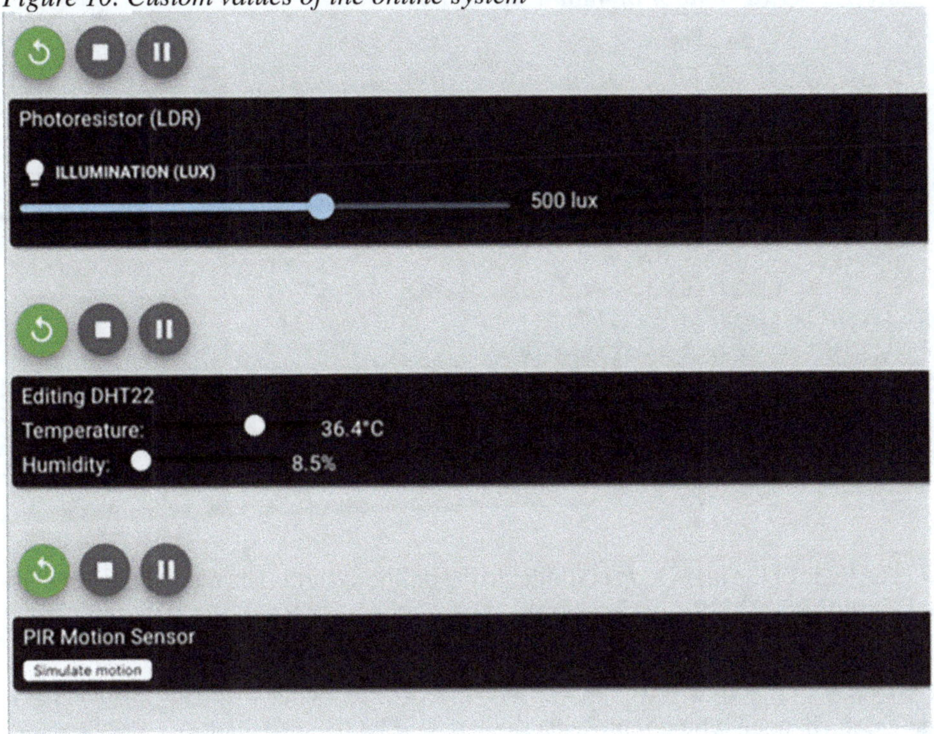

4.2 Physical System Simulation

The physical system implementation was conducted following the successful online simulation. The objective is to implement various functions in virtual devices in the real world, such as running automated logic and testing edge computing functions, to ensure the system functions correctly in the real world. Because physical tests use actual components, we can obtain more realistic performance data, such as response time, sensor accuracy and offline behavior.

After uploading the code to the ESP32, the ESP32 in Figure 11 first initializes the SHT4x sensor in high-precision mode. Then, it attempts to connect to the Wi-Fi network. If the connection succeeds, the local web server starts on port 80 and MQTT is initialized. The first data point will be recorded and saved to the circular buffer. The Serial Monitor will display the system status as shown in the image below. The objective is also to provide loading feedback to the user.

Figure 11. Serial monitor of initializing ESP32

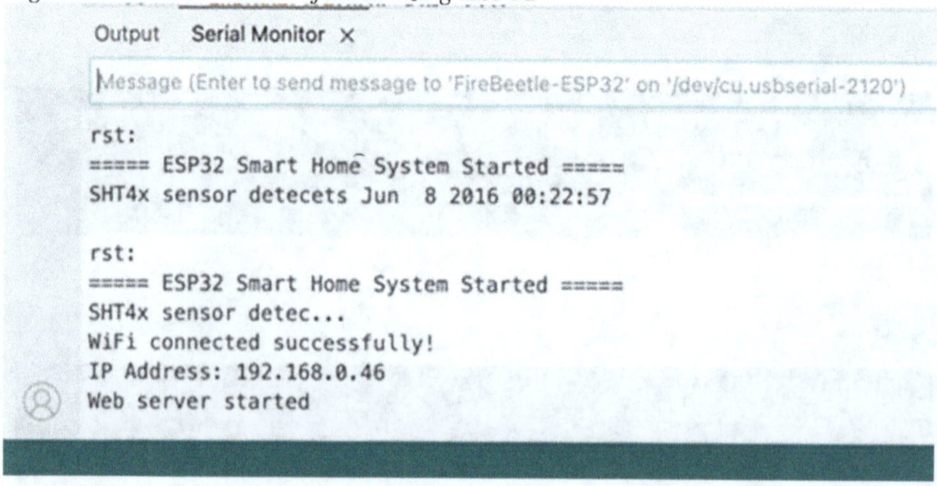

The logic of the main loop is similar to the online system. The main loop executes every 100 ms and all sensors read the data within each iteration. These real-time data are then used to evaluate the following automation logic:

If the temperature is between 24 °C and 30 °C and manual override is not active, the green channel, which represents the fan, will be lit.

If the temperature exceeds 31 °C and manual override is not active, the red channel, which represents the air conditioner, will be lit.

Otherwise, the LED will not be lit as shown in Figure 12.

Additionally, the LDR value is mapped to a PWM brightness level and applied to the LED, adjusting the brightness based on the ambient light conditions.

Figure 12. LED simulates the operation of electrical appliances

However, historical data is saved into the circular buffer every 10 seconds and Serial Monitor output is limited to every 5 seconds to prevent flooding. Differentiating the cycle times of different tasks according to their importance and requirements ensures that automated control can respond immediately but data will not overflow, thus achieving a balance between components.

To verify edge computing capability, the Wi-Fi SSID was input with an incorrect value. The ESP32 enters offline mode after failed Wi-Fi connections. If the Wi-Fi connection fails, the Serial Monitor will print a message to notify the user and the ESP32 will continue all critical functions locally. The result shows that even without a network, the system continues to read the latest data, control the LED based on the automation logic and record data in the circular buffer. Additionally, the device will continue to detect Wi-Fi and reconnect when Wi-Fi becomes available, regardless of how long the device has been offline. This confirms that all critical functions remain fully operational without internet dependency as shown Figure 13.

Figure 13. Wi-Fi connection failure notification

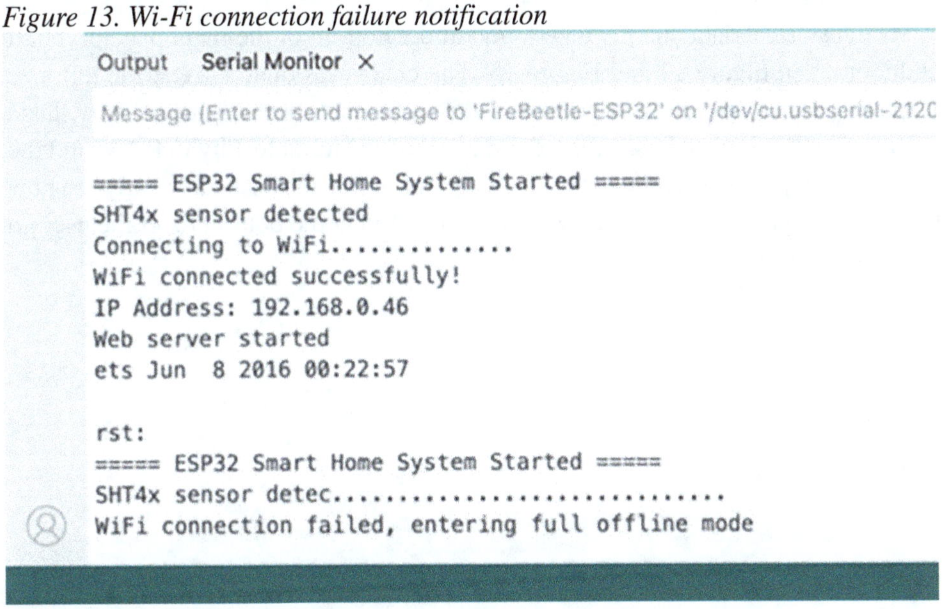

These results demonstrate the successful migration of the Chapter from virtual simulation to physical implementation and confirm the application of edge computing principles.

4.3 Webpage Testing

To evaluate the usability and real-time performance of the web server, a webpage test was conducted on the physical system. On the same Wi-Fi network, the webpage can be accessed by entering the ESP32's local IP address in the browser.

The main page serves as the central dashboard and displays the following real-time information at the top:

Under the real-time readings are four buttons allowing manual control. Clicking the buttons can trigger the corresponding endpoints:/fan_on, /fan_off, /ac_on, /ac_off, and/reset. This will instantly update the LED state. All manual override flags, such as Manual_FanOff and Manual_AcOff, are respected by the automation logic in loop(), ensuring the user can easily control electrical appliances. The result shows that the LED lights switched almost instantly after the button was pressed as in Figure 14 and Figure 15.

The historical data chart can be found at the bottom of the main page or /chart endpoint as in Figure 14 and Figure 15. The chart uses dual Y-axes: the left axis displays temperature and humidity and the right axis shows light intensity. Data points are pulled from the 48-record circular buffer stored locally in RAM and the X-axis displays the exact time of each data point so that the user can clearly see how the data changes over time. Lastly, there is a button at the bottom to let the user go back to the homepage.

Figure 14. Homepage of the web server

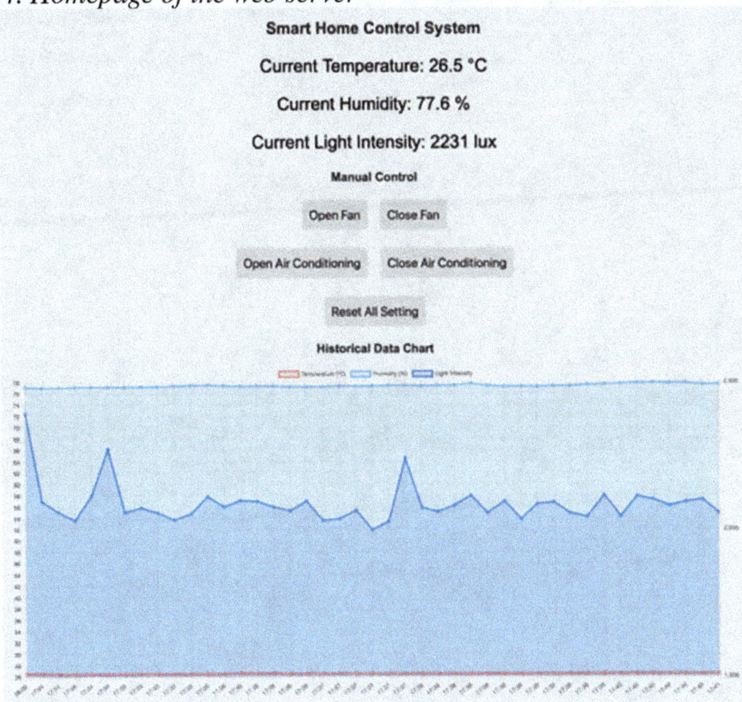

Additionally, a loading screen has been added to the webpage for manual control. When any control button is pressed by the user, a loading screen will appear to provide the user with loading feedback. After that, the webpage will automatically redirect to the homepage. This design directly follows Jakob Nielsen's first usability heuristic, Visibility of system status. It states that "the system should always keep users informed about what is going on, through appropriate feedback within a reasonable time" (Nielsen, 2006). This can prevent users from repeatedly clicking the same button because of feeling uncertain.

To conclude, the webpage testing shows that the system provides a reliable user interface for monitoring and control. At the same time, the webpage is very important for the following iOS Shortcuts app, laying the foundation for its development.

Figure 15. Chart endpoint

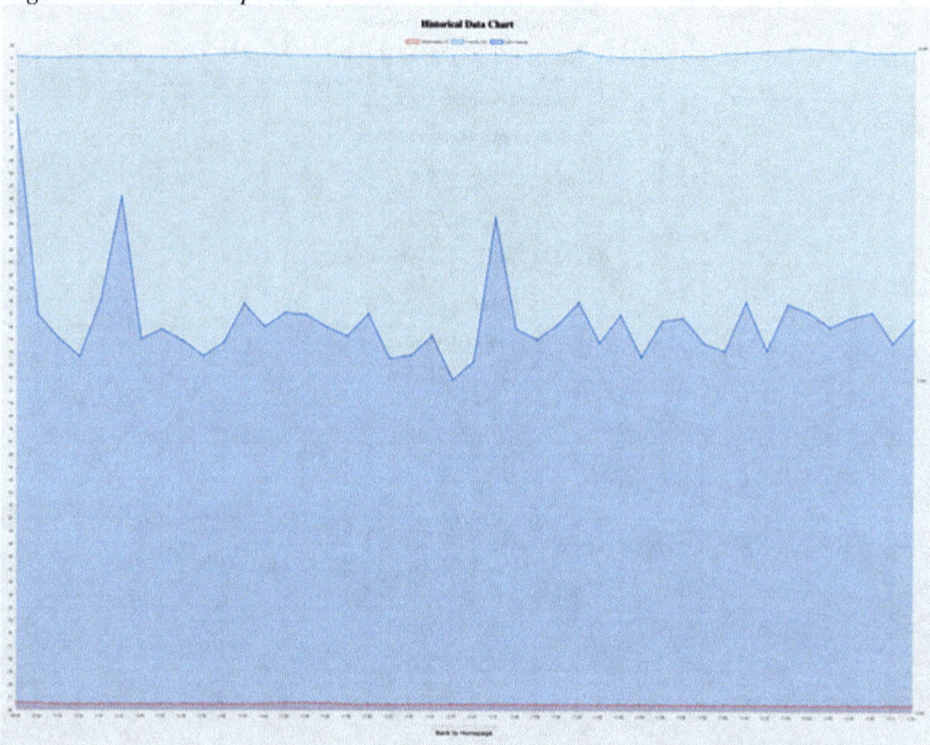

Figure 16. Loading feedback of the web server

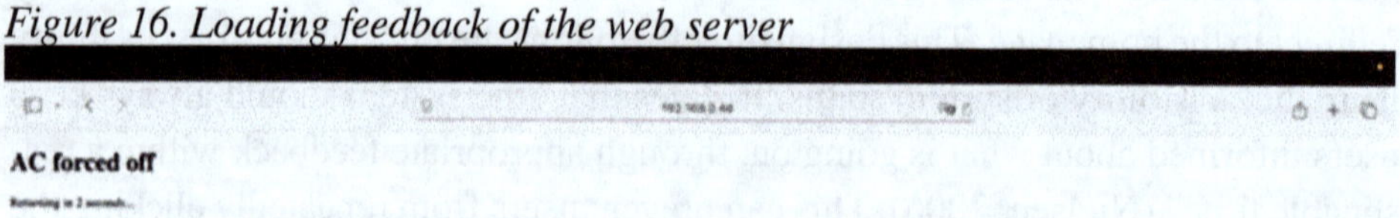

4.4 iOS Shortcuts Testing

After the webpage is set up, iOS Shortcuts can be developed and integrated with the ESP32 to perform different actions, such as viewing data or controlling appliances. The shortcuts use simple HTTP GET requests with existing endpoints, such as fan_on and /fan_off, shown in Figure 16, to perform the relevant action.

View real-time temperature, humidity and light intensity shown in Figure 17 and Figure 18: After clicking the shortcut, the real-time environmental data is successfully displayed on the phone.

Open/Close Fan: After clicking the Close Fan shortcut, the LED's green channel, which represents the fan, will close immediately until the user clicks the Open Fan shortcut. A notification will be sent to the user after completion.

Figure 17. Demonstration of viewing real-time data

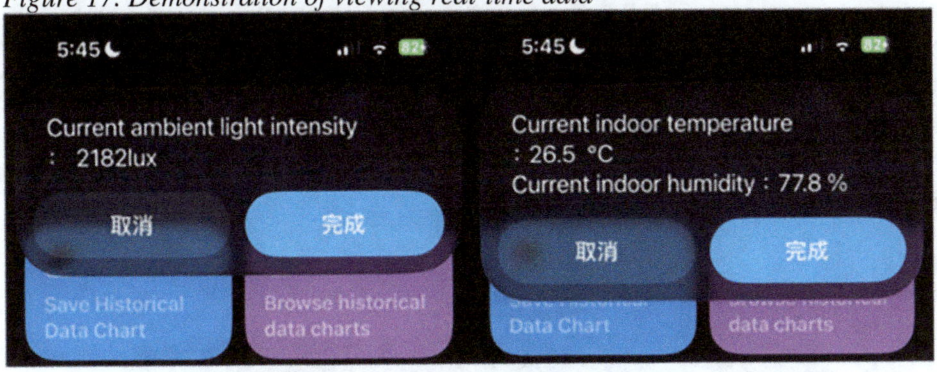

Figure 18. Notification of iOS Shortcuts

Open/Close Air Conditioning: After clicking the Air Conditioning shortcut, the LED's red channel, which represents the Air Conditioning, will close immediately until the user clicks the Open Air Conditioning shortcut. A notification will be sent to the user after completion.

Figure 19. Notification of iOS Shortcuts

Browse the historical data chart: After clicking this shortcut as shown in Figure 19, the browser successfully automatically opens and redirects to the webpage, http://[IP Address]/chart, which only contains the history charts.

Save the historical data chart as shown in Figure 20 and Figure 21: After clicking the shortcut, a notification will pop up after a few seconds indicating that the chart has been saved to the album. The newest chart in the open album will then appear in the newest position.

Figure 20. Open /chart endpoint by shortcuts

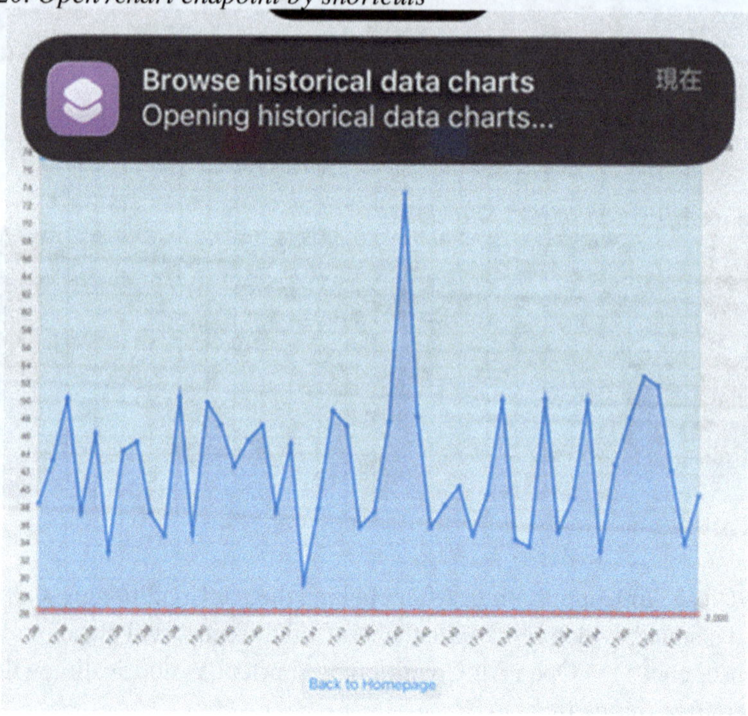

Figure 21. Successfully saved photos to the album

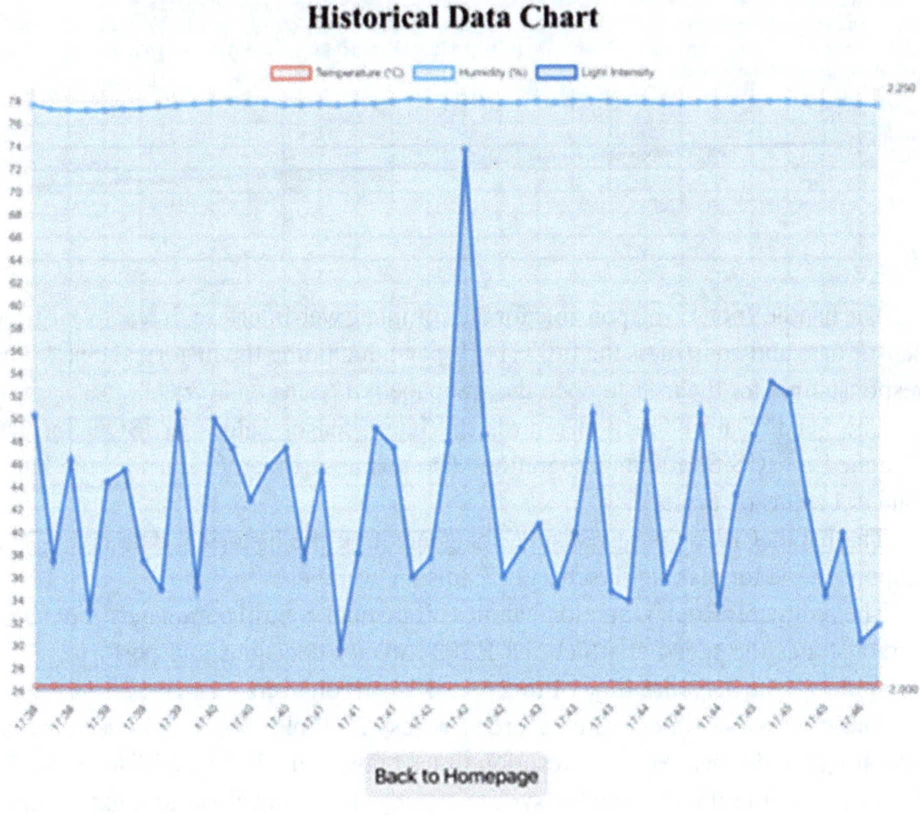

4.5 Reaction Time of Physical System

To demonstrate the device's real-time performance and edge computing speed, the response times of the three primary HTTP endpoints were systematically measured. This is achieved by using millis() function to record the duration time from the start to the server.send() command was executed. The following will describe the responsibilities of the three HTTP endpoints and their response times as shown in Figure 22:

Figure 22. The result of response time

```
HTTP Response Time (redirectToRoot): 2 ms
HTTP Response Time (redirectToRoot): 3 ms
HTTP Response Time (redirectToRoot): 2 ms
HTTP Response Time (redirectToRoot): 3 ms
HTTP Response Time (redirectToRoot): 3 ms
```

```
HTTP Response Time (handleChart): 7 ms
HTTP Response Time (handleChart): 8 ms
HTTP Response Time (handleChart): 8 ms
HTTP Response Time (handleChart): 8 ms
HTTP Response Time (handleChart): 8 ms
```

```
HTTP Response Time (handleData): 15 ms
HTTP Response Time (handleData): 14 ms
HTTP Response Time (handleData): 15 ms
HTTP Response Time (handleData): 15 ms
HTTP Response Time (handleData): 15 ms
```

```
HTTP Response Time (handleRoot): 18 ms
HTTP Response Time (handleRoot): 17 ms
HTTP Response Time (handleRoot): 18 ms
HTTP Response Time (handleRoot): 18 ms
HTTP Response Time (handleRoot): 17 ms
```

The handleRoot() is responsible for the complete web interface. It reads real-time sensor data and constructs the full HTML page, including the historical chart. The response time for the user to open the webpage is 17.6ms on average.

The handleData() function returns the latest sensor values in JSON format, designed for iOS Shortcuts integration. The response time of returning the latest data is 14.8ms on average.

The handleChart() is responsible for generating the historical data chart. The response time for making the chart is 7.8ms on average.

The redirectToRoot() function handles all manual control commands. This best demonstrates the speed at which the ESP32 makes decisions and performs edge computing. The response time of this part is 2.6ms on average.

These response times were recorded across multiple test runs under normal operating conditions. All of its reaction times range from 2 to 18 milliseconds. To compare with the cloud-based IoT system, a study shows that cloud solutions require longer response times. For example, controlling smart light required an average of 3.18 seconds for cloud-based mobile apps, while smart speaker took 4.53 seconds (Lee et al., 2020b). Both studies involve controlling home appliances. By comparing the test results of this Chapter, local computing is clearly much faster than the cloud system described in the research report.

5. RESULT ANALYSIS & DISCUSSION

5.1 Overall System Performance

To conclude, the simulation shows that all components of the implemented smart home system can be successfully integrated and demonstrate strong performance. In the online simulation, all sensor readings and actuator responses operated as expected. The physical system further validated these results in real-world conditions by using a portion of the sensors. These results confirm that the edge computing design can solve the limitations of the traditional cloud-based architecture. The three main problems currently facing cloud architecture mentioned above are privacy risks, heavy reliance on the network and higher power and bandwidth consumption, causing slower response times. How the device in this Chapter solves these problems through edge computing will be discussed as follows.

5.2 Privacy and Data Security

First of all, privacy and data security are substantially improved. Cloud-based architecture requires continuously transmitting data to third-party servers for processing and storage. This may increase the risk of data leakage through cyberattacks, unauthorized access or malicious actors. In contrast, ESP32 performs data processing and storing locally. No data will be uploaded to the external server unless the user requests it. Simulation clearly demonstrates this design. The data are processed directly by the ESP32 using if-else automation logic and stored in a circular buffer. The local web server and iOS Shortcuts communicate via HTTP GET requests to the ESP32's local IP address. This architecture eliminates the need for third-party cloud accounts, thereby removing the privacy risks associated with commercial servers.

5.3 Reliance on the Network

Another feature of the ESP32 is that it can maintain full functionality even without a network. This can directly address the limitation of reliance on the network of the cloud-based system. In cloud-based architectures, automation logic and data processing require data to be uploaded to the remote servers. This will result in an immediate loss of control and monitoring once the network is disconnected. However, the ESP32 demonstrates offline functionality in the testing. When Wi-Fi is turned off and the ESP32 is disconnected during testing, the ESP32 continues to read real-time data from the sensor and adjust LED color and brightness based on automation logic. Once Wi-Fi is turned back on, the device will automatically re-establish the connection and upload data to resume the local web server. This was repeatedly verified in physical tests by turning off the Wi-Fi for different durations, confirming that the system will not completely lose control even without a network.

5.4 Power and Bandwidth Consumption

One of the advantages of using edge computing architecture instead of cloud-based architecture is its significantly lower power consumption and bandwidth usage compared to cloud-based architecture. This is because cloud architecture requires data, such as sensor readings, to be periodically transmitted to the remote server for processing and storage, resulting in high bandwidth and power consumption. In contrast, ESP32 performs all critical functions, such as decision making and automation logic, locally on the device. It only sent data over the network if necessary.

Also, frequent cloud uploading typically requires higher power consumption because it requires remaining in transmit mode to send raw data to remote servers. However, the ESP32 consumes less power in RX mode and consumes higher power when TX mode is required for sending data. According to Espressif's official datasheet, Wi-Fi transmission in TX mode consumes 180–240 mA, while RX mode consumes only 95–100 mA (Espressif Systems, 2021). Compared to cloud architectures that force frequent TX transmissions, ESP32 can reduce overall power consumption.

5.5 Challenges Encountered and Solutions

During the development of the Chapter, some challenges were encountered and successfully solved. The first challenge occurred during the development of the Online System in Wokwi. When the system was just finished, it would suddenly stop during the simulation every time. After several tests, it was found that the system would stop after motion was detected, with the Red LED lit. It is because the Red LED that should light when detecting movement was initially set to remain on for 5 seconds using a simple delay(5000) command. However, this will block the entire main loop, causing the system to stop reading data or executing other automation logic during that period. As a result, the whole system became unresponsive whenever motion was detected. This seriously affects the performance and reliability of the entire system. To solve the problem, the code was redesigned by using millis() for a non-blocking design. Two variables, motionStartTime and motionActive, were also introduced for recording the start time of motion detection and the current state of the light. The five-second duration was calculated by millis() - motionStartTime. Finally, the non-blocking design was successfully implemented. The motion detection and red LED will not interrupt the main loop.

The second challenge was visual scaling of the historical data chart. In the beginning, the Y-axis was set with a scale of 5 degrees per major grid line when only temperature and humidity were displayed. Therefore, the plotted lines are always displayed between grid lines. It becomes difficult for users to read the values accurately. After the Light Intensity line was added to the graph, the chart became more crowded and the overall scale became smaller, causing the temperature and humidity lines to become harder to read. Additionally, the initial Y-axis scale of light intensity was too small, causing the line to exhibit excessive fluctuation. Considering that temperature and humidity are important parameters, such as providing meaningful daily weather references, while light intensity has comparatively lower reference for users, the scale of both Y-axis was adjusted for balance after testing different values. Finally, the temperature and humidity were adjusted to 1 degree per major grid line while the light intensity was adjusted to 500 lux per major grid line. This improved readability and allowed users to determine the exact values of important data more easily.

The most concerning challenge occurred when setting the iOS Shortcuts automated operation. Initially, I wanted the system to automatically execute all shortcuts that turn off appliances when the user leaves a certain area of their home. This can ensure that all appliances are turned off when no one is at home. This can prevent unnecessary expenses because of power usage and accidents caused by user negligence. However, I found that this approach is impractical because the ESP32 system relies on edge computing, which means everything runs locally on the same Wi-Fi network. Executing shortcuts requires all devices to be connected to the same network. However, when the user leaves home and disconnects from their home Wi-Fi, the shortcuts become unusable. The loss of Wi-Fi connectivity also prevents access to the device's local web pages, resulting in a total loss of control. Additionally, I recognized that iOS Shortcuts are exclusive to iOS devices. Given the diversity of smartphone brands available, this exclusivity limits the system's potential applicability across a broader range of homes and public spaces. Therefore, to overcome the limitation of the system, MQTT, which is called Message Queuing Telemetry Transport, was integrated into the physical ESP32 system as an additional communication layer. The details of MQTT will be discussed in Section 6.

6. EXPANSION AND FUTURE WORK

6.1 MQTT

Communication protocols are used in edge computing to facilitate communication between fog nodes, cloud servers, edge devices, and edge servers. Information may be sent between devices and network nodes in a safe, reliable, and effective manner thanks to these protocols. MQTT is one of the most popular communication protocols for edge computing (Pradeep, 2023). MQTT is a fast, lightweight and energy-efficient publish-subscribe protocol that uses a central broker as an intermediary. Devices can publish messages or subscribe to topics to receive commands (Collina et al., 2012c). It allows communication between the system and users even without the same network, as long as both devices have internet access.

Figure 23. MQTT WebSocket Client

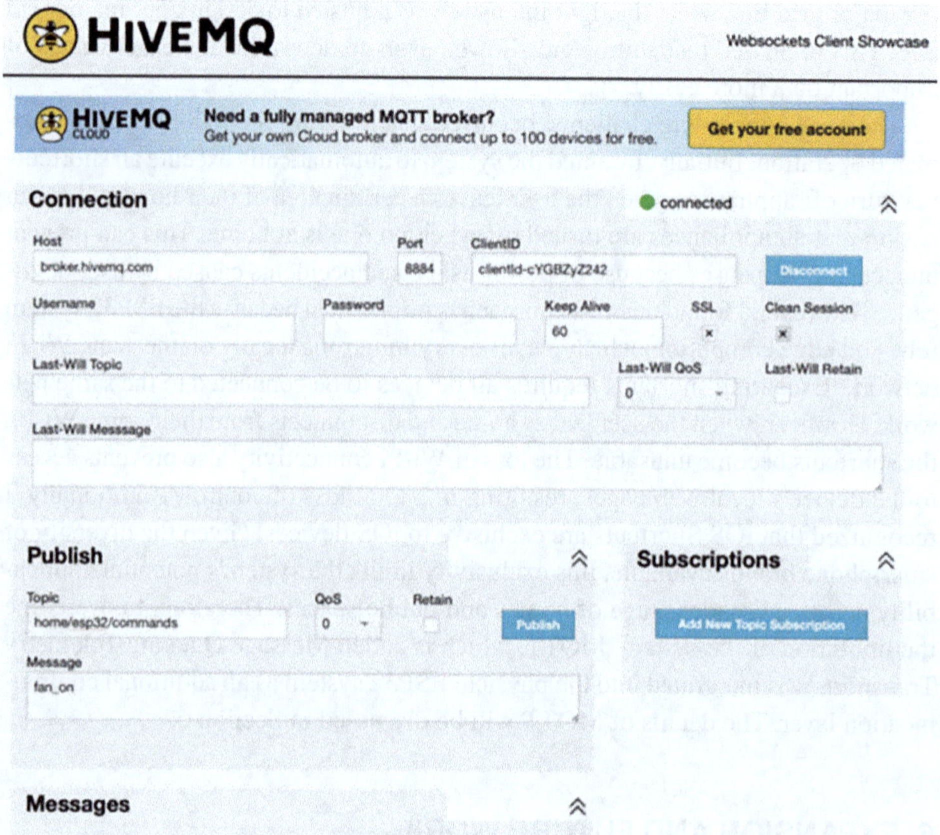

In this Chapter, MQTT was implemented by using the PubSubClient library as shown in Figure 23. The ESP32 publishes real-time data to the topic home/esp32/sensors every 5 seconds. Additionally, it subscribes to the topic home/esp32/commands to receive messages from external clients. The ESP32 will update the corresponding manual override flag and adjust the RGB LED after receiving a valid command, such as “fan_on” and “fan_off”.

Figure 22 shows the HiveMQ WebSocket Client, a web-based MQTT client interface provided by HiveMQ Cloud. This tool acts as an external MQTT client, allowing users to manually publish commands or subscribe to a specific topic. In this Chapter, it serves as a testing and demonstration platform to simulate remote control without the same network.

To compare with the current system, a study conducted a performance test by measuring the number of records successfully transmitted and inserted into the database over a 60-second period. The results showed that MQTT outperformed

HTTP, achieving up to 6 times the data transmission rate under the same test conditions. This performance is primarily attributed to MQTT's lightweight architecture (Atmoko et al., 2017).

By integrating MQTT, the requirement to use the same network has been successfully resolved. Users can now control appliances remotely from anywhere using an MQTT client application or a webpage. This hybrid approach retains the advantages of the original system's architecture and edge computing, while simultaneously expanding the system's future applicability and practicality by enabling true remote control independent of the local Wi-Fi network.

6.2 Future Improvement

Although the current system successfully demonstrates the automation logic control and core concepts of edge computing, several enhancements could further improve its functionality and practicality in the future.

Sensor Expansion: Currently, the physical device only implements two of the most commonly used sensors and data. Therefore, the scope of automation is limited by the number of sensors installed. In the future, more useful sensors such as PIR motion sensors and air quality monitors will be integrated. This expansion can make the device more versatile and applicable to a wider range of situations.

Customized Conditions Interface: Currently, all automation logic is fixed. For example, the air conditioner will only activate when the temperature is above 31 degrees Celsius. However, considering the different needs of different users and regions, a user-friendly interface will be developed on the web server. Users will be able to adjust the condition of the automation logic with a simple dashboard. This feature allows the system to adapt more effectively to varying home environments and seasonal changes. This feature can make the system more flexible.

Status Anomaly Alarm: An anomaly detection mechanism will be implemented. If the system detects abnormal conditions such as a sudden rapid increase in temperature, the ESP32 will immediately send a notification to the user's smartphone. This feature allows users to react promptly and avoid accidents that could result in loss of property or even life.

User Authentication and Security: The current web server does not require authentication, which may pose a security risk if the local network is compromised. To enhance system security, user authentication, such as using a username and password to log in, will be added to the web server. This upgrade will prevent unauthorized access to the control interface and data, enhancing system security.

Electricity Consumption Monitoring: By integrating a power sensing module, users will be able to monitor real-time and historical electricity consumption of the appliances directly through the webpage or iOS Shortcuts. This can help users to identify high consumption devices. Users can base on the data to optimize usage habits so that power consumption and electricity costs can be reduced.

These improvements can improve the current system into a more comprehensive smart automation system. Future work will focus on implementing these features to enhance the system reliability, security and applicability

7. CONCLUSION

This Chapter successfully developed a smart home automation system with edge computing using the ESP32 microcontroller. Online simulations have confirmed that it is feasible to use the ESP32 to control multiple sensors and electrical switches. It also demonstrates basic automation rules and non-blocking timing for detection. The physical devices demonstrated that real-time environmental monitoring and automated home appliances can be achieved using an ESP32, an SHT4X sensor and an LDR sensor, while overcoming the problems of traditional cloud platforms. It integrates with a local web server for data visualization and manual overrides. The most important is that it can still operate even when disconnected from the network, ensuring critical functions can run without internet dependency. This solves the problem of not being able to upload data to the cloud when there is no internet connection. By combining the results of online simulations and physical devices, a complete smart home system with edge computing can be created.

Overall, the system proves that edge computing enables privacy protection, low latency, power efficiency and reliability in network-disrupted environments. Compared to the commercial systems, this system offers a low-cost, open-source substitute for Hong Kong's market, where smart home adoption is limited by cost and privacy concerns. The Chapter's focus on local processing aligns with emerging IoT trends and showcases ESP32's versatility for embedded applications.

REFERENCES

Abdulsalam, Y. S., & Hedabou, M. (2021). Security and Privacy in Cloud Computing: Technical review. *Future Internet*, *14*(1), 11. DOI: 10.3390/fi14010011

Alzu'bi, A., Alomar, A., Alkhaza'leh, S., Abuarqoub, A., & Hammoudeh, M. (2024). A Review of Privacy and Security of Edge Computing in Smart Healthcare Systems: Issues, Challenges, and Research Directions. *Tsinghua Science and Technology*, *29*(4), 1152–1180. https://ieeexplore.ieee.org/document/10431732. DOI: 10.26599/TST.2023.9010080

Andriulo, F. C., Fiore, M., Mongiello, M., Traversa, E., & Zizzo, V. (2024). Edge computing and cloud computing for Internet of Things: A review. *Informatics (MDPI)*, *11*(4), 71. Advance online publication. DOI: 10.3390/informatics11040071

Atmoko, R. A., Riantini, R., & Hasin, M. K. (2017). IoT real time data acquisition using MQTT protocol. *Journal of Physics: Conference Series*, *853*, 012003. DOI: 10.1088/1742-6596/853/1/012003

Bajic, B., Cosic, I., Katalinic, B., Moraca, S., Lazarevic, M., & Rikalovic, A. (2019). Edge Computing vs. Cloud Computing: Challenges and Opportunities in Industry 4.0. In Annals of DAAAM for . . . & proceedings of the . . . International DAAAM Symposium (pp. 0864–0871). DOI: 10.2507/30th.daaam.proceedings.120

Cao, K., Liu, Y., Meng, G., & Sun, Q. (2020). An overview on edge computing research. *IEEE Access : Practical Innovations, Open Solutions*, *8*, 85714–85728. DOI: 10.1109/ACCESS.2020.2991734

Chun, B., Ihm, S., Maniatis, P., Naik, M., & Patti, A. (2011). CloneCloud: Elastic execution between mobile device and cloud. *EuroSys '11: Proceedings of the Sixth Conference on Computer Systems*, 301–314. DOI: 10.1145/1966445.1966473

Collina, M., Corazza, G. E., & Vanelli-Coralli, A. (2012c). Introducing the QEST broker: Scaling the IoT by bridging MQTT and REST. 2012 IEEE 23rd International Symposium on Personal, Indoor and Mobile Radio Communications, 36–41. https://ieeexplore.ieee.org/abstract/document/6362813

Espressif Systems. (2021). ESP32 Series Datasheet Version 5.2 [Report]. https://www.espressif.com/documentation/esp32_datasheet_en.pdf

Ficili, I., Giacobbe, M., Tricomi, G., & Puliafito, A. (2025). From Sensors to Data Intelligence: Leveraging IoT, Cloud, and Edge Computing with AI. *Sensors (Basel)*, *25*(6), 1763. DOI: 10.3390/s25061763 PMID: 40292910

Ghorbel, A., Ghorbel, M., & Jmaiel, M. (2017). Privacy in cloud computing environments: A survey and research challenges. *The Journal of Supercomputing*, *73*(6), 2763–2800. DOI: 10.1007/s11227-016-1953-y

Ha, K., Chen, Z., Hu, W., Richter, W., Pillai, P., & Satyanarayanan, M. (2014). Towards wearable cognitive assistance. *MobiSys '14: Proceedings of the 12th Annual International Conference on Mobile Systems, Applications, and Services*, 68–81. DOI: 10.1145/2594368.2594383

Lee, H., Mun, H., & Lee, Y. (2020b). Comparing Response Time of Home IoT Devices with or without Cloud. *2020 IEEE International Conference on Consumer Electronics*, 1–6. DOI: 10.1109/ICCE46568.2020.9043102

Nielsen, J. (2006). Ten usability heuristics. *Engineering in Life Sciences*, *20*(11), 485–493. DOI: 10.1002/elsc.202000024

Pradeep, A. (2023). Exploring the future of edge computing: advantages, limitations, and opportunities. In Communications in computer and information science (pp. 196–209). DOI: 10.1007/978-3-031-45124-9_15

Reis, M. J. C. S., & Serôdio, C. (2025). Edge AI for Real-Time Anomaly Detection in smart Homes. *Future Internet*, *17*(4), 179. DOI: 10.3390/fi17040179

Sajid, M., & Raza, Z. (2013). Cloud computing: issues & challenges. In International Conference on Cloud, Big Data and Trust 2013, Nov 13-15, RGPV, International Conference on Cloud, Big Data and Trust 2013, Nov 13-15, RGPV. https://wanivipin.wordpress.com/wp-content/uploads/2016/11/challenges-of-cc.pdf

Xiao, Z., & Xiao, Y. (2012). Security and privacy in cloud computing. *IEEE Communications Surveys and Tutorials*, *15*(2), 843–859. DOI: 10.1109/SURV.2012.060912.00182

Yu, W., Liang, F., He, X., Hatcher, W. G., Lu, C., Lin, J., & Yang, X. (2017). A survey on the edge computing for the internet of things. *IEEE Access : Practical Innovations, Open Solutions*, *6*, 6900–6919. https://ieeexplore.ieee.org/document/8123913. DOI: 10.1109/ACCESS.2017.2778504

Zwolenski, M., & Weatherill, L. (2014). The digital universe: Rich data and the increasing value of the internet of things. *Journal of Telecommunications and the Digital Economy*, *2*(3). Advance online publication. https://search.informit.org/doi/abs/10.3316/informit.678436300116927. DOI: 10.7790/ajtde.v2n3.47

Compilation of References

Abdelmoniem, A. M., Jaber, M., Anwar, A., Zhang, Y., & Gao, M. (2024). Towards a Decentralized Collaborative Framework for Scalable Edge AI. *Future Internet, 16*(11), 421. DOI: 10.3390/fi16110421

Abdulsalam, Y. S., & Hedabou, M. (2021). Security and Privacy in Cloud Computing: Technical review. *Future Internet, 14*(1), 11. DOI: 10.3390/fi14010011

Abujabal, N. A., Rabie, T., & Kamel, I. (2023). Path planning techniques for multi-robot systems: A systematic review. In 2023 15th International Conference on Innovations in Information Technology (IIT) (pp. 1-6). DOI: 10.1109/IIT59782.2023.10366472

Al-Fuqaha, A., Guizani, M., Mohammadi, M., Aledhari, M., & Ayyash, M. (2015). Internet of Things: A survey on enabling technologies, protocols, and applications. *IEEE Communications Surveys and Tutorials, 17*(4), 2347–2376. DOI: 10.1109/COMST.2015.2444095

Al-Garadi, M. A., Mohamed, A., Al-Ali, A. K., Du, X., Ali, I., & Guizani, M. (2020). A survey of machine and deep learning methods for Internet of Things (IoT) security. *IEEE Communications Surveys and Tutorials, 22*(3), 1646–1685. DOI: 10.1109/COMST.2020.2988293

Almarri, A., Hunaiti, Z., & Manivannan, N. (2025). A review of smart healthcare: Concept, drivers, characteristics, and challenges. *Hospitals, 2*(4), 26. DOI: 10.3390/hospitals2040026

Alzu'bi, A., Alomar, A., Alkhaza'leh, S., Abuarqoub, A., & Hammoudeh, M. (2024). A Review of Privacy and Security of Edge Computing in Smart Healthcare Systems: Issues, Challenges, and Research Directions. *Tsinghua Science and Technology, 29*(4), 1152–1180. https://ieeexplore.ieee.org/document/10431732. DOI: 10.26599/TST.2023.9010080

Amershi, S., Weld, D., Vorvoreanu, M., Fourney, A., Nushi, B., Collisson, P., & Horvitz, E. (2019). Guidelines for human-AI interaction. In *Proceedings of the 2019 CHI Conference on Human Factors in Computing Systems* (pp. 1-13). DOI: 10.1145/3290605.3300233

Anas, A. S., Alchamdani, , Syam, S., Arwan, , & Firmansyah, . (2025). The role of IoT and emerging technologies in shaping smart hospitals. *Medicor: Journal of Health Informatics and Health Policy, 3*(2), 96–109. DOI: 10.61978/medicor.v3i2.1083

Andriulo, F. C., Fiore, M., Mongiello, M., Traversa, E., & Zizzo, V. (2024). Edge computing and cloud computing for Internet of Things: A review. *Informatics (MDPI), 11*(4), 71. Advance online publication. DOI: 10.3390/informatics11040071

Arai, T., Pagello, E., & Parker, L. E. (2002). Editorial: Advances in multi-robot systems. *IEEE Transactions on Robotics and Automation, 18*(5), 655–661. DOI: 10.1109/TRA.2002.806024

Association of Hong Kong Nursing Staff. (2025, May). *Nursing manpower survey report 2025.* https://nurse.org.hk/storage/media/HA%E4%BA%BA%E6%89%8B%E5%95%8F%E5%8D%B7%E6%96%B0%E8%81%9E%E7%A8%BF_12052025%20(Final).pdf

Atmoko, R. A., Riantini, R., & Hasin, M. K. (2017). IoT real time data acquisition using MQTT protocol. *Journal of Physics: Conference Series, 853*, 012003. DOI: 10.1088/1742-6596/853/1/012003

Atzori, L., Iera, A., & Morabito, G. (2010). The Internet of Things: A survey. *Computer Networks, 54*(15), 2787–2805. DOI: 10.1016/j.comnet.2010.05.010

Bailey, J. (2022, December 7). Airlines burn $4 billion of untouched food and drink every year, here's why. Simple Flying. https://simpleflying.com/airlines-4-billion-untouched-food-drink/

Bajic, B., Cosic, I., Katalinic, B., Moraca, S., Lazarevic, M., & Rikalovic, A. (2019). Edge Computing vs. Cloud Computing: Challenges and Opportunities in Industry 4.0. In Annals of DAAAM for . . . & proceedings of the . . . International DAAAM Symposium (pp. 0864–0871). DOI: 10.2507/30th.daaam.proceedings.120

Banzi, M., & Shiloh, M. (2014). *Getting started with Arduino* (3rd ed.). Maker Media, Inc.

Baryannis, G., Validi, S., Dani, S., & Antoniou, G. (2019). Supply chain risk management and artificial intelligence: State of the art and future research directions. *International Journal of Production Research, 57*(7), 2179–2202. DOI: 10.1080/00207543.2018.1530476

Bédard, C., Lajoie, P.-Y., Beltrame, G., & Dagenais, M. (2023). Message flow analysis with complex causal links for distributed ROS 2 systems. *Robotics and Autonomous Systems*, *161*, 104361. DOI: 10.1016/j.robot.2022.104361

Bédard, C., Lütkebohle, I., & Dagenais, M. (2022). ros2_tracing: Multipurpose low-overhead framework for real-time tracing of ROS 2. *IEEE Robotics and Automation Letters*, *7*(3), 6511–6518. DOI: 10.1109/LRA.2022.3174346

Ben-Daya, M., Hassini, E., & Bahroun, Z. (2019). Internet of things and supply chain management: A literature review. *International Journal of Production Research*, *57*(15-16), 4719–4742. DOI: 10.1080/00207543.2017.1402140

Bergs, L., Tréheux, G., Schäper, L., Pinheiro de Souza, L. M., Münker, S., Göppert, A., & Schmitt, R. H. (2024). Edge computing framework for enhanced robotic adaptivity in line-less mobile assembly systems. *Procedia CIRP*, *129*, 26–31. DOI: 10.1016/j.procir.2024.07.006

Bewley, A., Ge, Z., Ott, L., Ramos, F., & Upcroft, B. (2016). *Simple online and realtime tracking*. arXiv. https://arxiv.org/abs/1602.00763 DOI: 10.1109/ICIP.2016.7533003

Bochkovskiy, A., Wang, C.-Y., & Liao, H.-Y. M. (2020). *YOLOv4: Optimal speed and accuracy of object detection*. arXiv. https://arxiv.org/abs/2004.10934

Bossard, L., Guillaumin, M., & Van Gool, L. (2014). Food-101: Mining discriminative components with random forests. In D. Fleet, T. Pajdla, B. Schiele, & T. Tuytelaars (Eds.), *Computer Vision - ECCV 2014* (pp. 446-461). Springer. DOI: 10.1007/978-3-319-10599-4_29

Bouman, R., Bukhsh, Z., & Heskes, T. (2024). Unsupervised anomaly detection algorithms on real-world data: How many do we need? *Journal of Machine Learning Research*, *25*(105), 1–34. https://arxiv.org/abs/2305.00735

Brambilla, M., Ferrante, E., Birattari, M., & Dorigo, M. (2013). Swarm robotics: A review from the swarm engineering perspective. *Swarm Intelligence*, *7*(1), 1–41. DOI: 10.1007/s11721-012-0075-2

Brintrup, A., Kito, T., & Wakolbinger, T. (2020). A review of machine learning in supply chain management: Applications, challenges and opportunities. *Computers & Industrial Engineering*, *139*, 105774. DOI: 10.1016/j.cie.2019.106764

Cadena, C., Carlone, L., Carrillo, H., Latif, Y., Scaramuzza, D., Neira, J., Reid, I., & Leonard, J. J. (2017). Past, present, and future of simultaneous localization and mapping: Toward the robust-perception age. *IEEE Transactions on Robotics*, *32*(6), 1309–1332. DOI: 10.1109/TRO.2016.2624754

Caesar, H., Bankiti, V., Lang, A. H., Vora, S., Liong, V. E., Xu, Q., Krishnan, A., Pan, Y., Baldan, G., & Beijbom, O. (2020). Nuscenes: A multimodal dataset for autonomous driving. *Proceedings of the IEEE/CVF Conference on Computer Vision and Pattern Recognition*, 11621–11631. DOI: 10.1109/CVPR42600.2020.01164

Campos, C., Elvira, R., Rodríguez, J. J. G., Montiel, J. M., & Tardós, J. D. (2021). Orb-slam3: An accurate open-source library for visual, visual–inertial, and multimap slam. *IEEE Transactions on Robotics*, *37*(6), 1874–1890. DOI: 10.1109/TRO.2021.3075644

Cantero, D., Esnaola-Gonzalez, I., Miguel-Alonso, J., & Jauregi, E. (2022). Benchmarking object detection deep learning models in embedded devices. *Sensors (Basel)*, *22*(11), 4205. DOI: 10.3390/s22114205

Cao, K., Liu, Y., Meng, G., & Sun, Q. (2020). An overview on edge computing research. *IEEE Access : Practical Innovations, Open Solutions*, *8*, 85714–85728. DOI: 10.1109/ACCESS.2020.2991734

Cao, Y. U., Fukunaga, A. S., & Kahng, A. B. (1997). Cooperative mobile robotics: Antecedents and directions. *Autonomous Robots*, *4*(1), 7–27. DOI: 10.1023/A:1008855018923

Card, S. K., Moran, T. P., & Newell, A. (1983). *The psychology of human-computer interaction.* Lawrence Erlbaum Associates.

Cheng, T., Song, L., Ge, Y., Liu, W., Wang, X., & Shan, Y. (2024). YOLO-World: Real-time open-vocabulary object detection. *Proceedings of the IEEE/CVF Conference on Computer Vision and Pattern Recognition*, 16901-16911. https://openaccess.thecvf.com/content/CVPR2024/html/Cheng_YOLO-World_Real-Time_Open-Vocabulary_Object_Detection_CVPR_2024_paper.html

Cheng, Y., Wang, D., Zhou, P., & Zhang, T. (2017). A survey of model compression and acceleration for deep neural networks. *arXiv Preprint arXiv:1710.09282*.

Chen, J., & Ran, X. (2019). Deep learning with edge computing: A review. *Proceedings of the IEEE, 107*(8), 1655–1674. DOI: 10.1109/JPROC.2019.2921977

Chen, J., & Yin, Y. (2019). Smart home architecture based on edge computing. *IEEE Access : Practical Innovations, Open Solutions*, *7*, 10245–10255.

Choi, H., Xiang, Y., & Kim, H. (2021). PiCAS: New design of priority-driven chain-aware scheduling for ROS2. In 2021 IEEE 27th Real-Time and Embedded Technology and Applications Symposium (RTAS) (pp. 251-263). DOI: 10.1109/RTAS52030.2021.00028

Chovet, L. P., Garcia, G. M., Bera, A., Richard, A., Yoshida, K., & Olivares-Mendez, M. A. (2025). Performance comparison of ROS2 middlewares for multi-robot mesh networks in planetary exploration. *Journal of Intelligent & Robotic Systems, 111*(1), 18. DOI: 10.1007/s10846-024-02211-2

Christopher, M., & Peck, H. (2004). Building the resilient supply chain. *International Journal of Logistics Management, 15*(2), 1–14. DOI: 10.1108/09574090410700275

Chun, B., Ihm, S., Maniatis, P., Naik, M., & Patti, A. (2011). CloneCloud: Elastic execution between mobile device and cloud. *EuroSys '11: Proceedings of the Sixth Conference on Computer Systems*, 301–314. DOI: 10.1145/1966445.1966473

Chung, S. J., Paranjape, A. A., Dames, P., Shen, S., & Kumar, V. (2018). A survey on aerial swarm robotics. *IEEE Transactions on Robotics, 34*(4), 837–855. DOI: 10.1109/TRO.2018.2857475

Collina, M., Corazza, G. E., & Vanelli-Coralli, A. (2012c). Introducing the QEST broker: Scaling the IoT by bridging MQTT and REST. 2012 IEEE 23rd International Symposium on Personal, Indoor and Mobile Radio Communications, 36–41. https://ieeexplore.ieee.org/abstract/document/6362813

Corsaro, A., Cominardi, L., Hecart, O., Baldoni, G., Avital, J. E. P., Loudet, J., Guimares, C., Ilyin, M., & Bannov, D. (2024). Zenoh: Unifying communication, storage and computation from the cloud to the microcontroller. In 2023 26th Euromicro Conference on Digital System Design (pp. 422-428). DOI: 10.1109/DSD60849.2023.00065

Costa, F., Melo, M. V., Gadelha, I., Fôlego, G., Gambaro, L., & Rodrigues, A. (2021). Self-portrait to ID document face matching: CNN-based face verification in cross-domain scenario. *Anais do XVII Workshop de Visão Computacional (WVC 2021)*, 31-36. DOI: 10.5753/wvc.2021.18885

Cunha, N. A., Bispo, B. C., Cavalcante, E. L., Inocêncio, A. V. M., Alves, G. J., & Rodrigues, M. A. B. (2022). Application of MQTT network for communication in healthcare establishment. In *XXVII Brazilian Congress on Biomedical Engineering (CBEB 2020)* (pp. 1465–1470). Springer. DOI: 10.1007/978-3-030-70601-2_216

Dai, X. (2024). Robust deep-learning-based refrigerator food recognition. *Frontiers in Artificial Intelligence*, *7*, 1442948. DOI: 10.3389/frai.2024.1442948 PMID: 39697798

Debnath, U., & Kim, S. (2025). A comprehensive review of heart rate measurement using remote photoplethysmography and deep learning. *Biomedical Engineering Online*, *24*(1), 73. Advance online publication. DOI: 10.1186/s12938-025-01405-5 PMID: 40542336

Dechouniotis, D., Spatharakis, D., & Papavassiliou, S. (2022). Edge robotics experimentation over next generation IIoT testbeds. In *2022 IEEE/IFIP Network Operations and Management Symposium* (pp. 1-3). DOI: 10.1109/NOMS54207.2022.9789889

Deng, J., Guo, J., Xue, N., & Zafeiriou, S. (2019). ArcFace: Additive angular margin loss for deep face recognition. *2019 IEEE/CVF Conference on Computer Vision and Pattern Recognition (CVPR)*, 4685-4694. DOI: 10.1109/CVPR.2019.00482

Dospinescu, O., Grigorcea, G.-L., & Lefter, B.-I. (2025). A mobile approach to food expiration date determination using OCR and on-cloud image classification. *Applied System Innovation*, *8*(6), 176. DOI: 10.3390/asi8060176

Dritsas, E., & Trigka, M. (2025). A Survey on Cybersecurity in IoT. *Future Internet*, *17*(1), 30. DOI: 10.3390/fi17010030

Dunbabin, M., & Marques, L. (2012). Robots for environmental monitoring: Significant advancements and applications. *IEEE Robotics & Automation Magazine*, *19*(1), 24–39. DOI: 10.1109/MRA.2011.2181683

Eclipse Foundation. (n.d.-a). *Eclipse Mosquitto*. https://mosquitto.org/

Eclipse Foundation. (n.d.-b). *Eclipse Paho MQTT Python Client*. https://eclipse.dev/paho/files/paho.mqtt.python/html/index.html

Engel, J., Koltun, V., & Cremers, D. (2017). Direct sparse odometry. *IEEE Transactions on Pattern Analysis and Machine Intelligence*, *40*(3), 611–625. DOI: 10.1109/TPAMI.2017.2658577 PMID: 28422651

Espressif Systems. (2021). ESP32 Series Datasheet Version 5.2 [Report]. https://www.espressif.com/documentation/esp32_datasheet_en.pdf

Espressif Systems. (n.d.-a). *ESP-IDF Programming Guide - ESP32-S3*. https://docs.espressif.com/projects/esp-idf/en/latest/esp32s3/

Espressif Systems. (n.d.-b). *Product overview - ESP32-S3*. https://docs.espressif.com/projects/esp-hardware-design-guidelines/en/latest/esp32s3/product-overview.html

Fall, K. (2003). A delay-tolerant network architecture for challenged internets. In *Proceedings of the 2003 Conference on Applications, Technologies, Architectures, and Protocols for Computer Communications (SIGCOMM)* (pp. 27-34). ACM. DOI: 10.1145/863955.863960

FastAPI. (n.d.). *Deployment*. https://fastapi.tiangolo.com/deployment/

Few, S. (2006). *Information dashboard design: The effective visual communication of data*. O'Reilly Media.

Feyerke, A. (2013). Designing offline-first web apps. A List Apart. https://alistapart .com/article/offline-first/

Ficili, I., Giacobbe, M., Tricomi, G., & Puliafito, A. (2025). From Sensors to Data Intelligence: Leveraging IoT, Cloud, and Edge Computing with AI. *Sensors (Basel)*, *25*(6), 1763. DOI: 10.3390/s25061763 PMID: 40292910

Florea, V., & Rebedea, T. (2020). Expiry date recognition using deep neural networks. *International Journal of User-System Interaction*, *13*(1), 1–17. DOI: 10.37789/ijusi.2020.13.1.1

Folego, G., Angeloni, M. A., Stuchi, J. A., Godoy, A., & Rocha, A. (2016). Cross-Domain Face Verification: Matching ID Document and Self-Portrait Photographs. /arXiv.1611.05755DOI: 10.48550

Francis-Coad, J., Hill, A.-M., Jacques, A., Chandler, A. M., Richey, P. A., Mion, L. C., & Shorr, R. I. (2020). Association between characteristics of injurious falls and fall preventive interventions in acute medical and surgical units. *The Journals of Gerontology. Series A, Biological Sciences and Medical Sciences*, *75*(10), e152–e158. DOI: 10.1093/gerona/glaa032 PMID: 31996903

Galliera, R., & Suri, N. (2022). Object detection at the edge: Off-the-shelf deep learning capable devices and accelerators. *Procedia Computer Science*, *205*, 239–248. DOI: 10.1016/j.procs.2022.09.025

Ganz, D. A., Huang, C., Saliba, D., Shier, V., Berlowitz, D., Lukas, C. V., Pelczarski, K., Schoelles, K., Wallace, L. C., & Neumann, P. (2013, January). *Preventing falls in hospitals: A toolkit for improving quality of care* (Prepared by RAND Corporation, Boston University School of Public Health, and ECRI Institute under Contract No. HHSA290201000017I TO #1; AHRQ Publication No. 13-0015-EF). Agency for Healthcare Research and Quality. https://www.ahrq.gov/sites/default/files/publications/files/fallpxtoolkit_0.pdf

Gautam, M. K., Pati, A., Mishra, S. K., Appasani, B., Kabalci, E., Bizon, N., & Thounthong, P. (2021). A comprehensive review of the evolution of networked control system technology and its future potentials. *Sustainability (Basel)*, *13*(5), 2962. DOI: 10.3390/su13052962

Geifman, Y., & El-Yaniv, R. (2019). SelectiveNet: A deep neural network with an integrated reject option. In *Proceedings of the 36th International Conference on Machine Learning* (pp. 2151-2159). PMLR. https://proceedings.mlr.press/v97/geifman19a.html

Gettens, S., Fulbrook, P., Jessup, M., & Low Choy, N. (2018). The patients' perspective of sustaining a fall in hospital: A qualitative study. *Journal of Clinical Nursing*, *27*(3–4), 743–752. DOI: 10.1111/jocn.14075 PMID: 28926152

Ghorbel, A., Ghorbel, M., & Jmaiel, M. (2017). Privacy in cloud computing environments: A survey and research challenges. *The Journal of Supercomputing*, *73*(6), 2763–2800. DOI: 10.1007/s11227-016-1953-y

Golshany, H., Ni, Y., Yu, Q., & Fan, L. (2025). IoT-enabled smart kitchen technologies and their impact on food storage, preparation, and culinary experiences: A systematic review. *Food Research International*, *213*, 116557. DOI: 10.1016/j.foodres.2025.116557 PMID: 40436590

Gong, D., Jiang, Y., Sun, L., Xu, J., & Wu, Y. (2021). A novel unified deep neural networks methodology for use-by-date recognition in retail food package image. *Signal, Image and Video Processing*, *15*(3), 449–457. DOI: 10.1007/s11760-020-01764-7

Grigorik, I. (2013). *High performance browser networking*. O'Reilly Media.

Grosso, S., Longhini, J., Tonet, S., Bernard, I., Corso, J., Marchi, D., Dorigo, L., Funes, G., Lussu, M., Oppio, N., Grassetti, L., Pais Dei Mori, L., & Palese, A. (2021). Prevalence and reasons for non-nursing tasks as perceived by nurses: Findings from a large cross-sectional study. *Journal of Nursing Management*, *29*(8), 2658–2673. DOI: 10.1111/jonm.13451 PMID: 34369615

Grous, A. (n.d.). Sky high economics: Quantifying the commercial opportunities of passenger connectivity. London School of Economics. https://www.lse.ac.uk/business/consulting/assets/documents/sky-high-economics-chapter-one.pdf

Gubbi, J., Buyya, R., Marusic, S., & Palaniswami, M. (2013). Internet of Things (IoT): A vision, architectural elements, and future directions. *Future Generation Computer Systems*, *29*(7), 1645–1660. DOI: 10.1016/j.future.2013.01.010

Gunasekaran, A., Papadopoulos, T., Dubey, R., Wamba, S. F., Childe, S. J., Hazen, B., & Akter, S. (2017). Big data and predictive analytics for supply chain and organizational performance. *Journal of Business Research*, *70*, 308–317. DOI: 10.1016/j.jbusres.2016.08.004

Guo, C., Pleiss, G., Sun, Y., & Weinberger, K. Q. (2017). On calibration of modern neural networks. In *Proceedings of the 34th International Conference on Machine Learning* (pp. 1321-1330). PMLR. https://proceedings.mlr.press/v70/guo17a.html

Gupta, M. K., Shah, R., Rathod, J., & Kumar, A. (2021). SmartIdOCR: Automatic detection and recognition of identity card number using deep networks. *2021 Sixth International Conference on Image Information Processing (ICIIP)*, 267-272. DOI: 10.1109/ICIIP53038.2021.9702703

Ha, K., Chen, Z., Hu, W., Richter, W., Pillai, P., & Satyanarayanan, M. (2014). Towards wearable cognitive assistance. *MobiSys '14: Proceedings of the 12th Annual International Conference on Mobile Systems, Applications, and Services*, 68–81. DOI: 10.1145/2594368.2594383

Hashimoto, M. (2016). Infrastructure as code. In *Proceedings of the 2016 ACM SIGPLAN International Conference on Object-Oriented Programming, Systems, Languages, and Applications (OOPSLA)*. ACM.

Hassan, M. (2021). Memory optimization techniques for embedded systems: A review. *. *Journal of Embedded Computing*, *5*(2), 112–125.

Healthcare Quality Improvement Partnership. (2015, October 8). *National audit of inpatient falls: Audit report 2015*. https://www.hqip.org.uk/resource/national-audit-of-inpatient-falls-audit-report-2015/

He, K., Zhang, X., Ren, S., & Sun, J. (2016). Deep residual learning for image recognition. In *Proceedings of the IEEE Conference on Computer Vision and Pattern Recognition* (pp. 770-778). DOI: 10.1109/CVPR.2016.90

Hospital Authority. (n.d.). *Hospital Authority data sharing portal*. https://www3.ha.org.hk/data/HAStatistics/MajorReport?language=en

Hsu, C., Yuan, E. J., Fen, J., Lee, C., Ming, J., Chen, T., Lee, W., & Chen, S. (2020). Deployment of a computerized ward visitor registration system in coronavirus disease 2019 epidemic: Experiences of a large academic medical center in Taiwan. *Journal of the Chinese Medical Association*, *83*(6), 561–565. DOI: 10.1097/JCMA.0000000000000327 PMID: 32502118

Huang, K., Wang, K., Lee, P. K. C., & Yeung, A. C. L. (2023). The impact of Industry 4.0 on supply chain capability and supply chain resilience: A dynamic resource-based view. *International Journal of Production Economics*, *262*, 108913. DOI: 10.1016/j.ijpe.2023.108913

Hunkeler, U., Truong, H. L., & Stanford-Clark, A. (2008). MQTT-S; A publish/subscribe protocol for Wireless Sensor Networks. *2008 3rd International Conference on Communication Systems Software and Middleware and Workshops (COMSWARE '08)*, 791-798.

IATA. (2019, August). IATA cabin waste handbook. https://www.iata.org/contentassets/821b593dd8cd4f4aa33b63ab9e35368b/iata-cabin-waste-handbook.pdf

Ivanov, D. (2022). Viable supply chain model: Integrating agility, resilience and sustainability perspectives-lessons from and thinking beyond the COVID-19 pandemic. *Annals of Operations Research, 319*(1), 1411–1431. DOI: 10.1007/s10479-020-03640-6 PMID: 32836614

Ivanov, D., & Dolgui, A. (2021). A digital supply chain twin for managing the disruption risks and resilience in the era of Industry 4.0. *Production Planning and Control, 32*(9), 775–788. DOI: 10.1080/09537287.2020.1768450

Jacob, B., Kligys, S., Chen, B., Zhu, M., Tang, M., Howard, A., Adam, H., & Kalenichenko, D. (2018). Quantization and training of neural networks for efficient integer-arithmetic-only inference. In *Proceedings of the IEEE/CVF Conference on Computer Vision and Pattern Recognition* (pp. 2704-2713). DOI: 10.1109/CVPR.2018.00286

Javaid, M., & Khan, I. H. (2021). Internet of things (IoT) enabled healthcare helps to take the challenges of COVID-19 pandemic. *Journal of Oral Biology and Craniofacial Research, 11*(2), 209–214. DOI: 10.1016/j.jobcr.2021.01.015 PMID: 33665069

Jayanth, R., Gupta, N., & Prasanna, V. (2024). Benchmarking edge AI platforms for high-performance ML inference. In *2024 IEEE High Performance Extreme Computing Conference*. DOI: 10.1109/HPEC62836.2024.10938499

Jocher, G., Chaurasia, A., & Qiu, J. (2023). *Ultralytics YOLOv8* (Version 8.0.0). https://github.com/ultralytics/ultralytics; GitHub.

Jonas, E., Schleier-Smith, J., Sreekanti, V., Tsai, C.-C., Khandelwal, A., Pu, Q., Shankar, V., Carreira, J., Kraska, T., Gonzalez, J. E., Popa, R. A., Stoica, I., & Patterson, D. A. (2019). Cloud programming simplified: A Berkeley view on serverless computing. arXiv preprint, arXiv:1902.03383. https://arxiv.org/abs/1902.03383

Joshi, S., Gupta, N., Burmi, S., & Sharma, Y. (2023, November). AI-IOT: Integrating artificial intelligence & Internet of Things for healthcare systems. In *2023 International Conference on Advances in Computation, Communication and Information Technology (ICAICCIT)* (pp. 281–285). IEEE. DOI: 10.1109/ICAICCIT60255.2023.10465828

Kang, P., & Somtham, A. (2022). An evaluation of modern accelerator-based edge devices for object detection applications. *Mathematics*, *10*(22), 4299. DOI: 10.3390/math10224299

Kapusi, T. P., Erdei, T. I., Husi, G., & Hajdu, A. (2022). Application of deep learning in the deployment of an industrial SCARA machine for real-time object detection. *Robotics (Basel, Switzerland)*, *11*(4), 69. DOI: 10.3390/robotics11040069

Karimi, K., & Atkinson, G. (2013). What the Internet of Things (IoT) needs to become a reality. *White Paper, FreeScale and ARM.*

Karpaga Priya, P. S., & Prabhune, G., A. (. (2024). Transforming hospital crowd management: A case study on Pravesh electronic visitor management system. *BOHR International Journal of Internet of things. Artificial Intelligence and Machine Learning*, *3*(1), 40–44. DOI: 10.54646/bijiam.2024.27

Kehl, P., Ansari, J., Jafari, M. H., Becker, P., Sachs, J., König, N., Göppert, A., & Schmitt, R. H. (2022). Prototype of 5G integrated with TSN for edge-controlled mobile robotics. *Electronics (Basel)*, *11*(11), 1666. DOI: 10.3390/electronics11111666

Khamis, A., Hussein, A., & Elmogy, A. (2015). Multi-robot task allocation: A review of the state-of-the-art. *Studies in Computational Intelligence*, *604*, 31–51. DOI: 10.1007/978-3-319-18299-5_2

Kido, S., Miyasaka, T., Tanaka, T., Shimizu, T., & Saga, T. (2009, January). Fall detection in toilet rooms using thermal imaging sensors. In *2009 IEEE/SICE International Symposium on System Integration (SII)* (pp. 83–88). IEEE. DOI: 10.1109/SI.2009.5384550

Kirkpatrick, J., Pascanu, R., Rabinowitz, N., Veness, J., Desjardins, G., Rusu, A. A., Milan, K., Quan, J., Ramalho, T., Grabska-Barwinska, A., Hassabis, D., Clopath, C., Kumaran, D., & Hadsell, R. (2017). Overcoming catastrophic forgetting in neural networks. *Proceedings of the National Academy of Sciences of the United States of America*, *114*(13), 3521–3526. DOI: 10.1073/pnas.1611835114 PMID: 28292907

Kittipanya-ngam, P., & Tan, K. H. (2020). A framework for food supply chain digitalization: Lessons from Thailand. *Production Planning and Control*, *31*(2-3), 158–172. DOI: 10.1080/09537287.2019.1631462

Kleppmann, M. (2017). *Designing data-intensive applications.* O'Reilly Media.

Kolling, A., Walker, P., Chakraborty, N., Sycara, K., & Lewis, M. (2016). Human interaction with robot swarms: A survey. *IEEE Transactions on Human-Machine Systems, 46*(1), 9–26. DOI: 10.1109/THMS.2015.2480801

Kong, L., Tan, J., Huang, J., Chen, G., Wang, S., Jin, X., Zeng, P., Khan, M., & Das, S. K. (2023). Edge-computing-driven Internet of Things: A survey. *ACM Computing Surveys, 55*(8), 174. Advance online publication. DOI: 10.1145/3555308

Krishnamoorthi, R. (2018). Quantizing deep convolutional networks for efficient inference: A whitepaper. *arXiv Preprint arXiv:1806.08342*.

Kronauer, T., Pohlmann, J., Matthé, M., Smejkal, T., & Fettweis, G. (2021). Latency analysis of ROS2 multi-node systems. In *2021 IEEE International Conference on Multisensor Fusion and Integration for Intelligent Systems (MFI)* (pp. 1-7). DOI: 10.1109/MFI52462.2021.9591166

Kumar, T. B., & Prashar, D. (2021). Review on Efficient Food Waste Management System Using Internet of Things. *International Journal of Current Research and Review, 13*(6), 143–150. DOI: 10.31782/IJCRR.2021.13603

Kupyn, O., Martyniuk, T., Wu, J., & Wang, Z. (2019). Deblurgan-v2: Deblurring (orders-of-magnitude) faster and better. *Proceedings of the IEEE/CVF International Conference on Computer Vision*, 8878–8887.

Lai, K. C., & Cetinkaya, A. (2026). Car-net: A cascade refinement network for rotational motion deblurring under angle information uncertainty. *Proceedings of the International Conference on Artificial Intelligence and Machine Learning (AAIML)*. DOI: 10.1109/AAIML67890.2026.11498173

Lai, K. C., & Mohammed, A. M. (2026). Robust visual perception for mobile robots: Integrating physics-informed deblurring in ROS2. *International Conference on Wireless Sensor Networks, Ubiquitous Computing and Applications (ICWSNUCA).*

Lam, G. (2022). Shortage of nurses in Hong Kong: The challenges ahead. *Asian Education and Development Studies, 12*(1), 89–102. DOI: 10.1108/AEDS-08-2021-0179

Laurenzi, A., Antonucci, D., Tsagarakis, N. G., & Muratore, L. (2023). The XBot2 real-time middleware for robotics. *Robotics and Autonomous Systems, 163*, 104379. DOI: 10.1016/j.robot.2023.104379

Lee, H., Mun, H., & Lee, Y. (2020b). Comparing Response Time of Home IoT Devices with or without Cloud. *2020 IEEE International Conference on Consumer Electronics*, 1–6. DOI: 10.1109/ICCE46568.2020.9043102

Lee, J., Bagheri, B., & Kao, H.-A. (2015). A cyber-physical systems architecture for Industry 4.0-based manufacturing systems. *Manufacturing Letters*, *3*, 18–23. DOI: 10.1016/j.mfglet.2014.12.001

Lee, Y. K. (2020). Mitigating servo jitter in microcontroller systems utilizing software UART interactions. *International Journal of Hardware Engineering*, *12*(3), 45–58.

Lema, D. G., Usamentiaga, R., & García, D. F. (2024). Quantitative comparison and performance evaluation of deep learning-based object detection models on edge computing devices. *Integration (Amsterdam)*, *95*, 102127. DOI: 10.1016/j.vlsi.2023.102127

Liang, J., Cao, J., Sun, G., Zhang, K., Van Gool, L., & Timofte, R. (2021). Swinir: Image restoration using swin transformer. *Proceedings of the IEEE/CVF International Conference on Computer Vision*, 1833–1844.

Liegeard, J., & Manning, L. (2020). Use of intelligent applications to reduce household food waste. *Critical Reviews in Food Science and Nutrition*, *60*(6), 1048–1061. DOI: 10.1080/10408398.2018.1556580 PMID: 30633547

Light, R. A. (2017). Mosquitto: Server and client implementation of the MQTT protocol. *Journal of Open Source Software*, *2*(13), 265. DOI: 10.21105/joss.00265

Lin, T.-Y., Maire, M., Belongie, S., Hays, J., Perona, P., Ramanan, D., Dollár, P., & Zitnick, C. L. (2014). Microsoft coco: Common objects in context. *European Conference on Computer Vision*, 740–755.

Liu, F. T., Ting, K. M., & Zhou, Z.-H. (2008, December). *Isolation forest*. In *2008 Eighth IEEE International Conference on Data Mining* (pp. 413–422). IEEE. DOI: 10.1109/ICDM.2008.17

Li, Z., Hasegawa, A., & Azumi, T. (2022). Autoware.perf: A tracing and performance analysis framework for ROS 2 applications. *Journal of Systems Architecture*, *123*, 102341. DOI: 10.1016/j.sysarc.2021.102341

Luan, S., Yang, C., Qin, X., Chen, D., & Sui, W. (2024). Towards robust visual odometry by motion blur recovery. *Frontiers in Signal Processing*, *4*, 1417363. DOI: 10.3389/frsip.2024.1417363

Lu, Y., Liu, C., Kevin, I., Wang, K., Huang, H., & Xu, X. (2020). Digital twin-driven smart manufacturing: Connotation, reference model, applications and research issues. *Robotics and Computer-integrated Manufacturing*, *61*, 101837. DOI: 10.1016/j.rcim.2019.101837

Macario Barros, A., Michel, M., Moline, Y., Corre, G., & Carrel, F. (2022). A comprehensive survey of visual slam algorithms. *Robotics (Basel, Switzerland), 11*(1), 24. DOI: 10.3390/robotics11010024

Macenski, S., Foote, T., Gerkey, B., Lalancette, C., & Woodall, W. (2022). Robot Operating System 2: Design, architecture, and uses in the wild. *Science Robotics, 7*(66), eabm6074. Advance online publication. DOI: 10.1126/scirobotics.abm6074

Mao, Y., You, C., Zhang, J., Huang, K., & Letaief, K. B. (2017). A survey on mobile edge computing: The communication perspective. *IEEE Communications Surveys and Tutorials, 19*(4), 2322–2358. DOI: 10.1109/COMST.2017.2745201

Maruyama, Y., Kato, S., & Azumi, T. (2016). Exploring the performance of ROS2. In *Proceedings of the 13th International Conference on Embedded Software* (pp. 1-10). DOI: 10.1145/2968478.2968502

Mastorakis, G., Kopanakis, I., Makridis, J., Chroni, C., Synani, K., Lasaridi, K., Abeliotis, K., Louloudakis, I., Daliakopoulos, I. N., & Manios, T. (2024). Managing household food waste with the FoodSaveShare mobile application. *Sustainability (Basel), 16*(7), 2800. DOI: 10.3390/su16072800

Mathisen, T. F., & Johansen, O. J. (2022). The impact of smartphone apps designed to reduce food waste on improving healthy eating, financial expenses and personal food waste: Crossover pilot intervention trial studying students' user experiences. *JMIR Formative Research, 6*(9), e38520. DOI: 10.2196/38520 PMID: 36053667

Mecocci, A., Micheli, F., Zoppetti, C., & Baghini, A. (2016, October). Automatic falls detection in hospital-room context. In *2016 7th IEEE International Conference on Cognitive Infocommunications (CogInfoCom)* (pp. 127–132). IEEE. DOI: 10.1109/CogInfoCom.2016.7804537

Mell, P., & Grance, T. (2011). The NIST definition of cloud computing. *National Institute of Standards and Technology*, Special Publication 800-145.

Merkel, D. (2014). Docker: Lightweight linux containers for consistent development and deployment. *Linux J., 2014*(239).

Metcalfe, R. M., & Boggs, D. R. (1976). Ethernet: Distributed packet switching for local computer networks. *Communications of the ACM, 19*(7), 395–404. DOI: 10.1145/360248.360253

Miller, T. (2019). Explanation in artificial intelligence: Insights from the social sciences. *Artificial Intelligence, 267*, 1–38. DOI: 10.1016/j.artint.2018.07.007

Ming, Z., Visani, M., Luqman, M. M., & Burie, J. (2020). A survey on anti-spoofing methods for facial recognition with RGB cameras of generic consumer devices. *Journal of Imaging*, *6*(12), 139. DOI: 10.3390/jimaging6120139 PMID: 34460536

Mohammadi, M., Al-Fuqaha, A., Sorour, S., & Guizani, M. (2018). Deep learning for IoT big data and streaming analytics: A survey. *IEEE Communications Surveys and Tutorials*, *20*(4), 2923–2960. DOI: 10.1109/COMST.2018.2844341

Mohammed, A., & Zangana, H. (Eds.). (2026). *Intelligent Construction Monitoring Systems: Real-Time Safety, Environmental Prediction, and Risk Management*. IGI Global Scientific Publishing., DOI: 10.4018/979-8-3373-9245-5

Monarch, R. M. (2021). *Human-in-the-loop machine learning: Active learning and annotation for human-centered AI*. Manning Publications.

Mosqueira-Rey, E., Hernández-Pereira, E., Alonso-Ríos, D., Bobes-Bascarán, J., & Fernández-Leal, Á. (2023). Human-in-the-loop machine learning: A state of the art. *Artificial Intelligence Review*, *56*(4), 3005–3054. DOI: 10.1007/s10462-022-10246-w

Mtowe, D. P., Long, L., & Kim, D. M. (2025). Low-latency edge-enabled digital twin system for multi-robot collision avoidance and remote control. *Sensors (Basel)*, *25*(15), 4666. DOI: 10.3390/s25154666

Murshed, M. G. S., Murphy, C., Hou, D., Khan, N., Ananthanarayanan, G., & Hussain, F. (2021). Machine learning at the network edge: A survey. *ACM Computing Surveys*, *54*(8), 170. Advance online publication. DOI: 10.1145/3469029

National Institute of Standards and Technology. (2019). *Considerations for managing Internet of Things (IoT) cybersecurity and privacy risks* (NIST IR 8228). DOI: 10.6028/NIST.IR.8228

Newsome, R., Balestrini, C. G., Baum, M. D., Corby, J., Fisher, W., Goodburn, K., Labuza, T. P., Prince, G., Thesmar, H. S., & Yiannas, F. (2014). Applications and perceptions of date labeling of food. *Comprehensive Reviews in Food Science and Food Safety*, *13*(4), 745–769. DOI: 10.1111/1541-4337.12086 PMID: 33412704

Nho, Y.-H., Lim, J. G., & Kwon, D.-S. (2020). Cluster-analysis-based user-adaptive fall detection using fusion of heart rate sensor and accelerometer in a wearable device. *IEEE Access : Practical Innovations, Open Solutions*, *8*, 40389–40401. DOI: 10.1109/ACCESS.2020.2969453

Nielsen, J. (1993). *Usability engineering*. Academic Press. DOI: 10.1016/B978-0-08-052029-2.50007-3

Nielsen, J. (2006). Ten usability heuristics. *Engineering in Life Sciences*, *20*(11), 485–493. DOI: 10.1002/elsc.202000024

NVIDIA Corporation. (2023). *NVIDIA TensorRT documentation*. NVIDIA Corporation. https://developer.nvidia.com/tensorrt

NVIDIA. (2025). *Jetson AGX Orin Developer Kit User Guide*. https://docs.nvidia.com/jetson/agx-orin-devkit/user-guide/latest/index.html

Nygard, M. T. (2018). *Release it! Design and deploy production-ready software* (2nd ed.). Pragmatic Bookshelf.

Open, O. A. S. I. S. (2019). *MQTT version 5.0*. https://docs.oasis-open.org/mqtt/mqtt/v5.0/mqtt-v5.0.pdf

Ozer, A. S., & Cinar, A. (2025). Real-time and fully automated robotic stacking system with deep learning-based visual perception. *Sensors (Basel)*, *25*(22), 6960. DOI: 10.3390/s25226960

Paho. (2024). *Eclipse Paho - MQTT Client Library for Python*. https://eclipse.dev/paho/index.php/page/clients/python/index.html

Palan, A., Nadoda, M., & Sajwan, P. (2025). A comprehensive framework for smart hospitals using IoT and artificial intelligence. *International Journal of Advanced Research in Computer and Communication Engineering*, *14*(12). Advance online publication. DOI: 10.17148/IJARCCE.2025.141204

Park, J., Delgado, R., & Choi, B. W. (2020). Real-time characteristics of ROS 2.0 in multiagent robot systems: An empirical study. *IEEE Access : Practical Innovations, Open Solutions*, *8*, 154637–154651. DOI: 10.1109/ACCESS.2020.3018122

Paszke, A., Gross, S., Massa, F., Lerer, A., Bradbury, J., Chanan, G., Killeen, T., Lin, Z., Gimelshein, N., Antiga, L., Desmaison, A., Kopf, A., Yang, E., DeVito, Z., Raison, M., Tejani, A., Chilamkurthy, S., Steiner, B., Fang, L., . . . Chintala, S. (2019). PyTorch: An imperative style, high-performance deep learning library. In H. Wallach, H. Larochelle, A. Beygelzimer, F. d'Alché-Buc, E. Fox, & R. Garnett (Eds.), *Advances in neural information processing systems* (Vol. 32). Curran Associates, Inc. https://proceedings.neurips.cc/paper_files/paper/2019/file/bdbca288fee7f92f2bfa9f7012727740-Paper.pdf

Peng, H., Bayon, J., Recas, J., & Guijarro, M. (2025). Efficient expiration date recognition in food packages for mobile applications. *Algorithms*, *18*(5), 286. DOI: 10.3390/a18050286

Perera, C., Zaslavsky, A., Christen, P., & Georgakopoulos, D. (2014). Context aware computing for the internet of things: A survey. *IEEE Communications Surveys and Tutorials*, *16*(1), 414–454. DOI: 10.1109/SURV.2013.042313.00197

Pradeep, A. (2023). Exploring the future of edge computing: advantages, limitations, and opportunities. In Communications in computer and information science (pp. 196–209). DOI: 10.1007/978-3-031-45124-9_15

PrimeTek. (n.d.). *PrimeVue introduction*. https://primevue.org/introduction/

Puck, L., Keller, P., Schnell, T., Plasberg, C., Tanev, A., Heppner, G., Rönnau, A., & Dillmann, R. (2021). Performance evaluation of real-time ROS2 robotic control in a time-synchronized distributed network. In 2021 IEEE 17th International Conference on Automation Science and Engineering (CASE) (pp. 1670-1676). DOI: 10.1109/CASE49439.2021.9551447

Python Software Foundation. (2023). *Tkinter; Python interface to Tcl/Tk*. https://docs.python.org/3/library/tkinter.html

Queiroz, M. M., Pereira, S. C. F., Telles, R., & Machado, M. C. (2021). Industry 4.0 and digital supply chain capabilities: A framework for understanding digitalisation challenges and opportunities. *Benchmarking*, *28*(5), 1761–1782. DOI: 10.1108/BIJ-12-2018-0435

Quigley, M., Gerkey, B., Conley, K., Faust, J., Foote, T., Leibs, J., Berger, E., Wheeler, R., & Ng, A. Y. (2009). ROS: An open-source robot operating system. *ICRA Workshop on Open Source Software*, *3*, 5.

Radford, A., Kim, J. W., Hallacy, C., Ramesh, A., Goh, G., Agarwal, S., Sastry, G., Askell, A., Mishkin, P., Clark, J., Krueger, G., & Sutskever, I. (2021). Learning transferable visual models from natural language supervision. In *Proceedings of the 38th International Conference on Machine Learning* (pp. 8748-8763). PMLR. https://proceedings.mlr.press/v139/radford21a.html

Rajan, S. S., & Wani, K. M. (2025). A review of smart food and packaging technologies: Revolutionizing nutrition and sustainability. *Food and Humanity*, *4*, 100593. DOI: 10.1016/j.foohum.2025.100593

Ramdhani, T. W., Budi, I., & Purwandari, B. (2021). Optical character recognition engines performance comparison in information extraction. *International Journal of Advanced Computer Science and Applications*, *12*(8). Advance online publication. DOI: 10.14569/IJACSA.2021.0120814

Raspberry Pi Ltd. (n.d.). *Raspberry Pi 5*. https://www.raspberrypi.com/documentation/computers/raspberry-pi-5.html

Redmon, J., Divvala, S., Girshick, R., & Farhadi, A. (2016). You only look once: Unified, real-time object detection. *Proceedings of the IEEE Conference on Computer Vision and Pattern Recognition*, 779-788. https://openaccess.thecvf.com/content_cvpr_2016/html/Redmon_You_Only_Look_CVPR_2016_paper.html

Reimers, N., & Gurevych, I. (2019). Sentence-BERT: Sentence embeddings using Siamese BERT-networks. In *Proceedings of the 2019 Conference on Empirical Methods in Natural Language Processing (EMNLP)* (pp. 3982-3992). https://arxiv.org/abs/1908.10084

Reis, M. J. C. S., & Serôdio, C. (2025). Edge AI for Real-Time Anomaly Detection in smart Homes. *Future Internet*, *17*(4), 179. DOI: 10.3390/fi17040179

Rewicki, F., Denzler, J., & Niebling, J. (2023). *Is it worth it? An experimental comparison of six deep- and classical machine learning methods for unsupervised anomaly detection in time series*. arXiv. https://arxiv.org/abs/2212.11080

Ricárdez-Estrada, J., Garay-Rondero, C. L., Romero, D., Wuest, T., & Pinto, R. (2024). Identifying digital supply chain capabilities. *Procedia Computer Science*, *232*, 1182–1191. DOI: 10.1016/j.procs.2024.01.116

Sajid, M., & Raza, Z. (2013). Cloud computing: issues & challenges. In International Conference on Cloud, Big Data and Trust 2013, Nov 13-15, RGPV, International Conference on Cloud, Big Data and Trust 2013, Nov 13-15, RGPV. https://wanivipin.wordpress.com/wp-content/uploads/2016/11/challenges-of-cc.pdf

Salman, O., Elhajj, I., Kayssi, A., & Chehab, A. (2015). Edge computing enabling the Internet of Things. *2015 IEEE 2nd World Forum on Internet of Things (WF-IoT)*, 603-608.

Sandler, M., Howard, A., Zhu, M., Zhmoginov, A., & Chen, L.-C. (2018). MobileNetV2: Inverted residuals and linear bottlenecks. In *Proceedings of the IEEE/CVF Conference on Computer Vision and Pattern Recognition* (pp. 4510-4520). DOI: 10.1109/CVPR.2018.00474

Sapkota, R., Cheppally, R. H., Sharda, A., & Karkee, M. (2025). YOLO26: Key architectural enhancements and performance benchmarking for real-time object detection. *arXiv Preprint arXiv:2509.25164*. https://arxiv.org/abs/2509.25164

Saputra, M. R. U., Markham, A., & Trigoni, N. (2018). Visual SLAM and structure from motion in dynamic environments: A survey. *ACM Computing Surveys*, *51*(2), 1–36. DOI: 10.1145/3177853

Satyanarayanan, M. (2017). The emergence of edge computing. *Computer*, *50*(1), 30–39. DOI: 10.1109/MC.2017.9

Satyanarayanan, M., Bahl, P., Caceres, R., & Davies, N. (2009). The case for VM-based cloudlets in mobile computing. *IEEE Pervasive Computing*, *8*(4), 14–23. DOI: 10.1109/MPRV.2009.82

Schwartz, M. (2016). *Internet of Things with Arduino Cookbook*. Packt Publishing Ltd.

Seker, A. C., & Ahn, Y. H. (2022). A generalized framework for recognition of expiration dates on product packages using fully convolutional networks. *Expert Systems with Applications*, *204*, 117310. DOI: 10.1016/j.eswa.2022.117310

Seol, Y., Hyeon, D., Min, J., Kim, Y., Kim, J., Lee, S., & Kim, H. (2021). Timely survey of time-sensitive networking: Past and future directions. *IEEE Access : Practical Innovations, Open Solutions*, *9*, 142506–142527. DOI: 10.1109/ACCESS.2021.3120769

Sethi, P., & Sarangi, S. R. (2017). Internet of Things: Architectures, protocols, and applications. *Journal of Electrical and Computer Engineering*, *2017*, 1–25. DOI: 10.1155/2017/9324035

Shahrad, M., Fonseca, R., Goiri, I., Chaudhry, G., Bancy, P., Cober, T., Fontoura, M., Fonseca, R., & Bianchini, R. (2020). Serverless in the wild: Characterizing and optimizing the serverless workload at a large cloud provider. In Proceedings of the 2020 USENIX Annual Technical Conference (ATC). USENIX.

Shapiro, M., Preguica, N., Baquero, C., & Zawirski, M. (2011). Conflict-free replicated data types. In *Proceedings of the 13th International Conference on Stabilization, Safety, and Security of Distributed Systems (SSS)* (pp. 386-400). Springer. DOI: 10.1007/978-3-642-24550-3_29

Shata, E., Chen, B., Hu, L., Seskar, I., Guo, Y., Mahmoudi, C., Shekhar, S., & Zou, Q. (2024). 5G-cloud-based real-time robotic part repairing for advanced manufacturing via computer vision. *Manufacturing Letters*, *41*, 1398–1404. DOI: 10.1016/j.mfglet.2024.09.166

Shi, W., Cao, J., Zhang, Q., Li, Y., & Xu, L. (2016). Edge computing: Vision and challenges. *IEEE Internet of Things Journal*, *3*(5), 637–646. DOI: 10.1109/JIOT.2016.2579198

Sixsmith, A., & Johnson, N. (2004). A smart sensor to detect the falls of the elderly. *IEEE Pervasive Computing*, *3*(2), 42–47. DOI: 10.1109/MPRV.2004.1316817

South China Morning Post. (2023, April 18). *Call for monitoring devices for Hong Kong patients after missing woman found dead*. https://www.scmp.com/news/hong-kong/health-environment/article/3217390/woman-58-chronic-illness-found-dead-hong-kong-hospital-toilet-after-waiting-more-12-hours-emergency

SQLite Consortium. (n.d.). *Database file format.* https://sqlite.org/fileformat.html

Stallings, W. (2017). *Cryptography and Network Security: Principles and Practice* (7th ed.). Pearson.

Stefanou, E. E., Bitilis, P., Bouchouras, G., & Kotis, K. (2025). Collecting, Integrating and Processing IoT Sensor Data on Edge Devices for PD Monitoring: A Scoping Review. *Applied Sciences (Basel, Switzerland), 15*(19), 10541. DOI: 10.3390/app151910541

Su, M.-C., Liao, J.-W., Wang, P.-C., & Wang, C.-H. (2017, June). A smart ward with a fall detection system. In *2017 IEEE International Conference on Environment and Electrical Engineering and 2017 IEEE Industrial and Commercial Power Systems Europe (EEEIC/I&CPS Europe)* (pp. 1–4). IEEE. DOI: 10.1109/EEEIC.2017.7977515

Szenderák, J., Fróna, D., & Rákos, M. H. (2025). Expiration date label knowledge and food waste reduction: A synthesis and critical evaluation of policy recommendations. *Environmental Challenges, 20*, 101269. DOI: 10.1016/j.envc.2025.101269

Sze, V., Chen, Y.-H., Yang, T.-J., & Emer, J. S. (2017). Efficient processing of deep neural networks: A tutorial and survey. *Proceedings of the IEEE, 105*(12), 2295–2329. DOI: 10.1109/JPROC.2017.2761740

Tahir, N.. (2025). Edge computing and its application in robotics: A survey. *Robotics (Basel, Switzerland), 14*(4), 65. DOI: 10.3390/robotics14040065

Tao, F., Zhang, H., Liu, A., & Nee, A. Y. C. (2019). Digital twin in industry: State-of-the-art. *IEEE Transactions on Industrial Informatics, 15*(4), 2405–2415. DOI: 10.1109/TII.2018.2873186

Tao, X., Gao, H., Shen, X., Wang, J., & Jia, J. (2018). Scale-recurrent network for deep image deblurring. *Proceedings of the IEEE Conference on Computer Vision and Pattern Recognition*, 8174–8182.

Tipsuwan, Y., & Chow, M.-Y. (2003). Control methodologies in networked control systems. *Control Engineering Practice, 11*(10), 1099–1111. DOI: 10.1016/S0967-0661(03)00036-4

United Nations. (n.d.). *World population projected to reach 9.8 billion in 2050, and 11.2 billion in 2100.* https://www.un.org/en/desa/world-population-projected-reach-98-billion-2050-and-112-billion-2100

Uvicorn. (n.d.). *Deployment.* https://www.uvicorn.org/deployment/ Vue.js. (n.d.). *Introduction.* https://vuejs.org/guide/introduction

Van Mourik, M. S., Perencevich, E. N., Gastmeier, P., & Bonten, M. J. (2017). Designing surveillance of healthcare-associated infections in the era of automation and reporting mandates. *Clinical Infectious Diseases*, *66*(6), 970–976. DOI: 10.1093/cid/cix835 PMID: 29514241

Waller, M. A., & Fawcett, S. E. (2013). Data science, predictive analytics, and big data: A revolution that will transform supply chain design and management. *Journal of Business Logistics*, *34*(2), 77–84. DOI: 10.1111/jbl.12010

Wamba, S. F., Gunasekaran, A., Akter, S., Ren, S. J.-F., Dubey, R., & Childe, S. J. (2017). Big data analytics and firm performance: Effects of dynamic capabilities. *Journal of Business Research*, *70*, 356–365. DOI: 10.1016/j.jbusres.2016.08.009

Wang, C., Huang, C., Zhu, X., Li, Z., & Zhao, L. (2023). Smart retail SKUs checkout using improved residual network. *Scientific Reports*, *13*(1), 22512. DOI: 10.1038/s41598-023-49543-x PMID: 38110520

Wang, S., Qu, Y., Shen, J., & Araki, T. (2023). Entity alignment with semantic embeddings: A survey. *ACM Computing Surveys*, *56*(2), 1–38. DOI: 10.1145/3604560

Wang, S., Tuor, T., Salonidis, T., Leung, K. K., Makaya, C., He, T., & Chan, K. (2019). Adaptive federated learning in resource constrained edge computing systems. *IEEE Journal on Selected Areas in Communications*, *37*(6), 1205–1221. DOI: 10.1109/JSAC.2019.2904348

Wang, T., Guo, J., Zhang, B., Yang, G., & Li, D. (2025). Deploying AI on edge: Advancement and challenges in edge intelligence. *Mathematics*, *13*(11), 1878. DOI: 10.3390/math13111878

Wang, X.-F., He, Z.-H., Wang, K., Wang, Y.-F., Zou, L., & Wu, Z.-Z. (2023). A survey of text detection and recognition algorithms based on deep learning technology. *Neurocomputing*, *556*, 126702. DOI: 10.1016/j.neucom.2023.126702

Wang, Z., Bovik, A. C., Sheikh, H. R., & Simoncelli, E. P. (2004). Image quality assessment: From error visibility to structural similarity. *IEEE Transactions on Image Processing*, *13*(4), 600–612. DOI: 10.1109/TIP.2003.819861 PMID: 15376593

Wei, X.-S., Cui, Q., Yang, L., Wang, P., Liu, L., & Yang, J. (2022). RPC: A large-scale and fine-grained retail product checkout dataset. *Science China. Information Sciences*, *65*(9), 197101. Advance online publication. DOI: 10.1007/s11432-022-3513-y

Wickham, H. (2014). Tidy data. *Journal of Statistical Software*, *59*(10), 1–23. DOI: 10.18637/jss.v059.i10

Wild, D., Nayak, U. S., & Isaacs, B. (1981). How dangerous are falls in old people at home? *British Medical Journal (Clinical Research Ed.)*, *282*(6260), 266–268. DOI: 10.1136/bmj.282.6260.266 PMID: 6779979

Wilson, N. L. W., & Miao, R. (2025). Food waste, date labels, and risk preferences: An experimental exploration. *Applied Economic Perspectives and Policy*, *47*(3), 1029–1057. Advance online publication. DOI: 10.1002/aepp.13507

Wong, S., Leung, M., Tong, D. W., Lee, L. L., Leung, W. L., Chan, F. W., Chen, J. H., Hung, I. F., Yuen, K., Yeung, D. T., Chung, K., & Cheng, V. C. (2020). Infection control challenges in setting up community isolation and treatment facilities for patients with coronavirus disease 2019 (COVID-19): Implementation of directly observed environmental disinfection. *Infection Control and Hospital Epidemiology*, *42*(9), 1037–1045. DOI: 10.1017/ice.2020.1355 PMID: 33280617

World Health Organization. (2021, April 27). *Step safely: Strategies for preventing and managing falls across the life-course*. https://www.who.int/publications/i/item/978924002191-4

Xiao, Z., & Xiao, Y. (2012). Security and privacy in cloud computing. *IEEE Communications Surveys and Tutorials*, *15*(2), 843–859. DOI: 10.1109/SURV.2012.060912.00182

Yang, Y., & Azumi, T. (2020). Exploring real-time executor on ROS 2. In *2020 IEEE International Conference on Embedded Software and Systems (ICESS)* (pp. 1-8). DOI: 10.1109/ICESS49830.2020.9301530

Yanto, W., Nayan, N. M., & Sulaiman, R. S. (2023). Implementation of optical character recognition (Ocr) based on text image using raspberry Pi with Arduino to write text. DOI: 10.2139/ssrn.4500046

Yu, W., Liang, F., He, X., Hatcher, W. G., Lu, C., Lin, J., & Yang, X. (2018). A Survey on the Edge Computing for the Internet of Things. *IEEE Access : Practical Innovations, Open Solutions*, *6*, 6900–6919. DOI: 10.1109/ACCESS.2017.2778504

Zamir, S. W., Arora, A., Khan, S., Hayat, M., Khan, F. S., & Yang, M.-H. (2022). Restormer: Efficient transformer for high-resolution image restoration. *Proceedings of the IEEE/CVF Conference on Computer Vision and Pattern Recognition*, 5728–5739. DOI: 10.1109/CVPR52688.2022.00564

Zanella, A., Bui, N., Castellani, A., Vangelista, L., & Zorzi, M. (2014). Internet of things for smart cities. *IEEE Internet of Things Journal*, *1*(1), 22–32. DOI: 10.1109/JIOT.2014.2306328

Zhang, Y., Sun, P., Jiang, Y., Yu, D., Weng, F., Yuan, Z., Luo, P., Liu, W., & Wang, X. (2022). ByteTrack: Multi-object tracking by associating every detection box. In *Computer Vision - ECCV 2022*. European Conference on Computer Vision. https://www.ecva.net/papers/eccv_2022/papers_ECCV/html/315_ECCV_2022_paper.php

Zhang, C., Zhou, G., Li, J., Chang, F., Ding, K., & Ma, D. (2023a). A multi-access edge computing enabled framework for the construction of a knowledge-sharing intelligent machine tool swarm in Industry 4.0. *Journal of Manufacturing Systems*, *66*, 56–70. DOI: 10.1016/j.jmsy.2022.11.015

Zhang, H., Zhang, R., & Sun, J. (2025). Developing real-time IoT-based public safety alert and emergency response systems. *Scientific Reports*, *15*(1), 29056. DOI: 10.1038/s41598-025-13465-7 PMID: 40781521

Zhang, J., Yu, X., Ha, S., Queralta, J. P., & Westerlund, T. (2024a). Comparison of middlewares in edge-to-edge and edge-to-cloud communication for distributed ROS 2 systems. *Journal of Intelligent & Robotic Systems*, *110*(4), 162. DOI: 10.1007/s10846-024-02187-z

Zhang, T., Wang, G., Xue, C., Wang, J., Nixon, M., & Han, S. (2024b). Time-sensitive networking (TSN) for industrial automation: Current advances and future directions. *ACM Computing Surveys*. Advance online publication. DOI: 10.1145/3695248

Zhang, Y., Wurll, C., & Hein, B. (2023b). KubeROS: A unified platform for automated and scalable deployment of ROS2-based multi-robot applications. In *2023 IEEE International Conference on Robotics and Automation (ICRA)* (pp. 9097-9103). DOI: 10.1109/ICRA48891.2023.10160632

Zhou, Z., Chen, X., Li, E., Zeng, L., Luo, K., & Zhang, J. (2019). Edge intelligence: Paving the last mile of artificial intelligence with edge computing. *Proceedings of the IEEE*, *107*(8), 1738–1762. DOI: 10.1109/JPROC.2019.2918951

Zhu, Y., Yin, X., Liew, A. W.-C., & Tian, H. (2024). *Privacy-preserving in medical image analysis: A review of methods and applications*. arXiv. https://arxiv.org/abs/2412.03924

Zwolenski, M., & Weatherill, L. (2014). The digital universe: Rich data and the increasing value of the internet of things. *Journal of Telecommunications and the Digital Economy*, *2*(3). Advance online publication. https://search.informit.org/doi/abs/10.3316/informit.678436300116927. DOI: 10.7790/ajtde.v2n3.47

[illegible] Bolt, F. [illegible] (2022). [illegible] based [illegible] association [illegible] detection [illegible] papers [illegible] CVPR [illegible] 2022 [illegible]

Zhu, [illegible] (2022). [illegible] methods and applications. [illegible] 2112.05[illegible]

[illegible] (2024). The digital [illegible] and [illegible] the [illegible] internet of things. [illegible] Online publication. [illegible] DOI: 10.[illegible]

About the Contributors

Aquil Mirza Mohammed is a senior member of IEEE, a researcher and educator specializing in AI, AIoT, robotics, sensor networks, and medical imaging. He is a Dean and Professor at Liaoning Technical University, China and a Senior Lecturer at The Hong Kong Polytechnic University, Hong Kong. His research has been funded by major grants, and his work has earned several national and international awards such as the IEEE Best Student Paper awards, Young Investigator Award, etc. With a strong academic background and innovative research, he continues to inspire future engineers.

Cong Wu received his PhD from Dongbei University of Finance and Economics and holds dual Master's degrees in Engineering (Computer Application Technology) and Management (Management Science and Engineering). He is a council member of the Higher Education Branch of the Chinese Association for Optimization, Overall Planning and Economic Mathematics, a JD Cloud & AI certified engineer, and an AI engineering assessor for the Ministry of Industry; he is also an Outstanding Worker of the China Coal Industry Association, an Advanced Individual in Academic and Postgraduate Education in Liaoning Province, and a recipient of the Fuxin City May Day Labor Medal. Currently, he is the Vice Dean of the Innovation and Practice College, and his research focuses on multimodal learning and large-scale models, data analysis and intelligent decision-making, big data and cloud computing, and intelligent sensing in the Internet of Things.

***

Mohammad Al-Khaldy is an Assistant professor of Business Intelligence and Data Analytics at University of Petra, Jordan. He completed his PhD in Atrificial Intelligence and Data Science for the University of Hull -United Kingdom - 20017. Alkhaldy's research interests include data analytics, machine learning, predictive analytics, NLP, and decision support systems. His research has been published extensively in academic journals and conferences. Alkhaldy taught a variety of courses in Artificial intelligence and business intelligence, including Data Mining, Business Analytics, Computer Programming, Intelligent business systems, Data Visualization, and machine learning. Alkhaldy is a member of Jordan Computers Society, and Arab Robotics & AI Association. He also a reviewer for several other academic conferences and journals.

Hewa Zangana is an Assistant Professor currently affiliated with Duhok Polytechnic University (DPU) in Iraq. He holds a Doctorate in Philosophy (PhD) degree in ITM, which he is currently pursuing at DPU. He has also held administrative positions such as Curriculum Division Director - Presidency of DPU, Head of the Computer Science Department at Nawroz University, and Acting Dean of the College of Computer and IT at Nawroz University. Hewa Majeed Zangana's research interests span a wide range of topics in computer science, including network systems, information security, mobile communication, data communication, and intelligent systems. He has published extensively in peer-reviewed journals and has published many academic books with IGI Global, several of which are indexed in Scopus and Web of Science (Clarivate). In addition to his research contributions, Hewa Majeed Zangana has been actively involved in editorial roles. He has also been a member of various scientific committees and administrative bodies at Nawroz University.

Index

A

B

D

E

F

H

I

J

L

M

O

P

R

S

T

Y

www.ingramcontent.com/pod-product-compliance
Lightning Source LLC
LaVergne TN
LVHW081535300926
849161LV00006B/1433